W9-AFL-711

The Music of the Primes

The Music of the Primes

Searching to Solve the Greatest Mystery in Mathematics

Marcus du Sautoy

HarperCollins*Publishers*

FIRST EDITION

Designed by Geoff Green Book Design

Printed on acid-free paper

Library of Congress Cataloging-in-Publication Data is available upon request.

ISBN 0-06-621070-4

03 04 05 06 07 ❖/RRD 10 9 8 7 6 5 4 3 2

For the memory of

Yonathan du Sautoy

October 21, 2000

Contents

The Music of the Primes

CHAPTER ONE

Who Wants To Be a Millionaire?

'Do we know what the sequence of numbers is? Okay, here, we can do it in our heads . . . fifty-nine, sixty-one, sixty-seven . . . seventy-one . . . Aren't these all prime numbers?' A little buzz of excitement circulated through the control room. Ellie's own face momentarily revealed a flutter of something deeply felt, but this was quickly replaced by a sobriety, a fear of being carried away, an apprehension about appearing foolish, unscientific. Carl Sagan, *Contact*

One hot and humid morning in August 1900, David Hilbert of the University of Göttingen addressed the International Congress of Mathematicians in a packed lecture hall at the Sorbonne, Paris. Already recognised as one of the greatest mathematicians of the age, Hilbert had prepared a daring lecture. He was going to talk about what was unknown rather than what had already been proved. This went against all the accepted conventions, and the audience could hear the nervousness in Hilbert's voice as he began to lay out his vision for the future of mathematics. 'Who of us would not be glad to lift the veil behind which the future lies hidden; to cast a glance at the next advances of our science and at the secrets of its development during future centuries?' To herald the new century, Hilbert challenged the audience with a list of twenty-three problems that he believed should set the course for the mathematical explorers of the twentieth century.

The ensuing decades saw many of the problems answered, and those who discovered the solutions make up an illustrious band of mathematicians known as 'the honours class'. It includes the likes of Kurt Gödel and Henri Poincaré, along with many other pioneers whose ideas have transformed the mathematical landscape. But there was one problem, the eighth on Hilbert's list, which looked as if it would survive the century without a champion: the Riemann Hypothesis.

Of all the challenges that Hilbert had set, the eighth had a special place in his heart. There is a German myth about Frederick Barbarossa, a much-loved German emperor who died during the Third Crusade. A legend grew that he was still alive, asleep in a cavern in the Kyffhäuser Mountains. He would awake only when Germany needed him. Somebody allegedly asked Hilbert, 'If you were to be revived like Barbarossa, after five hundred years, what would you do?' His reply: 'I would ask, "Has someone proved the Riemann Hypothesis?"'

As the twentieth century drew to a close, most mathematicians had resigned themselves to the fact that this jewel amongst all of Hilbert's problems was not only likely to outlive the century, but might still be unanswered when Hilbert awoke from his five-hundred-year slumber. He had stunned the first International Congress of the twentieth century with his revolutionary lecture full of the unknown. However, there turned out to be a surprise in store for those mathematicians who were planning to attend the last Congress of the century.

On April 7, 1997, computer screens across the mathematical world flashed up some extraordinary news. Posted on the website of the International Congress of Mathematicians that was to be held the following year in Berlin was an announcement that the Holy Grail of mathematics had finally been claimed. The Riemann Hypothesis had been proved. It was news that would have a profound effect. The Riemann Hypothesis is a problem which is central to the whole of mathematics. As they read their email, mathematicians were thrilling to the prospect of understanding one of the greatest mathematical mysteries.

The announcement came in a letter from Professor Enrico Bombieri. One could not have asked for a better, more respected source. Bombieri is one of the guardians of the Riemann Hypothesis and is based at the prestigious Institute for Advanced Study in Princeton, once home to Einstein and Gödel. He is very softly spoken, but mathematicians always listen carefully to anything he has to say.

Bombieri grew up in Italy, where his prosperous family's vineyards gave him a taste for the good things in life. He is fondly referred to by colleagues as 'the Mathematical Aristocrat'. In his youth he always cut a dashing figure at conferences in Europe, often arriving in a fancy sports car. Indeed, he was quite happy to fuel a rumour that he'd once come sixth in a twenty-four-hour rally in Italy. His successes on the mathematical circuit were more concrete and led to an invitation in the 1970s to go to Princeton, where he has remained ever since. He has replaced his enthusiasm for rallying with a passion for painting, especially portraits.

But it is the creative art of mathematics, and in particular the challenge of the Riemann Hypothesis, that gives Bombieri the greatest buzz. The Riemann Hypothesis had been an obsession for Bombieri ever since he first read about it at the precocious age of fifteen. He had always been fascinated by properties of numbers as he browsed through the mathematics books his father, an economist, had collected in his extensive library. The Riemann Hypothesis, he discovered, was regarded as the deepest and most fundamental problem in number theory. His passion for the problem was further fuelled when his father offered to buy him a Ferrari if he ever solved it – a desperate attempt on his father's part to stop Enrico driving his own model.

According to his email, Bombieri had been beaten to his prize. 'There are fantastic developments to Alain Connes's lecture at IAS last Wednesday,' Bombieri began. Several years previously, the mathematical world had been set alight by the news that Alain Connes had turned his attention to trying to crack the Riemann Hypothesis. Connes is one of the revolutionaries of the subject, a benign Robespierre of mathematics to Bombieri's Louis XVI. He is an extraordinarily charismatic figure whose fiery style is far from the image of the staid, awkward mathematician. He has the drive of a fanatic convinced of his world-view, and his lectures are mesmerising. Amongst his followers he has almost cult status. They will happily join him on the mathematical barricades to defend their hero against any counter-offensive mounted from the *ancien régime*'s entrenched positions.

Connes is based at France's answer to the Institute in Princeton, the Institut des Hautes Études Scientifiques in Paris. Since his arrival there in

Alain Connes, professor at the Institut des Hautes Études Scientifiques and at the Collège de France.

1979, he has created a completely new language for understanding geometry. He is not afraid to take the subject to the extremes of abstraction. Even the majority of the mathematical ranks who are usually at home with their subject's highly conceptual approach to the world have balked at the abstract revolution Connes is proposing. Yet, as he has demonstrated to those who doubt the necessity for such stark theory, his new language for geometry holds many clues to the real world of quantum physics. If it has instilled terror in the hearts of the mathematical masses, then so be it.

Connes's audacious belief that his new geometry could unmask not only the world of quantum physics but explain the Riemann Hypothesis – the greatest mystery about numbers – was met with surprise and even shock. It reflected his disregard for conventional boundaries that he dare venture into the heart of number theory and confront head-on the most difficult outstanding problem in mathematics. Since his arrival on the scene in the mid-nineties, there had been an expectancy in the air that if anyone had the resources to conquer this notoriously difficult problem, it was Alain Connes.

But it was not Connes who appeared to have found the last piece in the complex jigsaw. Bombieri went on to explain that a young physicist in the audience had seen 'in a flash' how to use his bizarre world of 'supersymmetric fermionic–bosonic systems' to attack the Riemann Hypothesis. Not many mathematicians knew quite what this cocktail of buzzwords meant, but Bombieri explained that it described 'the physics corresponding to a near-absolute zero ensemble of a mixture of anyons and morons with opposite spins'. It still sounded rather obscure, but then this was after all the solution to the most difficult problem in the history of mathematics, so no one was expecting an easy solution. According to Bombieri, after six days of uninterrupted work and with the help of a new computer language called MISPAR, the young physicist had finally cracked mathematics' toughest problem.

Bombieri concluded his email with the words, 'Wow! Please give this the highest diffusion.' Although it was extraordinary that a young physicist had ended up proving the Riemann Hypothesis, it came as no great surprise. Much of mathematics had found itself entangled with physics over the past few decades. Despite being a problem with its heart in the theory of numbers, the Riemann Hypothesis had for some years been showing unexpected resonances with problems in particle physics.

Mathematicians were changing their travel plans to fly in to Princeton to share the moment. Memories were still fresh with the excitement of a few years earlier when an English mathematician, Andrew Wiles, had announced a proof of Fermat's Last Theorem in a lecture delivered in Cambridge in June 1993. Wiles had proved that Fermat had been right in his claim that the equation $x^n + y^n = z^n$ has no solutions when n is bigger than 2. As Wiles laid down his chalk at the end of the lecture, the champagne bottles started popping and the cameras began flashing.

Mathematicians knew, however, that proving the Riemann Hypothesis would be of far greater significance for the future of mathematics than knowing that Fermat's equation has no solutions. As Bombieri had discovered at the tender age of fifteen, the Riemann Hypothesis seeks to understand the most fundamental objects in mathematics – prime numbers.

Prime numbers are the very atoms of arithmetic. The primes are those indivisible numbers that cannot be written as two smaller numbers multiplied together. The numbers 13 and 17 are prime, whilst 15 is not since it can be written as 3 times 5. The primes are the jewels studded throughout the vast expanse of the infinite universe of numbers that mathematicians have explored down the centuries. For mathematicians they instil a sense of wonder: 2, 3, 5, 7, 11, 13, 17, 19, 23, . . . – timeless numbers that exist in some world independent of our physical reality. They are Nature's gift to the mathematician.

Their importance to mathematics comes from their power to build all other numbers. Every number that is not a prime can be constructed by multiplying together these prime building blocks. Every molecule in the physical world can be built out of atoms in the periodic table of chemical elements. A list of the primes is the mathematician's own periodic table. The prime numbers 2, 3 and 5 are the hydrogen, helium and lithium in the mathematician's laboratory. Mastering these building blocks offers the mathematician the hope of discovering new ways of charting a course through the vast complexities of the mathematical world.

Yet despite their apparent simplicity and fundamental character, prime numbers remain the most mysterious objects studied by mathematicians. In a subject dedicated to finding patterns and order, the primes offer the ultimate challenge. Look through a list of prime numbers, and you'll find that it's impossible to predict when the next prime will appear. The list seems chaotic, random, and offers no clues as to how to determine the next number. The list of primes is the heartbeat of mathematics, but it is a pulse wired by a powerful caffeine cocktail:

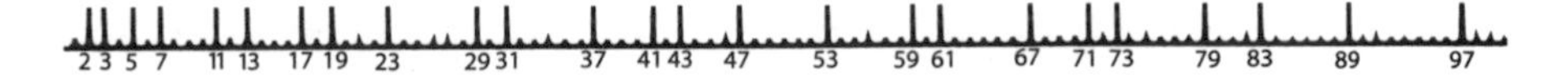

The prime numbers up to 100 – mathematics' irregular heartbeat.

Can you find a formula that generates the numbers in this list, some magic rule that will tell you what the 100th prime number is? This question has been plaguing mathematical minds down the ages. Despite over two thousand years of endeavour, prime numbers seem to defy attempts to fit them into a straightforward pattern. Generations have sat listening to the rhythm of the prime-number drum as it beats out its sequence of numbers: two beats, followed by three beats, five, seven, eleven. As the beat goes on, it becomes easy to believe that random white noise, without any inner logic, is responsible. At the centre of mathematics, the pursuit of order, mathematicians could only hear the sound of chaos.

Mathematicians can't bear to admit that there might not be an explanation for the way Nature has picked the primes. If there were no structure to mathematics, no beautiful simplicity, it would not be worth studying. Listening to white noise has never caught on as an enjoyable pastime. As the French mathematician Henri Poincaré wrote, 'The scientist does not study Nature because it is useful; he studies it because he delights in it, and he delights in it because it is beautiful. If Nature were not beautiful, it would not be worth knowing, and if Nature were not worth knowing, life would not be worth living.'

One might hope that the prime-number heartbeat settles down after a jumpy start. Not so – things just seem to get worse the higher you count. Here are the primes amongst the 100 numbers either side of 10,000,000. First, those below 10,000,000:

9,999,901 9,999,907, 9,999,929, 9,999,931, 9,999,937, 9,999,943, 9,999,971, 9,999,973, 9,999,991

But look now at how few there are in the 100 numbers above 10,000,000:

10,000,019, 10,000,079.

It is hard to guess at a formula that could generate this kind of pattern. In fact, this procession of primes resembles a random succession of numbers much more than it does a nice orderly pattern. Just as knowing the first 99 tosses of a coin won't help you much in guessing the result of the 100th toss, so do the primes seem to defy prediction.

Prime numbers present mathematicians with one of the strangest tensions in their subject. On the one hand a number is either prime or it isn't. No flip of a coin will suddenly make a number divisible by some smaller number. Yet there is no denying that the list of primes looks like a randomly chosen sequence of numbers. Physicists have grown used to the idea that a quantum die decides the fate of the universe, randomly choosing at each throw where scientists will find matter. But it is something of an embarrassment to have to admit that these fundamental numbers on which mathematics is based appear to have been laid out by Nature flipping a coin, deciding at each toss the fate of each number. Randomness and chaos are anathema to the mathematician.

Despite their randomness, prime numbers – more than any other part of our mathematical heritage – have a timeless, universal character. Prime numbers would be there regardless of whether we had evolved sufficiently to recognise them. As the Cambridge mathematician G.H. Hardy said in his famous book *A Mathematician's Apology*, '317 is a prime not because we think so, or because our minds are shaped in one way or another, but *because it is so*, because mathematical reality is built that way.'

Some philosophers might take issue with such a Platonist view of the world – this belief in an absolute and eternal reality beyond human existence – but to my mind that is what makes them philosophers and not mathematicians. There is a fascinating dialogue between Alain Connes, the mathematician who featured in Bombieri's email, and the neurobiologist Jean-Pierre Changeux in *Conversations on Mind, Matter and Mathematics*. The tension in this book is palpable as the mathematician argues for the existence of mathematics outside the mind, and the neurologist is determined to refute any such idea: 'Why wouldn't we see "$\pi = 3.1416$" written in gold letters in the sky or "6.02×10^{23}" appear in the reflections of a crystal ball?' Changeux declares his frustration at Connes's insistence that 'there exists, independently of the human mind, a raw and immutable mathematical reality' and at the heart of that world we find the unchanging list of primes. Mathematics, Connes declares, 'is unquestionably the only *universal* language'. One can imagine a different chemistry or biology on the other side of the universe, but prime numbers will remain prime whichever galaxy you are counting in.

In Carl Sagan's classic novel *Contact*, aliens use prime numbers to contact life on earth. Ellie Arroway, the book's heroine, has been working at SETI, the Search for Extraterrestrial Intelligence, listening to the crackle of the cosmos. One night, as the radio telescopes are turned towards Vega, they suddenly pick up strange pulses through the background noise. It takes Ellie no time to recognise the drumbeat in this radio signal. Two pulses are followed by a pause, then three pulses, five, seven, eleven, and so on through all the prime numbers up to 907. Then it starts all over again.

This cosmic drum was playing a music that earthlings couldn't fail to recognise. Ellie is convinced that only intelligent life could generate this beat: 'It's hard to imagine some radiating plasma sending out a regular set of mathematical signals like this. The prime numbers are there to attract our attention.' Had the alien culture transmitted the previous ten years of alien winning lottery numbers, Ellie couldn't have distinguished them from the background noise. Even though the list of primes looks as random a list as the lottery winnings, its universal constancy has determined the choice of each number in this alien broadcast. It is this structure that Ellie recognises as the sign of intelligent life.

Communicating using prime numbers is not just science fiction. Oliver Sacks in his book *The Man Who Mistook His Wife for a Hat* documents twenty-six-year-old twin brothers, John and Michael, whose deepest form of communication was to swap six-digit prime numbers. Sacks tells of when he first discovered them secretly exchanging numbers in the corner of a room: 'they looked, at first, like two connoisseurs wine-tasting, sharing rare tastes, rare appreciations'. At first, Sacks can't figure out what the twins are up to. But as soon as he cracks their code, he memorises some eight-digit primes which he drops surreptitiously into the conversation at their next meeting. The twins' surprise is followed by deep concentration which turns to jubilation as they recognise another prime number. Whilst Sacks had resorted to prime number tables to find his primes, how the twins were generating their primes is a tantalising puzzle. Could it be that these autistic-savants were in possession of some secret formula that generations of mathematicians had missed?

The story of the twins is a favourite of Bombieri's.

> It is hard for me to hear this story without feeling awe and astonishment at the workings of the brain. But I wonder: Do my non-mathematical friends have the same response? Do they have any inkling how bizarre, how prodigious and even other-worldly was the singular talent the twins so naturally enjoyed? Are they aware that mathematicians have been struggling for centuries to come up with a way to do what John and Michael did spontaneously: to generate and recognize prime numbers?

Before anyone could find out how they were doing it, the twins were separated at the age of thirty-seven by their doctors, who believed that their private numerological language had been hindering their development. Had they listened to the arcane conversations that can be heard in the common rooms of university maths departments, these doctors would probably have recommended closing them down too.

It's likely that the twins were using a trick based on what's called Fermat's Little Theorem to test whether a number is prime. The test is similar to the way in which autistic-savants can quickly identify that April 13, 1922, for instance, was a Thursday – a feat the twins performed regularly on TV chat shows. Both tricks depend on doing something called clock or modular arithmetic. Even if they lacked a magic formula for the primes, their skill was still extraordinary. Before they were separated they had reached twenty-digit numbers, well beyond the upper limit of Sacks's prime number tables.

Like Sagan's heroine listening to the cosmic prime number beat and Sacks eavesdropping on the prime number twins, mathematicians for centuries had been straining to hear some order in this noise. Like Western ears listening to the music of the East, nothing seemed to make sense. Then, in the middle of the nineteenth century, came a major breakthrough. Bernhard Riemann began to look at the problem in a completely new way. From his new perspective, he began to understand something of the pattern responsible for the chaos of the primes. Underlying the outward noise of the primes was a subtle and unexpected harmony. Despite this great step forward, this new music kept many of its secrets out of earshot. Riemann, the Wagner of the mathematical world, was undaunted. He made a bold prediction about the mysterious music that he had discovered. This prediction is what has become known as the Riemann Hypothesis. Whoever proves that Riemann's intuition about the nature of this music was right will have explained why the primes give such a convincing impression of randomness.

Riemann's insight followed his discovery of a mathematical looking-glass through which he could gaze at the primes. Alice's world was turned upside down when she stepped through her looking-glass. In contrast, in the strange mathematical world beyond Riemann's glass, the chaos of the primes seemed to be transformed into an ordered pattern as strong as any mathematician could hope for. He conjectured that this order would be maintained however far one stared into the never-ending world beyond the glass. His prediction of an inner harmony on the far side of the mirror would explain why outwardly the primes look so chaotic. The metamorphosis provided by Riemann's mirror, where chaos turns to order, is one which most mathematicians find almost miraculous. The challenge that Riemann left the mathematical world was to prove that the order he thought he could discern was really there.

Bombieri's email of April 7, 1997, promised the beginning of a new era. Riemann's vision had not been a mirage. The Mathematical Aristocrat had offered mathematicians the tantalising possibility of an explanation for the apparent chaos in the primes. Mathematicians were keen to loot the many other treasures they knew should be unearthed by the solution to this great problem.

A solution of the Riemann Hypothesis will have huge implications for many other mathematical problems. Prime numbers are so fundamental to the working mathematician that any breakthrough in understanding their nature will have a massive impact. The Riemann Hypothesis seems unavoidable as a problem. As mathematicians navigate their way across the mathematical terrain, it is as though all paths will necessarily lead at some point to the same awesome vista of the Riemann Hypothesis.

Many people have compared the Riemann Hypothesis to climbing Mount Everest. The longer it remains unclimbed, the more we want to conquer it. And the mathematician who finally scales Mount Riemann will certainly be remembered longer than Edmund Hillary. The conquest of Everest is marvelled at not because the top is a particularly exciting place to be, but because of the challenge it poses. In this respect the Riemann Hypothesis differs significantly from the ascent of the world's tallest peak. Riemann's peak is a place we all want to sit upon because we already know the vistas that will open up to us should we make it to the top. The person who proves the Riemann Hypothesis will have made it possible to fill in the missing gaps in thousands of theorems that rely on it being true. Many mathematicians have simply had to assume the truth of the Hypothesis in reaching their own goals.

The dependence of so many results on Riemann's challenge is why mathematicians refer to it as a hypothesis rather than a conjecture. The word 'hypothesis' has the much stronger connotation of a necessary assumption that a mathematician makes in order to build a theory. 'Conjecture', in contrast, represents simply a prediction of how mathematicians believe their world behaves. Many have had to accept their inability to solve Riemann's riddle and have simply adopted his prediction as a working hypothesis. If someone can turn the hypothesis into a theorem, all those unproven results would be validated.

By appealing to the Riemann Hypothesis, mathematicians are staking their reputations on the hope that one day someone will prove that Riemann's intuition was correct. Some go further than just adopting it as a working hypothesis. Bombieri regards it as an article of faith that the primes behave as Riemann's Hypothesis predicts. It has become virtually a cornerstone in the pursuit of mathematical truth. If, however, the Riemann Hypothesis turns out to be false, it will completely destroy the faith we have in our intuition to sniff out the way things work. So convinced have we become that Riemann was right that the alternative will require a radical revision of our view of the mathematical world. In particular, all the results that we believe exist beyond Riemann's peak would disappear in a puff of smoke.

Most significantly, a proof of the Riemann Hypothesis would mean that mathematicians could use a very fast procedure guaranteed to locate a prime number with, say, a hundred digits or any other number of digits you care to choose. You might legitimately ask, 'So what?' Unless you are a mathematician such a result looks unlikely to have a major impact on your life.

Finding hundred-digit primes sounds as pointless as counting angels on a pinhead. Although most people recognise that mathematics underlies the construction of an aeroplane or the development of electronics technology, few would expect the esoteric world of prime numbers to have much impact on their lives. Indeed, even in the 1940s G.H. Hardy was of the same mind: 'both Gauss and lesser mathematicians may be justified in rejoicing that here is one science [number theory] at any rate whose very remoteness from ordinary human activities should keep it gentle and clean'.

But a more recent turn of events has seen prime numbers take centre stage in the rough and dirty world of commerce. No longer are prime numbers confined to the mathematical citadel. In the 1970s, three scientists, Ron Rivest, Adi Shamir and Leonard Adleman, turned the pursuit of prime numbers from a casual game played in the ivory towers of academia into a serious business application. By exploiting a discovery made by Pierre de Fermat in the seventeenth century, these three found a way to use the primes to protect our credit card numbers as they travel through the electronic shopping malls of the global marketplace. When the idea was first proposed in the 1970s, no one had any idea how big e-business would turn out to be. But today, without the power of prime numbers there is no way this business could exist. Every time you place an order on a website, your computer is using the security provided by the existence of prime numbers with a hundred digits. The system is called RSA after its three inventors. So far, over a million primes have already been put to use to protect the world of electronic commerce.

Every business trading on the Internet therefore depends on prime numbers with a hundred digits to keep their business transactions secure. The expanding role of the Internet will ultimately lead to each of us being uniquely identified by our very own prime numbers. Suddenly there is a commercial interest in knowing how a proof of the Riemann Hypothesis might help in understanding how primes are distributed throughout the universe of numbers.

The extraordinary thing is that although the *construction* of this code depends on discoveries about primes made by Fermat over three hundred years ago, to *break* this code depends on a problem that we still can't answer. The security of RSA depends on our inability to answer basic questions about prime numbers. Mathematicians know enough about the primes to build these Internet codes, but not enough to break them. We can understand one half of the equation but not the other. The more we demystify the primes, however, the less secure these Internet codes are becoming. These numbers are the keys to the locks that protect the world's electronic secrets. This is why companies such as AT&T and Hewlett-Packard are ploughing money into endeavours to understand the subtleties of prime numbers and the Riemann Hypothesis. The insights gained could help to break these prime number codes, and all companies with an Internet presence want to be the first to know when their codes become insecure. And this is the reason why number theory and business have become such strange bedfellows. Business and security agencies are keeping a watchful eye on the blackboards of the pure mathematicians.

So it wasn't only the mathematicians who were getting excited about Bombieri's announcement. Was this solution of the Riemann Hypothesis going to cause a meltdown of e-business? Agents from the NSA, the US National Security Agency, were dispatched to Princeton to find out. But as mathematicians and security agents made their way to New Jersey, a number of people began to smell something fishy in Bombieri's email. Fundamental particles have been given some crazy names – gluons, cascade hyperons, charmed mesons, quarks, the last of these courtesy of James Joyce's *Finnegans Wake*. But 'morons'? Surely not! Bombieri has an unrivalled reputation for appreciating the ins and outs of the Riemann Hypothesis, but those who know him personally are also aware of his wicked sense of humour.

Fermat's Last Theorem had fallen foul of an April Fool prank that emerged just after a gap had appeared in the first proof that Andrew Wiles had proposed in Cambridge. With Bombieri's email, the mathematical community had been duped again. Eager to relive the buzz of seeing Fermat proved, they had grabbed the bait that Bombieri had thrown at them. And the delights of forwarding email meant that the first of April had disappeared from the original source as it rapidly disseminated. This, combined with the fact that the email was read in countries with no concept of April Fool's Day, made the prank far more successful than Bombieri could have imagined. He finally had to own up that his email was a joke. As the twenty-first century approached, we were still com-

Enrico Bombieri, professor at the Institute
for Advanced Study, Princeton.

pletely in the dark as to the nature of the most fundamental numbers in mathematics. It was the primes that had the last laugh.

Why had mathematicians been so gullible that they believed Bombieri? It's not as though they give up their trophies lightly. The stringent tests that mathematicians require to be passed before a result can be declared proven far exceed those deemed sufficient in other subjects. As Wiles realised when a gap appeared in his first proof of Fermat's Last Theorem, completing 99 per cent of the jigsaw is not enough: it would be the person who put in the last piece who would be remembered. And the last piece can often remain hidden for years.

The search for the secret source that fed the primes had been going on for over two millennia. The yearning for this elixir had made mathematicians all too susceptible to Bombieri's ruse. For years, many had simply been too frightened to go anywhere near this notoriously difficult problem. But it was striking how, as the century drew to a close, more and more mathematicians were prepared to talk about attacking it. The proof of Fermat's Last Theorem only helped to fuel the expectation that great problems could be solved.

Mathematicians had enjoyed the attention that Wiles's solution to Fermat had brought them as mathematicians. This feeling undoubtedly contributed to the desire to believe Bombieri. Suddenly, Andrew Wiles was being asked to model chinos for Gap. It felt good. It felt almost sexy to be a mathematician. Mathematicians spend so much time in a world that fills them with excitement and pleasure. Yet it is a pleasure they rarely have the opportunity to share with the rest of the world. Here was a chance to flaunt a trophy, to show off the treasures that their long, lonely journeys had uncovered.

A proof of the Riemann Hypothesis would have been a fitting mathematical climax to the twentieth century. The century had opened with Hilbert's direct challenge to the world's mathematicians to crack this enigma. Of the twenty-three problems on Hilbert's list, the Riemann Hypothesis was the only problem to make it into the new century unvanquished.

On May 24, 2000, to mark the 100th anniversary of Hilbert's challenge, mathematicians and the press gathered in the Collège de France in Paris to hear the announcement of a fresh set of seven problems to challenge the mathematical community for the new millennium. They were proposed by a small group of the world's finest mathematicians, including Andrew Wiles and Alain Connes. The seven problems were new except for one that had appeared on Hilbert's list: the Riemann Hypothesis. In obeisance to the capitalist ideals that shaped the twentieth century, these challenges come with some extra spice. The Riemann Hypothesis and the other six problems now have a price tag of one million dollars apiece. Incentive indeed for Bombieri's fictional young physicist – if glory weren't enough.

The idea for the Millennium Problems was the brainchild of Landon T. Clay, a Boston businessman who made his money in trading mutual funds on a buoyant stock market. Despite dropping out of mathematics at Harvard he has a passion for the subject, a passion he wants to share. He realises that money is not the motivating force for mathematicians: 'It's the desire for truth and the response to the beauty and power and elegance of mathematics that drive mathematicians.' But Clay is not naive, and as a businessman he knows how a million dollars might inspire another Andrew Wiles to join the chase for the solutions of these great unsolved problems. Indeed, the Clay Mathematics Institute's website, where the Millennium Problems were posted, was so overwhelmed by hits the day after the announcement that it collapsed under the strain.

The seven Millennium Problems are different in spirit to the twenty-three problems chosen a century before. Hilbert had set a new agenda

The Collège de France.

for mathematicians in the twentieth century. Many of his problems were original and encouraged a significant shift in attitudes towards the subject. Rather than focusing on the particular, like Fermat's Last Theorem, Hilbert's twenty-three problems inspired the community to think more conceptually. Instead of picking over individual rocks in the mathematical landscape, Hilbert offered mathematicians the chance of a balloon flight high above their subject to encourage them to understand the overarching lay of the land. This new approach owes a lot to Riemann, who fifty years before had begun this revolutionary shift from mathematics as a subject of formulas and equations to one of ideas and abstract theory.

The choice of the seven problems for the new millennium was more conservative. They are the Turners in the mathematical gallery of problems, whereas Hilbert's questions were a more modernist, avant-garde collection. The conservatism of the new problems was partly because their solutions were expected to be sufficiently clear cut for their solvers to be awarded the million-dollar prize. The Millennium Problems are questions that mathematicians have known about for some decades, and in the case of the Riemann Hypothesis, over a century. They are a classic selection.

Clay's seven million dollars is not the first time that money has been offered for solutions to mathematical problems. In 1997 Wiles picked up 75,000 Deutschmarks for his proof of Fermat's Last Theorem, thanks to a prize offered in 1908 by Paul Wolfskehl. The story of the Wolfskehl Prize is what had brought Fermat to Wiles's attention at the impressionable age of ten. Clay believes that if he can do the same for the Riemann Hypothesis, it will be a million dollars well spent. More recently, two publishing houses, Faber & Faber in the UK and Bloomsbury in the USA, offered a million dollars for a proof of Goldbach's Conjecture as a publicity stunt to launch their publication of Apostolos Doxiadis's novel *Uncle Petros and Goldbach's Conjecture*. To earn the money you had to explain why every even number can be written as the sum of two prime numbers. However, the publishers didn't give you much time to crack it. The solution had to be submitted before midnight on March 15, 2002, and was bizarrely open only to US and UK residents.

Clay believes that mathematicians receive little reward or recognition for their labours. For example, there is no Nobel Prize for Mathematics that they can aspire to. Instead, the award of a Fields Medal is considered the ultimate prize in the mathematical world. In contrast to Nobel prizes, which tend to be awarded to scientists at the end of their careers for achievements long past, Fields Medals are restricted to mathematicians below the age of forty. This is not because of the generally held belief that mathematicians burn out at an early age. John Fields, who conceived of the idea and provided funds for the prize, wanted its award to spur on the most promising mathematicians to even greater achievements. The medals are awarded every four years on the occasion of the International Congress of Mathematicians. The first ones were awarded in Oslo in 1936.

The age limit is strictly adhered to. Despite Andrew Wiles's extraordinary achievement in proving Fermat's Last Theorem, the Fields Medal committee weren't able to award him a medal at the Congress in Berlin in 1998, the first opportunity after the final proof was accepted, for he was born in 1953. They did have a special medal struck to honour Wiles's achievement. But it still does not compare to being a member of the illustrious club of Fields Medal winners. The recipients include many of the key players in our drama: Enrico Bombieri, Alain Connes, Atle Selberg, Paul Cohen, Alexandre Grothendieck, Alan Baker, Pierre Deligne. Those names account for nearly a fifth of the medals ever awarded.

But it is not for the money that mathematicians aspire to these medals. In contrast to the big bucks behind the Nobel prizes, the purse that accompanies a Fields Medal contains a modest 15,000 Canadian dollars. So Clay's millions will help compete with the monetary kudos of the Nobel prizes. In contrast to Fields Medals and the Faber–Bloomsbury Goldbach prize, the money is there regardless of age or nationality, and with no time limits for a solution, except for the ticking clock of inflation.

However, the greatest incentive for the mathematician chasing one of the Millennium problems is not the monetary reward but the intoxicating prospect of the immortality that mathematics can bestow. Solving one of Clay's problems may earn you a million dollars, but that is nothing compared with carving your name on civilisation's intellectual map. The Riemann Hypothesis, Fermat's Last Theorem, Goldbach's Conjecture, Hilbert space, the Ramanujan tau function, Euclid's algorithm, the

Hardy–Littlewood Circle Method, Fourier series, Gödel numbering, a Siegel zero, the Selberg trace formula, the sieve of Eratosthenes, Mersenne primes, the Euler product, Gaussian integers – these discoveries have all immortalised the mathematicians who have been responsible for unearthing these treasures in our exploration of the primes. Those names will live on long after we have forgotten the likes of Aeschylus, Goethe and Shakespeare. As G.H. Hardy explained, 'languages die and mathematical ideas do not. "Immortality" may be a silly word, but probably a mathematician has the best chance of whatever it may mean.'

Those mathematicians who have laboured long and hard on this epic journey to understand the primes are more than just names set in mathematical stone. The twists and turns that the story of the primes has taken are the products of real lives, of a dramatis personae rich and varied. Historical figures from the French revolution and friends of Napoleon give way to modern-day magicians and Internet entrepreneurs. The stories of a clerk from India, a French spy spared execution and a Jewish Hungarian fleeing the persecution of Nazi Germany are bound together by an obsession with the primes. All these characters bring a unique perspective in their attempt to add their name to the mathematical roll call. The primes have united mathematicians across many national boundaries: China, France, Greece, America, Norway, Australia, Russia, India and Germany are just a few of the countries from which have come prominent members of the nomadic tribe of mathematicians. Every four years they converge to tell the stories of their travels at an International Congress.

It is not only the desire to leave a footprint in the past which motivates the mathematician. Just as Hilbert dared to look forward into the unknown, a proof of the Riemann Hypothesis would be the start of a new journey. When Wiles addressed the press conference at the announcement of the Clay prizes he was keen to stress that the problems are not the final destination:

> There is a whole new world of mathematics out there, waiting to be discovered. Imagine if you will, the Europeans in 1600. They know that across the Atlantic there is a New World. How would they have assigned prizes to aid in the discovery and development of the United States? Not a prize for inventing the airplane, not a prize for inventing the computer, not a prize for founding Chicago, not a prize for machines that would harvest areas of wheat. These things have become a part of America, but such things could not have been imagined in 1600. No, they would have given a prize for solving such problems as the problem of longitude.

The Riemann Hypothesis is the longitude of mathematics. A solution to the Riemann Hypothesis offers the prospect of charting the misty waters of the vast ocean of numbers. It represents just a beginning in our understanding of Nature's numbers. If we can only find the secret of how to navigate the primes, who knows what else lies out there, waiting for us to discover?

CHAPTER TWO

The Atoms of Arithmetic

When things get too complicated, it sometimes makes sense to stop and wonder: Have I asked the right question? Enrico Bombieri, 'Prime Territory' in *The Sciences*

Two centuries before Bombieri's April Fool had teased the mathematical world, equally exciting news was being trumpeted from Palermo by another Italian, Giuseppe Piazzi. From his observatory Piazzi had detected a new planet that orbited the Sun somewhere between the orbits of Mars and Jupiter. Christened Ceres, it was much smaller than the seven major planets then known, but its discovery on January 1, 1801, was regarded by everyone as a great omen for the future of science in the new century.

Excitement turned to despair a few weeks later as the small planet disappeared from view as its orbit took it around the other side of the Sun, where its feeble light was drowned out by the Sun's glare. It was now lost to the night sky, hidden once again amongst the plethora of stars in the firmament. Nineteenth-century astronomers lacked the mathematical tools for calculating its complete path from the short trajectory they had been able to track during the first few weeks of the new century. It seemed that they had lost the planet and had no way of predicting where it would next appear.

However, nearly a year after Piazzi's planet had vanished, a twenty-four-year-old German from Brunswick announced that he knew where astronomers should find the missing object. With no alternative prediction to hand, astronomers aimed their telescopes at the region of the night sky to which the young man had pointed. As if by magic, there it was. This unprecedented astronomical prediction was not, however, the mysterious magic of an astrologer. The path of Ceres had been worked out by a mathematician who had found patterns where others had only seen a tiny, unpredictable planet. Carl Friedrich Gauss had taken the minimal data that had been recorded for the planet's path and applied a new method he had recently developed to estimate where Ceres could be found at any future date.

The discovery of Ceres' path made Gauss an overnight star within the scientific community. His achievement was a symbol of the predictive power of mathematics in the burgeoning scientific age of the early nineteenth century. Whereas the astronomers had discovered the planet by chance, it was a mathematician who had brought to bear the necessary analytic skills to explain what was going to happen next.

Although Gauss's name was new to the astronomical fraternity, he had already made his mark as a formidable new voice in the mathematical world. He had successfully plotted the trajectory of Ceres, but his real passion was for finding patterns in the world of numbers. For Gauss, the universe of numbers presented the ultimate challenge: to find structure and order where others could only see chaos. 'Child prodigy' and 'mathematical genius' are titles that are bandied about far too often, but there are few mathematicians who would argue with these labels being attached to Gauss. The sheer number of new ideas and discoveries that he produced before he was even twenty-five seems to defy explanation.

Gauss was born into a labourer's family in Brunswick, Germany, in 1777. At the age of three he was correcting his father's arithmetic. At the age of nineteen, his discovery of a beautiful geometric construction of a 17-sided shape convinced him that he should dedicate his life to mathematics. Before Gauss, the Greeks had shown how to use a compass and straight edge to construct a perfect pentagon. No one since had been able to show how to use this simple equipment to construct other perfect, so-called regular polygons with a prime number of sides. The excitement that Gauss experienced when he found a way to build this perfect 17-sided shape prompted him to start a mathematical diary which he would keep for the next eighteen years. This diary, which remained in the family's hands until 1898, has become one of the most important documents in the history of mathematics, not least because it confirmed that Gauss had proved, but failed to publish, many results that it took other mathematicians well into the nineteenth century to rediscover.

One of Gauss's greatest early contributions was the invention of the clock calculator. This was an idea, rather than a physical machine, that unleashed the possibility of doing arithmetic with numbers that had previously been considered too unwieldy. The clock calculator works on exactly the same principle as a conventional clock. If your clock says it's 9 o'clock, and you add 4 hours, the hour hand moves round to 1 o'clock. Gauss's clock calculator would therefore return the answer 1 rather than 13. If Gauss wants to do a more complicated calculation such as 7×7, the clock calculator would come up with the remainder that is left after dividing $49 = 7 \times 7$ by 12. The result would again be 1 o'clock.

It is when Gauss wants to calculate the value of $7 \times 7 \times 7$ that the power and speed of the clock calculator begins to emerge. Instead of multiplying 49 by 7 again, Gauss can just multiply the last answer (which

Carl Friedrich Gauss (1777–1855).

was 1) by 7 to get the answer 7. So without having to calculate what 7 × 7 × 7 was (which happens to be 343), he still knew with little effort that it gave remainder 7 on division by 12. The power of the calculator came into its own when Gauss started exploring big numbers that lay beyond his computational reach. Although he had no idea what 7^{99} was, his clock calculator told him that the number gave remainder 7 on division by 12.

Gauss saw that there was nothing special about clocks with 12 hours on their face. He introduced the idea of doing clock arithmetic, sometimes called modular arithmetic, with any number of hours on the clock face. So, for example, if you enter 11 into a clock calculator divided into 4 hours, the answer is 3 o'clock since 11 leaves remainder 3 on division by 4. Gauss's account of this new sort of arithmetic revolutionised mathematics at the turn of the nineteenth century. Just as the telescope had allowed astronomers to see new worlds, the development of the clock calculator helped mathematicians to discover in the universe of numbers new patterns which had been hidden from view for generations. Even today, Gauss's clocks are central to the security of the Internet, which utilises calculators whose clock faces bear more hours than there are atoms in the observable universe.

Gauss, the child of a poor family, was lucky to get the chance to capitalise on his mathematical talent. He was born into an age when mathematics was still a privileged pursuit funded by noble courts and patrons, or practised by amateurs such as Pierre de Fermat in their spare time. Gauss's patron was the Duke of Brunswick, Carl Wilhelm Ferdinand. Ferdinand's family had always supported the culture and economy of their dukedom. Indeed, his father had founded the Collegium Carolinum, one of the oldest technical universities in Germany. Ferdinand was imbued with his father's ethos that education was the foundation of Brunswick's commercial successes, and he was always on the lookout for talent deserving of support. Ferdinand first came across Gauss in 1791, and was so impressed with his abilities that he offered to finance the young man to attend the Collegium Carolinum so that he could realise his obvious potential.

It was with much gratitude that Gauss dedicated his first book to the duke in 1801. This book, entitled *Disquisitiones Arithmeticae*, collected together many of Gauss's discoveries about the properties of numbers that he had recorded in his diaries. It is generally acknowledged as the book that heralded the birth of number theory as a subject in its own right, not just a ragbag collection of observations about numbers. Its publication is responsible for making the subject of number theory, as Gauss always liked to call it, 'the Queen of Mathematics'. For Gauss, the jewels in the crown were the primes, numbers which had fascinated and teased generations of mathematicians.

The first tentative evidence that humankind knew about the special qualities of prime numbers is a bone that dates from 6500 BC. Called the Ishango bone, it was discovered in 1960 in the mountains of central equatorial Africa. Marked on it are three columns of four notches. In one of the columns we find 11, 13, 17 and 19 notches, a list of all the primes between 10 and 20. The other columns do seem to be of a mathematical nature. It is unclear whether this bone, housed in Belgium's Royal Institute for Natural Sciences in Brussels, truly represents our ancestors' first attempts to understand the primes or whether the carvings are a random selection of numbers which just happen to be prime. Nevertheless, this ancient bone is perhaps intriguing and tantalising evidence for the first foray into the theory of prime numbers.

Some believe that the Chinese were the first culture to hear the beating of the prime number drum. They attributed female characteristics to even numbers and male to odd numbers. In addition to this straight divide they also regarded those odd numbers that are not prime, such as 15, as effeminate numbers. There is evidence that by 1000 BC they had evolved a very physical way of understanding what it is, amongst all the numbers, that makes prime numbers special. If you take 15 beans, you can arrange them in a neat rectangular array made up of three rows of five beans. Take 17 beans, though, and the only rectangle you can make is one with a single row of 17 beans. For the Chinese, the primes were macho numbers which resisted any attempt to break them down into a product of smaller numbers.

The ancient Greeks also liked to attribute sexual qualities to numbers, but it was they who first discovered, in the 4th century BC, the primes' true potency as the building blocks for all numbers. They saw that every number could be constructed by multiplying prime numbers together. Whilst the Greeks mistakenly believed fire, air, water and earth to be the building blocks of matter, they were spot on when it came to identifying the atoms of arithmetic. For many centuries, chemists strove to identify the basic constituents of their subject, and the Greeks' intuition finally culminated in Dmitri Mendeleev's Periodic Table, a complete description of the elements of chemistry. In contrast to the Greeks' head start in identifying the building blocks of arithmetic, mathematicians are still floundering in their attempts to understand their own table of prime numbers.

The librarian of the great ancient Greek research institute in Alexandria was the first person we know of to have produced tables of primes. Like some ancient mathematical Mendeleev, Eratosthenes in the third century BC discovered a reasonably painless procedure for determining which numbers are prime in a list of, say, the first 1,000 numbers. He began by writing out all the numbers from 1 to 1,000. He then took the first prime, 2, and struck off every second number in the list. Since all these numbers were divisible by 2, they weren't prime. He then moved to the next number that hadn't been struck off, namely 3. He then stuck off every third number after 3. Since these were all divisible by 3, they weren't prime either. He kept doing this, just picking up the next number which hadn't already been struck from the list and striking off all the numbers divisible by the new prime. By this systematic process he produced tables of primes. The procedure was later christened *the sieve of Eratosthenes*. Each new prime creates a 'sieve' which Eratosthenes uses to eliminate non-primes. The size of the sieve changes at each stage, but by the time he reaches 1,000 the only numbers to have made it through all the sieves are prime numbers.

When Gauss was a young boy he was given a present – a book containing a list of the first several thousand prime numbers which had probably been constructed using these ancient number sieves. To Gauss, these numbers just tumbled around randomly. Predicting the elliptical path of Ceres would be difficult enough. But the challenge posed by the primes had more in common with the near-impossible task of analysing the rotation of bodies such as Hyperion, one of Saturn's satellites, which is shaped like a hamburger. In contrast to the Earth's Moon, Hyperion is far from gravitationally stable and spins chaotically. Even though the spinning of Hyperion and the orbits of some asteroids are chaotic, at least it is known that their behaviour is determined by the gravitational pull of the Sun and the planets. But for the primes, no one had the faintest idea what was pulling and pushing these numbers around. As he gazed at his table of numbers, Gauss could see no rule that told him how far to jump to find the next prime. Were mathematicians just going to have to accept these numbers as determined by Nature, set like stars in the night sky with no rhyme or reason? Such a position was unacceptable to Gauss. The primary drive for the mathematician's existence is to find patterns, to discover and explain the rules underlying Nature, to predict what will happen next.

The search for patterns

The mathematician's quest for primes is captured perfectly by one of the tasks we have all faced at school. Given a list of numbers, find the next number. For example, here are three challenges:

1, 3, 6, 10, 15, . . .
1, 1, 2, 3, 5, 8, 13, . . .
1, 2, 3, 5, 7, 11, 15, 22, 30, . . .

Numerous questions spring to the mathematical mind when faced with such lists. What is the rule behind the creation of each list? Can you predict the next number on the list? Can you find a formula that will produce the 100th number on the list without having to calculate the first 99 numbers?

The first sequence of numbers above consists of what are called the *triangular numbers*. The tenth number on the list is the number of beans required to build a triangle with ten rows, starting with one bean in the first row and ending with ten beans in the last row. So the Nth triangular number is got by simply adding the first N numbers: $1 + 2 + 3 + \ldots + N$.

If you want to find the 100th triangular number, there is a long laborious method in which you attack the problem head on and add up the first 100 numbers.

Indeed, Gauss's schoolteacher liked to set this problem for his class, knowing that it always took his students so long that he could take forty winks. As each student finished the task they were expected to come and place their slate tablets with their answer written on it in a pile in front of the teacher. While the other students began labouring away, within seconds the ten-year-old Gauss had laid his tablet on the table. Furious, the teacher thought that the young Gauss was being cheeky. But when he looked at Gauss's slate, there was the answer – 5,050 – with no steps in the calculation. The teacher thought that Gauss must have cheated somehow, but the pupil explained that all you needed to do was put $N = 100$ into the formula $\frac{1}{2} \times (N + 1) \times N$ and you will get the 100th number in the list without having to calculate any other numbers on the list on the way.

Rather than tackling the problem head on, Gauss had thought laterally. He argued that the best way to discover how many beans there were in a triangle with 100 rows was to take a second similar triangle of beans which could be placed upside down on top of the first triangle. Now Gauss had a rectangle with 101 rows each containing 100 beans. Calculating the total number of beans in this rectangle built from the two triangles was easy: there are in total $101 \times 100 = 10{,}100$ beans. So one triangle must contain half this number, namely $\frac{1}{2} \times 101 \times 100 = 5{,}050$. There is nothing special here about 100. Replace it by N and you get the formula $\frac{1}{2} \times (N + 1) \times N$.

The picture overleaf illustrates the argument for the triangle with 10 rows instead of 100.

Instead of directly attacking his teacher's problem, Gauss had found a different angle from which to view the calculation. Lateral thinking, turning the problem upside down or inside out to see it from a new perspective, is an immensely important theme in mathematical discovery and is one reason why people who can think like the young Gauss make good mathematicians.

The second challenge sequence, 1, 1, 2, 3, 5, 8, 13, . . ., consists of the so-called *Fibonacci numbers*. The rule behind this sequence is that each new number is calculated by adding the two previous ones, for example, $13 = 5 + 8$. Fibonacci, a mathematician in the thirteenth-century court in Pisa, had struck upon the sequence in relation to the mating habits of rabbits. He had tried to bring European mathematics out of the Dark Ages by proselytising the discoveries of Arabic mathematicians. He failed.

An illustration of Gauss's proof of his formula for the triangular numbers.

Instead, it was the rabbits that immortalised him in the mathematical world. His model of the propagation of rabbits predicted that each new season would see the number of pairs of rabbits grow in a certain pattern. This pattern was based on two rules: each mature pair of rabbits will produce a new pair of rabbits each season, and each new pair will take one season to reach sexual maturity.

But it is not only in the rabbit world that these numbers prevail. This sequence of numbers crops up in all manner of natural ways. The number of petals on a flower invariably is a Fibonacci number, as is the number of spirals in a fir cone. The growth of a seashell over time reflects the progression of the Fibonacci numbers.

Is there a fast formula like Gauss's formula for the triangular numbers that will produce the 100th Fibonacci number? Again, at first sight it looks as though we might have to calculate all the previous 99 numbers since the way to get the 100th number is to add together the two previous ones. Is it possible that there is a formula that could give us this 100th number just by plugging the number 100 into the formula? This turns out be much trickier, despite the simplicity of the rule for generating these numbers.

The formula for generating the Fibonacci numbers is based upon a special number called the *golden ratio*, a number which begins 1.618 03. . . Like the number π, the golden ratio is a number whose decimal expansion continues without end, demonstrating no patterns. Yet it encapsulates what many people down the centuries have regarded as perfect proportions. If you examine the canvases in the Louvre or the Tate Gallery, you'll find that very often the artist will have chosen a rectangle whose sides are in a ratio of 1 to 1.618 03 . . . Experiment reveals that a person's height when compared to the distance from their feet to their belly button favours the same ratio. The golden ratio is a number which appears in Nature in an uncanny fashion. Despite its chaotic decimal expansion, this number also holds the key to generating the Fibonacci numbers. The Nth Fibonacci number can be expressed by a formula built from the Nth power of the golden ratio.

I will leave the third sequence of numbers, 1, 2, 3, 5, 7, 11, 15, 22, 30, . . ., as a teasing challenge which I will return to later. Its properties helped cement the fame of one of the most intriguing mathematicians of the twentieth century, Srinivasa Ramanujan, who had an extraordinary ability to discover new patterns and formulas in areas of mathematics where others had tried and failed.

It is not just Fibonacci numbers that one finds in Nature. The animal kingdom also knows about prime numbers. There are two species of cicada called *Magicicada septendecim* and *Magicicada tredecim* which often live in the same environment. They have a life cycle of exactly 17 and 13 years, respectively. For all but their last year they remain in the ground feeding on the sap of tree roots. Then, in their last year, they metamorphose from nymphs into fully formed adults and emerge en masse from the ground. It is an extraordinary event as, every 17 years, *Magicicada septendecim* takes over the forest in a single night. They sing loudly, mate, eat, lay eggs, then die six weeks later. The forest goes quiet for another 17 years. But why has each species chosen a prime number of years as the length of their life cycle?

There are several possible explanations. Since both species have evolved prime number life cycles, they will be synchronised to emerge in the same year very rarely. In fact they will have to share the forest only every $221 = 17 \times 13$ years. Imagine if they had chosen cycles which weren't prime, for example 18 and 12. Over the same period they would have been in synch 6 times, namely in years 36, 72, 108, 144, 180 and 216. These are the years which share the prime building blocks of both 18 and 12. The prime numbers 13 and 17, on the other hand, allow the two species of cicada to avoid too much competition.

Another explanation is that a fungus developed which emerged simultaneously with the cicadas. The fungus was deadly for the cicadas, so they evolved a life cycle which would avoid the fungus. By changing to a prime number cycle of 17 or 13 years, the cicadas ensured that they emerged in the same years as the fungus less frequently than if they had a non-prime life cycle. For the cicadas, the primes weren't just some abstract curiosity but the key to their survival.

Evolution might be uncovering primes for the cicadas, but mathematicians wanted a more systematic way to find these numbers. Of all the number challenges it was the list of primes above all others for which mathematicians sought some secret formula. One has to be careful, though, about expecting patterns and order to be everywhere in the mathematical world. Many people throughout history have got lost in the vain attempt to find structure hidden in the decimal expansion of π, one of the most important numbers in mathematics. But its importance has fuelled desperate attempts to discover messages buried in its chaotic decimal expansion. Whilst alien life had used the primes to catch Ellie Arroway's attention at the beginning of Carl Sagan's book *Contact*, the ultimate message of the book is buried deep in the expansion of π, in which a series of 0's and 1's suddenly appears, mapping out a pattern that is meant to reveal 'there is an intelligence that antedates the universe'. Darren Aronofsky's film 'π' also plays on this popular cultural image.

As a warning to those captivated by the idea of uncovering hidden messages in numbers such as π, mathematicians have been able to prove that most decimal numbers have hidden somewhere in their infinite expansions any sequence of numbers you might be looking for. So there is a good chance that π will contain the computer code for the book of Genesis if you search for long enough. One has to find the right viewpoint from which to look for patterns. π is an important number not because its decimal expansion contains hidden messages. Its importance becomes apparent when it is examined from a different perspective. The same was true of the primes. Armed with his table of primes and his knack for lateral thinking, Gauss was on the lookout for the right angle and viewpoint from which to stare at the primes so that some previously hidden order might emerge from behind the façade of chaos.

Proof, the mathematician's travelogue

Although finding patterns and structure in the mathematical world is one part of what a mathematician does, the other part is *proving* that a pattern will persist. The concept of proof perhaps marks the true beginning of mathematics as the art of deduction rather than just numerological observation, the point at which mathematical alchemy gave way to mathematical chemistry. The ancient Greeks were the first to understand that it was possible to prove that certain facts would remain true however far you counted, however many instances you examined.

The mathematical creative process starts with a guess. Often, the guess emerges from the intuition that the mathematician develops after years of exploring the mathematical world, cultivating a feel for its many twists and turns. Sometimes simple numerical experiments reveal a pattern which one might guess will persist for ever. Mathematicians during the seventeenth century, for example, discovered what they believed might be a fail-safe method to test if a number N was prime: calculate 2 to the power N and divide by N – if the remainder is 2 then the number N is a prime. In terms of Gauss's clock calculator, these mathematicians were trying to calculate 2^N on a clock with N hours. The challenge then is to prove whether this guess is right or wrong. It is these mathematical guesses or predictions that the mathematician calls a 'conjecture' or 'hypothesis'.

A mathematical guess only earns the name of 'theorem' once a proof has been provided. It is this movement from 'conjecture' or 'hypothesis' to 'theorem' that marks the mathematical maturity of a subject. Fermat left mathematics with a whole slew of predictions. Subsequent generations of mathematicians have made their mark by proving Fermat right or wrong. Admittedly, Fermat's Last Theorem was always called a theorem and never a conjecture. But that is unusual, and probably came about because Fermat claimed in notes that he scribbled in his copy of Diophantus's *Arithmetica* that he had a marvellous proof that was unfortunately too large to write in the margin of the page. Fermat never recorded his supposed proof anywhere, and his marginal comments became the biggest mathematical tease in the history of the subject. Until Andrew Wiles provided an argument, a proof of why Fermat's equations really had no interesting solutions, it actually remained a hypothesis – merely wishful thinking.

Gauss's schoolroom episode encapsulates the movement from guess via proof to theorem. Gauss had produced a formula which he predicted would produce any number you wanted on the list of triangular numbers.

How could he guarantee that it would work every time? He certainly couldn't test every number on the list to see whether his formula gave the correct answer, since the list is infinitely long. Instead, he resorted to the powerful weapon of mathematical proof. His method of combining two triangles to make a rectangle guaranteed, without the need for an infinite number of calculations, that the formula would always work. In contrast, the seventeenth-century prime number test based on 2^N was finally thrown out of the mathematical court in 1819. The test works correctly for all numbers up to 340, but then declares that 341 is prime. This is where the test fails, since $341 = 11 \times 31$. This exception wasn't discovered until Gauss's clock calculator with 341 hours on the clock face could be used to simplify the analysis of a number like 2^{341}, which on a conventional calculator stretches to over a hundred digits.

The Cambridge mathematician G.H. Hardy, author of *A Mathematician's Apology*, used to describe the process of mathematical discovery and proof in terms of mapping out distant landscapes: 'I have always thought of a mathematician as in the first instance an observer, a man who gazes at a distant range of mountains and notes down his observations.' Once the mathematician has observed a distant mountain, the second task is then to describe to people how to get there.

You begin in a place where the landscape is familiar and there are no surprises. Within the boundaries of this familiar land are the axioms of mathematics, the self-evident truths about numbers, together with those propositions that have already been proved. A proof is like a pathway from this home territory leading across the mathematical landscape to distant peaks. Progress is bound by the rules of deduction, like the legitimate moves of a chess piece, prescribing the steps you are permitted to take through this world. At times you arrive at what looks like an impasse, and need to take that characteristic lateral step, moving sideways or even backwards to find a way around. Sometimes you need to wait for new tools, like Gauss's clock calculators, to be invented, so that you can continue your ascent.

In Hardy's words, the mathematical observer

> sees A sharply, while of B he can obtain only transitory glimpses. At last he makes out a ridge which leads from A, and following it to its end he discovers that it culminates in B. If he wishes someone else to see it, he points to it, either directly or through the chain of summits which led him to recognise it himself. When his pupil also sees it, the research, the argument, the proof is finished.

The proof is the story of the trek and the map charting the coordinates of that journey – the mathematician's log. Readers of the proof will experience the same dawning realisation as its author. Not only do they finally see the way to the peak, but also they understand that no new development will undermine the new route. Very often a proof will not seek to dot every i and cross every t. It is a description of the journey and not necessarily the re-enactment of every step. The arguments that mathematicians provide as proofs are designed to create a rush in the mind of the reader. Hardy used to describe the arguments we give as '*gas*, rhetorical flourishes designed to affect psychology, pictures on the board in the lecture, devices to stimulate the imagination of pupils'.

The mathematician is obsessed with proof, and will not be satisfied simply with experimental evidence for a mathematical guess. This attitude is often marvelled at and even ridiculed in other scientific disciplines. Goldbach's Conjecture has been checked for all numbers up to 400,000,000,000,000 but has not been accepted as a theorem. Most other scientific disciplines would be happy to accept this overwhelming numerical data as a convincing enough argument, and move on to other things. If, at a later date, new evidence were to crop up which required a reassessment of the mathematical canon, then fine. If it is good enough for the other sciences, why is mathematics any different?

Most mathematicians would quiver at the thought of such heresy. As the French mathematician André Weil expressed it, 'Rigour is to the mathematician what morality is to men.' Part of the reason is that evidence is often quite hard to assess in mathematics. More than any other part of mathematics, the primes take a long time to reveal their true colours. Even Gauss was taken in by overwhelming data in support of a hunch he had about prime numbers, but theoretical analysis later revealed that he had been duped. This is why a proof is essential: first appearances can be deceptive. While the ethos of every other science is that experimental evidence is all that you can truly rely on, mathematicians have learnt never to trust numerical data without proof.

In some respects, the ethereal nature of mathematics as a subject of the mind makes the mathematician more reliant on providing proof to lend some feeling of reality to this world. Chemists can happily investigate the structure of a solid buckminsterfullerene molecule; sequencing the genome presents the geneticist with a concrete challenge; even the physicists can sense the reality of the tiniest subatomic particle or a distant black hole. But the mathematician is faced with trying to understand objects with no obvious physical reality such as shapes in eight dimensions, or prime numbers so large they exceed the number of atoms in the physical universe. Given a palette of such abstract concepts the mind can play strange tricks, and without proof there is a danger of creating a house of cards. In the other scientific disciplines, physical observation and experiment provide some reassurance of the reality of a subject. While other scientists can use their eyes to see this physical reality, mathematicians rely on mathematical proof, like a sixth sense, to negotiate their invisible subject.

Searching for proofs of patterns that have already been spotted is also a great catalyst for further mathematical discovery. Many mathematicians feel that it may be better if these defining problems never get solved because of the wonderful new mathematics encountered along the way. The problems allow for exploration of a kind which forces mathematical pioneers to pass through lands they could never have envisaged at the outset of their journey.

But perhaps the most convincing argument for why the culture of mathematics places such stock in proving that a statement is true is that, unlike the other sciences, there is the luxury of being able to do so. In how many other disciplines is there anything that parallels the statement that Gauss's formula for triangular numbers will *never* fail to give the right answer? Mathematics may be an ethereal subject confined to the mind, but its lack of tangible reality is more than compensated for by the certitude that proof provides.

Unlike the other sciences, in which models of the world can crumble between one generation and the next, proof in mathematics allows us to establish with 100 per cent certainty that facts about prime numbers will not change in the light of future discoveries. Mathematics is a pyramid where each generation builds on the achievements of the last without fear of any collapse. This durability is what is so addictive about being a mathematician. For no science other than mathematics can we say that what the ancient Greeks established in their subject holds true today. We may scoff now at the Greeks' belief that matter was made from fire, air, water and earth. Will future generations look back on the list of 109 atoms that make up Mendeleev's Periodic Table of elements with as much disdain as we view the Greek model of the chemical world? In contrast, all mathematicians begin their mathematical education by learning what the ancient Greeks proved about prime numbers.

The certainty that proof gives to the mathematician is something that is envied by members of other university departments as much as it is jeered at. The permanence created by mathematical proof leads to the genuine immortality to which Hardy referred. This is often why people surrounded by a world of uncertainty are drawn to the subject. Time after time has the mathematical world offered a refuge for young minds yearning to escape from a real world they cannot cope with.

Our faith in the durability of a proof is reflected in the rules governing the award of Clay's Millennium Prizes. The prize money is released two years after publication of the proof and with the general acceptance of the mathematical community. Of course, this is no guarantee that there isn't a subtle error, but it does recognise that we generally believe that errors can be spotted in proofs without waiting many years for new evidence. If there is an error, it must be there on the page in front of us.

Are mathematicians arrogant in believing that they have access to absolute proof? Can one argue that a proof that all numbers are built from primes is as likely to be overthrown as the theory of Newtonian physics or the theory of an indivisible atom? Most mathematicians believe that the axioms that are taken as self-evident truths about numbers will never crumble under future scrutiny. The laws of logic used to build upon these foundations, if applied correctly, will in their view produce proofs of statements about numbers that will never be overturned by new insights. Maybe this is philosophically naive, but it is certainly the central tenet of the sect of mathematics.

There is also the emotional buzz the mathematician experiences in charting new pathways across the mathematical landscape. There is an amazing feeling of exhilaration at discovering a way to reach the summit of some distant peak which has been visible for generations. It is like creating a wonderful story or a piece of music which truly transports the mind from the familiar to the unknown. It is great to make that first sighting of the possible existence of a far-off mountain like Fermat's Last Theorem or the Riemann Hypothesis. But it doesn't compare to the satisfaction of navigating the land in between. Even those who follow in the trail of that first pioneer will experience something of the sense of spiritual elevation that accompanied the first moment of epiphany at discovering a new proof. And this is why mathematicians continue to value the pursuit of proof even if they are utterly convinced that something like the Riemann Hypothesis is true. Because mathematics is as much about travelling as it is about arrival.

Is mathematics an act of creation or an act of discovery? Many mathematicians fluctuate between feeling they are being creative and a sense they are discovering absolute scientific truths. Mathematical ideas can often appear very personal and dependent on the creative mind that conceived them. Yet that is balanced by the belief that its logical character means that every mathematician is living in the same mathematical world that is full of immutable truths. These truths are simply waiting to be unearthed, and no amount of creative thinking will undermine their existence. Hardy encapsulates perfectly this tension between creation and discovery that every mathematician battles with: 'I believe that mathematical reality lies outside us, that our function is to discover or observe it and that the theorems which we prove and which we describe grandiloquently as our "creations" are simply our notes of our observations.' But at other times he favoured a more artistic description of the process of doing mathematics: 'Mathematics is not a contemplative but a creative subject,' he wrote in *A Mathematician's Apology*, a book Graham Greene ranked with Henry James's notebooks as the best account of what it is like to be a creative artist.

Although the primes, and other aspects of mathematics, transcend cultural barriers, much of mathematics is creative and a product of the human psyche. Proofs, the stories mathematicians tell about their subject, can often be narrated in different ways. It is likely that Wiles's proof of Fermat's Last Theorem would be as mysterious to aliens as listening to Wagner's *Ring* cycle. Mathematics is a creative art under constraints – like writing poetry or playing the blues. Mathematicians are bound by the logical steps they must take in crafting their proofs. Yet within such constraints there is still a lot of freedom. Indeed, the beauty of creating under constraints is that you get pushed in new directions and find things you might never have expected to discover unaided. The primes are like notes in a scale, and each culture has chosen to play these notes in its own particular way, revealing more about historical and social influences than one might expect. The story of the primes is a social mirror as much as the discovery of timeless truths. The burgeoning love of machines in the seventeenth and eighteenth centuries is reflected in a very practical, experimental approach to the primes; in contrast, Revolutionary Europe created an atmosphere where new abstract and daring ideas were brought to bear on their analysis. The choice of how to narrate the journey through the mathematical world is something which is specific to each individual culture.

Euclid's fables

The first to start telling these stories were the ancient Greeks. They realised the power of proof to forge permanent pathways to mountains in the mathematical world. Once they were reached, no longer was there the fear that these mountains were some distant mathematical mirage. For example, how can we be really sure that there aren't some rogue numbers out there which can't actually be built by multiplying together prime numbers? The Greeks were the first to come up with an argument that would leave no doubt in their minds or in the minds of future generations that no such rogue numbers could ever turn up.

Mathematicians often discover proofs by taking a particular instance of the general theory they are trying to prove, and begin by trying to understand why the theory is true for this example. They hope that the argument or recipe that was successful when applied to the example will work regardless of the particular case they chose to analyse. For instance, to prove that every number is a product of primes, start by considering the particular case of the number 140. Suppose you had checked that every number below 140 is either a prime number or the product of prime numbers multiplied together. What about the number 140 itself? Is it possible that this is a rogue number which is neither prime nor equal to a product of prime numbers? First, you would discover that the number is not prime. How would you do this? By showing it could be written as two smaller numbers multiplied together. For example, 140 is 4×35. Now we are 'in' because we have already confirmed that 4 and 35, numbers lower than our first candidate rogue, 140, can be written as primes multiplied together: 4 is 2×2 and 35 is 5×7. Piecing this information together, we see that 140 is in fact the product $2 \times 2 \times 5 \times 7$. So 140 is not a rogue after all.

The Greeks understood how they could translate this particular example into a general argument that would apply to all numbers. Curiously, their argument begins by asking us to imagine that there *are* such rogue numbers – ones that are neither prime nor can be written as prime numbers multiplied together. If there are such rogues, then, as we count through the sequence of all the numbers, we must eventually encounter the first of these rogue numbers. We shall call it N (it is sometimes referred to as the *minimal criminal*). Since this hypothetical number N isn't a prime number, we must be able to write it as two smaller numbers, A and B, multiplied together. After all, if that weren't possible, N would be prime.

Since A and B are smaller than N, our choice of N implies that A and B can be written as products of primes. So if we multiply together all the primes coming from A and all the primes coming from B, then we must get the original number, N. We have now shown that N can be written as prime numbers multiplied together, which contradicts our original choice of N. So our original assumption that there were rogue numbers can't be tenable. Hence every number must be prime or built by multiplying primes.

When I tried this argument out on friends, they felt as if they had been cheated somewhere along the way. There is something slightly slippery about our opening gambit: assume the things you don't want to exist do exist, and end up proving they don't. This strategy of thinking the unthinkable became a powerful tool in the Greeks' construction of proofs. It relies on the logical fact that a statement has to be either true or false. If we assume the statement is false and we get a contradiction, we can infer that our assumption was wrong and deduce that the statement must have been true after all.

The Greek proof appeals to the lazy side of most mathematicians. Instead of being faced with the impossible task of doing an infinite number of explicit calculations to prove that all numbers can be built from primes, the abstract argument captures the essence of every such computation. It's like knowing how to climb an infinite ladder without physically having to perform the task.

More than any other Greek mathematician, Euclid is regarded as the father of the art of proof. He was part of the research institute that the Greek leader Ptolemy I established in Alexandria around 300 BC. There, Euclid wrote one of the most influential textbooks in all of recorded history: *The Elements*. In the first part of this book he set down axioms for geometry describing the relationship between points and lines. These axioms were put forward as self-evident truths about the objects of geometry, so that geometry would then act as a mathematical description of the physical world. He went on to use the rules of deduction to produce five hundred theorems of geometry.

The middle part of Euclid's *Elements* deals with the properties of numbers, and it is here that we find what many regard as the first truly brilliant piece of mathematical reasoning. In Proposition 20, Euclid explains a simple but fundamental truth about prime numbers: there are infinitely many of them. He begins with the fact that every number can be built by multiplying prime numbers together. On top of this he constructs his next proof. If these prime numbers are the building blocks of all numbers, is it possible, he asks himself, for there to be only a finite number of these building blocks? The Periodic Table of the chemical elements was constructed by Mendeleev, and in its present form it classifies 109 different atoms from which we can build all matter. Could the same be true for prime numbers? What if a mathematical Mendeleev presented Euclid with

Euclid (*c*.350–300 BC).

a list of 109 primes and challenged Euclid to prove that some primes were missing from the list?

Why, for example, can't all numbers be built simply by multiplying together different combinations of the primes 2, 3, 5 and 7? Euclid thought about how you might look for numbers that aren't built from any of these primes. You might say, 'Well that's easy – just take the next prime, 11.' This certainly can't be built from 2, 3, 5 and 7. But sooner or later this strategy is going to fail precisely because, even today, we have no clue about how to guarantee finding where the next prime will be. And because of this unpredictability, Euclid had to try a different tack in his search for a method that would work, regardless of how long the list of primes was.

Whether it was truly Euclid's own idea or whether he was simply recording ideas that others had dreamt up in Alexandria, we have no way of knowing. Whichever it was, he was able to show how to build a number that couldn't be built from any finite list of primes that he might be given. Take the primes 2, 3, 5 and 7. Euclid multiplied them together to get $2 \times 3 \times 5 \times 7 = 210$, then – and this is his act of genius – he added 1 to this product to get 211. Euclid had constructed this number, 211, in such a way that none of the primes in the list, 2, 3, 5 and 7, would divide into it exactly. By adding 1 to the product, he could guarantee that dividing by a prime on the list would always leave remainder 1.

Now, Euclid knew that all numbers were built by multiplying together primes. So what about 211? Since it can't be divided by 2, 3, 5 or 7, there had to be some other primes not on the list that built the number 211. In this particular example, 211 is itself a prime. Euclid was not claiming that the number he built would always be prime – only that it was a number that was built out of primes that were not on the list that our mathematical Mendeleev was offering us.

For example, what if one claimed that all numbers could be built from the finite list of primes 2, 3, 5, 7, 11 and 13. Euclid's number built from these primes is $2 \times 3 \times 5 \times 7 \times 11 \times 13 + 1 = 30{,}031$. This number is not a prime. All Euclid was saying was that, given any list containing finitely many primes, he could produce a number that had to be built out of primes that were not on the list. In this particular case the primes you need are 59 and 509. But in general, Euclid had no way of knowing how to find the precise value of these new primes. He knew only that they must exist.

It was a wonderful argument. Euclid had no idea how to produce primes explicitly, but he could prove why they would never run dry. It is striking that we do not know whether infinitely many of Euclid's numbers themselves are prime, even though they are sufficient to prove that there must be an infinite number of primes. With Euclid's proof, gone was the chance of fitting together a Periodic Table listing all the primes or of discovering some prime number genome coding billions of them. No simple butterfly collecting would ever allow us to understand these numbers. Here, then, was the ultimate challenge: the mathematician, armed with a limited weaponry, pitched against the infinite expanse of prime numbers. How could we possibly chart a path through such an infinite chaotic jumble and find some pattern which might predict their behaviour?

Hunting for primes

Generations have striven without success to improve on Euclid's understanding of the primes, and there have been many intriguing speculations. But as Cambridge don Hardy liked to say, 'Every fool can ask questions about prime numbers that the wisest man cannot answer.' The Twin Primes Conjecture, for example, asks whether there are infinitely many primes p such that the number $p + 2$ is also prime. An example of such a pair is 1,000,037 and 1,000,039. (Note that this is the closest that two primes numbers can be, since N and $N + 1$ cannot both be prime – except when $N = 2$ – because at least one of these numbers is divisible by 2.) Might Sacks's autistic-savant twins have had an extra facility for finding these twin primes? Euclid proved two millennia ago that there are infinitely many primes, but no one knows whether there might be some number beyond which there are no longer such close primes. We believe that there are infinitely many twin primes. Guesses are one thing, but proof remains the ultimate goal.

Mathematicians tried, with varying degrees of success, to come up with formulas that, even if they don't generate all prime numbers, do produce a list of primes. Fermat thought he had one. He guessed that if you raise 2 to the power 2^N and add 1, the resulting number $2^{2^N} + 1$ is a prime. This number is called the Nth *Fermat number*. For example, taking $N = 2$ and raising 2 to the power $2^2 = 4$, you get 16. Add 1 and you get the prime number 17, the second Fermat number. Fermat thought that his formula would always give him a prime number, but it turned out to be one of the few guesses that he got wrong. The Fermat numbers get very large very quickly. Even the fifth Fermat number has 10 digits, and was out of Fermat's computational reach. It is the first Fermat number which is not prime, being divisible by 641.

Fermat's numbers were very dear to Gauss's heart. The fact that 17 is one of Fermat's primes is the key to why Gauss could construct his perfect 17-sided shape. In his great treatise *Disquisitiones Arithmeticae*, Gauss shows why it is that, if the Nth Fermat number is a prime, you can make a geometric construction of an N-sided shape only using a straight edge and compass. The fourth Fermat number, 65,537, is prime, so with these very basic instruments it is possible to construct a perfect 65,537-sided figure.

Fermat's numbers have failed to throw up more than four primes to date, but he had more success in uncovering some of the very special properties that prime numbers have. Fermat discovered a curious fact about those prime numbers that leave remainder 1 on division by 4 – examples are 5, 13, 17 and 29. Such prime numbers can always be written as the sum of two squares – for example, $29 = 2^2 + 5^2$. This was another of Fermat's teases. Although he claimed to have a proof, he failed to record much of the details.

On Christmas Day, 1640, Fermat wrote of his discovery – that certain primes could be expressed as the sum of two squares – in a letter to a French monk called Marin Mersenne. Mersenne's interests were not confined to liturgical matters. He loved music and was the first to develop a coherent theory of harmonics. He also loved numbers. Mersenne and Fermat corresponded regularly about their mathematical discoveries, and Mersenne broadcast many of Fermat's claims to a wider audience. Mersenne became renowned for his role as an international scientific clearing house through which mathematicians could disseminate their ideas.

Just as generations had been captivated by the search for order in the primes, Mersenne too had caught the bug. Although he couldn't see a way to find a formula that would produce all the primes, he did come across a formula that in the long run has proved far more successful at finding primes than Fermat's formula has. Like Fermat, he started by considering powers of 2. But instead of adding 1, as Fermat had, Mersenne decided to subtract 1 from the answer. So, for example, $2^3 - 1 = 8 - 1 = 7$, a prime number. Maybe Mersenne's musical intuition was coming to his aid. Doubling the frequency of a note takes the note up an octave, so powers of 2 produce harmonic notes. You might expect a shift of 1 to sound a very dissonant note, not compatible with any previous frequency – a 'prime note'.

Mersenne quickly discovered that his formula wasn't going to yield a prime every time. For example, $2^4 - 1 = 15$. Mersenne realised that if n was not prime then there was no chance that $2^n - 1$ was going to be prime. But now he boldly claimed that, for n up to 257, $2^n - 1$ would be prime precisely if n was one of the following numbers: 2, 3, 5, 7, 13, 19, 31, 67, 127, 257. He had discovered that even if n was prime, it still annoyingly didn't guarantee that his number $2^n - 1$ would be prime. He could calculate $2^{11} - 1$ by hand and get 2,047, which is 23×89. Generations of mathematicians marvelled at Mersenne's ability to assert that a number as large as $2^{257} - 1$ was prime. This number has 77 digits. Did the monk have access to some mystical arithmetic formula that told him why this number, beyond any human computational abilities, was prime?

Mathematicians believe that if one continues Mersenne's list, there will be infinitely many choices for n which will make Mersenne's numbers $2^n - 1$ into prime numbers. But we are still missing a proof that this guess is true. We are still waiting for a modern day Euclid to prove that Mersenne's primes never run dry. Or perhaps this far-off peak is just a mathematical mirage.

Many mathematicians of Fermat and Mersenne's generation had played around with interesting numerological properties of the primes, but their methods did not match up to the ancient Greek ideal of proof. This explains in part why Fermat gave no details of many of the proofs he claimed to have discovered. There was a distinct lack of interest during this period in providing such logical explanations. Mathematicians were quite content with a more experimental approach to their subject, where in an increasingly mechanised world results were justified by their practical applications. In the eighteenth century, however, there arrived a mathematician who would rekindle a sense of the value of proof in mathematics. The Swiss mathematician Leonhard Euler, born in 1707, came up with explanations for many of the patterns that Fermat and Mersenne had discovered but failed to account for. Euler's methods would later play a significant role in opening new theoretical windows onto our understanding of the primes.

Euler, the mathematical eagle

The mid-eighteenth century was a time of court patronage. This was pre-Revolutionary Europe, when countries were ruled by enlightened despots: Frederick the Great in Berlin, Peter the Great and Catherine the Great in St Petersburg, Louis XV and Louis XVI in Paris. Their patronage supported the academies that drove the intellectual development of the Enlightenment, and indeed they saw it as a mark of their standing that they be surrounded in their courts by intellectuals. And they were well aware of the potential of the sciences and mathematics to boost the military and industrial capabilities of their countries.

Euler was the son of a clergyman who hoped that his son would join him in the church. The young Euler's precocious mathematical talents, however, had brought him to the notice of the powers that be. Euler was soon being courted by the academies throughout Europe. He had been tempted to join the Academy in Paris, which by this time had become the world's centre of mathematical activity. He chose instead to accept an offer made to him in 1726 to join the Academy of Sciences in St Petersburg, the capstone for Peter the Great's campaign to improve education in Russia. He would be joining friends from Basel who had stimulated his interest in mathematics as a child. They wrote to Euler from St Petersburg asking whether he could bring from Switzerland fifteen pounds of coffee, one pound of the best green tea, six bottles of brandy, twelve dozen fine tobacco pipes and a few dozen packs of playing cards. Laden down with gifts, it took the young Euler seven weeks to complete the long journey by boat, on foot and by post wagon, and in May 1727 he

Leonhard Euler (1707–83).

finally arrived in St Petersburg to pursue his mathematical dreams. His subsequent output was so extensive that the St Petersburg Academy was still publishing material that had been housed in their archives some fifty years after Euler's death in 1783.

The role of the court mathematician is perfectly illustrated by a story that was told of Euler's time in St Petersburg. Catherine the Great was hosting the famous French philosopher and atheist Denis Diderot. Diderot was always rather damning of mathematics, declaring that it added nothing to experience and served only to draw a veil between human beings and nature. Catherine, though, quickly tired of her guest, not because of Diderot's disparaging views on mathematics but rather his tiresome attempts to rattle the religious faith of her courtiers. Euler was promptly called to court to assist in silencing the insufferable atheist. In appreciation of her patronage, Euler duly consented and addressed Diderot in serious tones before the assembled court: 'Sir, $(a + b^n)/n = x$, hence God exists; reply.' Diderot is reported to have retreated in the light of such a mathematical onslaught.

This anecdote, told by the famous English mathematician Augustus De Morgan in 1872, had probably been embroidered for popular consumption and is a reflection more of the fact that most mathematicians enjoy putting down philosophers. But it does show how the royal courts of Europe had not considered themselves complete without a smattering of mathematicians amongst the ranks of astronomers, artists and composers.

Catherine the Great was interested not so much in mathematical proofs of the existence of God, but rather in Euler's work on hydraulics, ship construction and ballistics. The Swiss mathematician's interests ranged far and wide over the mathematics of the day. As well as military mathematics, Euler also wrote on the theory of music, but ironically his treatise was regarded as too mathematical for musicians and too musical for mathematicians.

One of his popular triumphs was the solution of the Problem of the Bridges of Königsberg. The River Pregel, known now as the Pregolya, runs through Königsberg, which in Euler's day was in Prussia (it's now in Russia, and called Kaliningrad). The river divides, creating two islands in the centre of the town, and the Königsbergers had built seven bridges to span it (see overleaf).

It had become a challenge amongst the citizens to see if anyone could walk around the town, crossing each bridge once and only once, and return to their starting point. It was Euler who eventually proved in 1735 that the task was impossible. His proof is often cited as the beginning of topology, where the actual physical dimensions of the problem are not relevant. It was the network of connections between different parts of the town that was important to Euler's solution, and not their actual locations or distances apart – the map of the London Underground illustrates this principle.

It was numbers above all that captivated Euler's heart. As Gauss would write:

> The peculiar beauties of these fields have attracted all those who have been active there; but none has expressed this so often as Euler, who, in almost every one of his many papers on number-theory, mentions again and again his delight in such investigations, and the welcome change he finds there from tasks more directly related to practical applications.

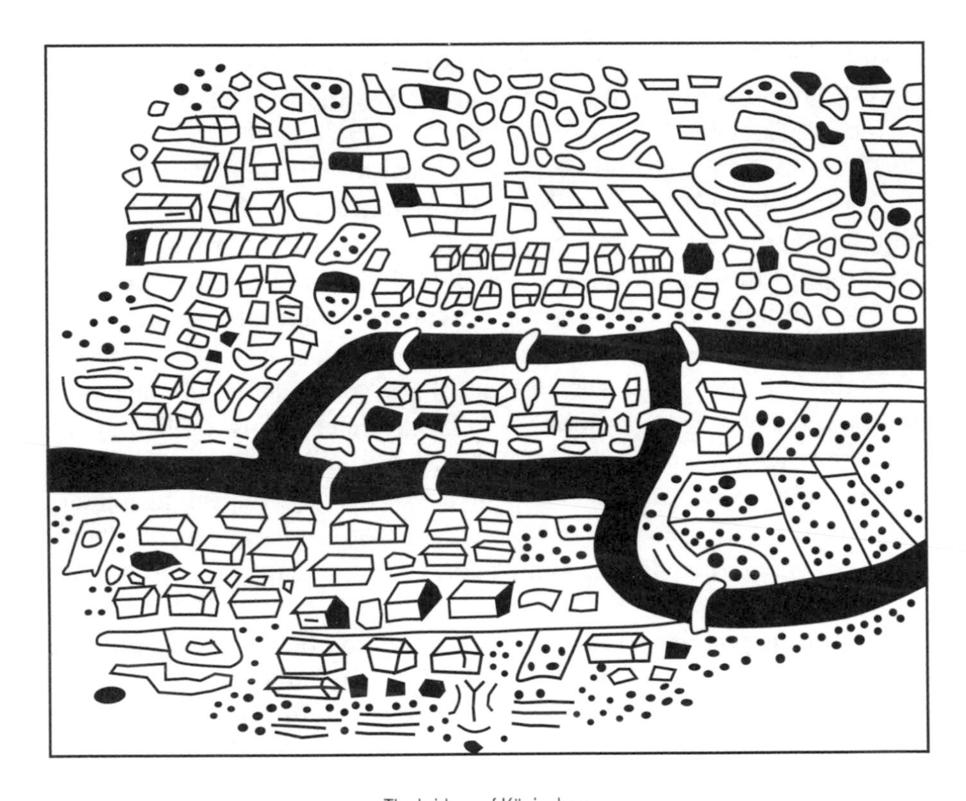

The bridges of Königsberg.

Euler's passion for number theory had been stimulated by correspondence with Christian Goldbach, an amateur German mathematician who was living in Moscow and unofficially employed as secretary of the Academy of Sciences in St Petersburg. Like the amateur mathematician Mersenne before him, Goldbach was fascinated by playing around with numbers and doing numerical experiments. It was to Euler that Goldbach communicated his conjecture that every even number could be written as a sum of two primes. Euler in return would write to Goldbach to try out many of the proofs he had constructed to confirm Fermat's mysterious catalogue of discoveries. In contrast to Fermat's reticence in keeping his supposed proofs a secret from the world, Euler was happy to show off to Goldbach his proof of Fermat's claim that certain primes can be written as the sum of two squares. Euler even managed to prove an instance of Fermat's Last Theorem.

Despite his passion for proof, Euler was still very much an experimental mathematician at heart. Many of his arguments flew close to the mathematical wind, containing steps that weren't completely rigorous.

That did not concern him if it led to interesting new discoveries. He was a mathematician of exceptional computational skill and very adept at manipulating mathematical formulas until strange connections emerged. As the French academician François Arago observed, 'Euler calculated without apparent effort, as men breathe, or eagles sustain themselves in the wind.'

Above all else, Euler loved calculating prime numbers. He produced tables of all the primes up to 100,000 and a few beyond. In 1732, he was also the first to show that Fermat's formula for primes, $2^{2^N} + 1$, broke down when $N = 5$. Using new theoretical ideas, he managed to show how to crack this ten-digit number into a product of two smaller numbers. One of his most curious discoveries was a formula that seemed to generate an uncanny number of primes. In 1772, he calculated all the answers that you get when you feed the numbers from 0 to 39 into the formula $x^2 + x + 41$. He got the following list:

41, 43, 47, 53, 61, 71, 83, 97, 113, 131, 151, 173, 197, 223, 251, 281, 313, 347, 383, 421, 461, 503, 547, 593, 641, 691, 743, 797, 853, 911, 971, 1,033, 1,097, 1,163, 1,231, 1,301, 1,373, 1,447, 1,523, 1,601

It seemed bizarre to Euler that you could generate so many primes with this formula. He realised that the process would have to break down at some point. It might already be clear to you that when you input 41, the output has to be divisible by 41. Also, for $x = 40$ you get a number which is not prime.

Nonetheless, Euler was quite struck by his formula's ability to produce so many primes. He began to wonder what other numbers might work instead of 41. He discovered that in addition to 41 you could also choose $q = 2, 3, 5, 11, 17$, and the formula $x^2 + x + q$ would spit out primes when fed numbers from 0 to $q - 2$.

But finding such a simple formula for generating all the primes was beyond even the great Euler. As he wrote in 1751, 'There are some mysteries that the human mind will never penetrate. To convince ourselves we have only to cast a glance at tables of primes and we should perceive that there reigns neither order nor rule.' It seems paradoxical that the fundamental objects on which we build our order-filled world of mathematics should behave so wildly and unpredictably.

It would turn out that Euler had been sitting on an equation that would break the prime number deadlock. But it would take another hundred years, and another great mind, to show what Euler could not.

That mind belonged to Bernhard Riemann. It was Gauss, though, who by initiating another of his classic lateral moves, would eventually inspire Riemann's new perspective.

Gauss's guess

If centuries of searching had failed to unearth some magic formula which would generate the list of prime numbers, perhaps it was time to adopt a different strategy. This was what the fifteen-year-old Gauss was thinking in 1792. He had been given a book of logarithms as a present the previous year. Until a few decades ago, tables of logarithms were familiar to every teenager doing calculations in the schoolroom. With the advent of pocket calculators, they lost their place as an essential tool in everyday life, but several hundred years ago every navigator, banker and merchant would have been exploiting these tables to turn difficult multiplication into simple addition. Included at the back of Gauss's new book was a table of prime numbers. It was uncanny that primes and logarithms should appear together, because Gauss noticed after extensive calculations that there seemed to be a connection between these two seemingly unrelated topics.

The first table of logarithms was conceived in 1614 in an age when sorcery and science were bedfellows. Their creator, the Scottish Baron John Napier, was regarded by local residents as a magician who dealt in the dark arts. He skulked around his castle dressed in black, a jet-black cock perched on his shoulder, muttering that his apocalyptic algebra foretold that the Last Judgement would fall between 1688 and 1700. But as well as applying his mathematical skills to the practice of the occult, he also uncovered the magic of the logarithm function.

If you feed a number, say 100, into your calculator and then press the 'log' button, the calculator spits out a second number, the logarithm of 100. What your calculator has done is to solve a little puzzle: it has looked for the number x that makes the equation $10^x = 100$ correct. In this case the calculator outputs the answer 2. If we input 1,000, a number ten times larger than 100, then the new answer output by your calculator is 3. The logarithm goes up by 1. Here is the essential character of the logarithm: it turns multiplication into addition. Each time we *multiply* the input by 10, we get the new output by *adding* 1 to the previous answer.

It was a fairly major step for mathematicians to realise that they could talk about logarithms of numbers which weren't whole-number powers of 10. For example, Gauss would have been able to look up in his tables the logarithm of 128 and find that raising 10 to the power 2.107 21 would get him pretty close to 128. These calculations are what Napier had collected together in the tables that he had produced in 1614.

Tables of logarithms helped to accelerate the world of commerce and navigation that was blossoming in the seventeenth century. Because of the dialogue that logarithms created between multiplication and addition, the tables helped to convert a complicated problem of multiplying together two large numbers into the simpler task of adding their logarithms. To multiply together large numbers, merchants would add together the logarithms of the numbers and then use the log tables in reverse to find the result of the original multiplication. The increase in speed that the sailor or seller would gain via these tables might save the wrecking of a ship or the collapse of a deal.

But it was the supplementary table of prime numbers at the back of his book of logarithms that fascinated the young Gauss. In contrast to the logarithms, these tables of primes were nothing more than a curiosity to those interested in the practical application of mathematics. (Tables of primes constructed in 1776 by Antonio Felkel were considered so useless that they ended up being used for cartridges in Austria's war with Turkey!) The logarithms were very predictable; the primes were completely random. There seemed no way to predict when to expect the first prime after 1,000, for example.

The important step Gauss took was to ask a different question. Rather than attempting to predict the precise location of the next prime, he tried instead to see whether he could at least predict how many primes there were in the first 100 numbers, the first 1,000 numbers, and so on. If one took any number N, was there a way to estimate how many primes one would expect to find amongst the numbers from 1 to N? For example, there are twenty-five prime numbers up to 100. So you have a one in four chance of getting a prime if you choose a number at random between 1 and 100. How does this proportion change if we look at the primes from 1 to 1000, or 1 to 1,000,000? Armed with his prime number tables, Gauss began his quest. As he looked at the proportion of numbers that were prime, he found that when he counted higher and higher a pattern started to emerge. Despite the randomness of these numbers, a stunning regularity seemed to be looming out of the mist.

If we look at the table overleaf of values of the number of primes up to various powers of ten, based on more modern calculations, this regularity becomes apparent.

This table, which contains much more information than was available to Gauss, shows us more clearly the regularity that Gauss discovered. It is

N	Number of primes from 1 to N, often referred to as $\pi(N)$	On average, how many numbers you need to count before you reach a prime number
10	4	2.5
100	25	4.0
1,000	168	6.0
10,000	1,229	8.1
100,000	9,592	10.4
1,000,000	78,498	12.7
10,000,000	664,579	15.0
100,000,000	5,761,455	17.4
1,000,000,000	50,847,534	19.7
10,000,000,000	455,052,511	22.0

in the last column that the pattern manifests itself. This column represents the proportion of prime numbers amongst all the numbers being considered. For example, 1 in 4 numbers are prime counting up to 100, so in that interval you will need to count, on average, 4 to get from one prime to the next. Of the numbers up to 10 million, 1 in 15 are prime. (So, for example, there is a 1 in 15 chance that a seven-digit telephone number is a prime.) For N greater than 10,000, this last column seems to be just increasing by about 2.3 each time.

So every time Gauss multiplied by 10, he had to add about 2.3 to the ratio of the primes to all the numbers. This link between multiplication and addition is precisely the relationship embodied in a logarithm. Gauss, with his book of logarithms, would have found this connection staring him in the face.

The reason why the proportion of primes was increasing by 2.3 rather than 1 every time Gauss multiplied by 10 is because primes favour logarithms based not on powers of 10 but on powers of a different number. Pressing the 'log' button on your calculator when you input 100 produced the answer 2, which is the solution to the equation $10^x = 100$. But there is nothing that says we have to have 10 as the number to raise to the power x. It is our obsession with our ten fingers which makes 10 so appealing. The choice of the number 10 is called the *base* of the logarithm. We can talk about the logarithm of a number to a base other than 10. For example, the logarithm of 128 to base 2 rather than base 10 requires us to solve a different puzzle, to find a number x so that $2^x = 128$. If we had a 'log to base 2' button on our calculator, we could press it and get the answer 7, because we need to raise 2 to the power of 7 to get up to 128: $2^7 = 128$.

What Gauss discovered is that primes can be counted using logarithms to the base of a special number, called e, which to twelve decimal places is 2.718 281 828 459 . . . (like π, it has an infinite decimal expansion with no repeating patterns). e turns out to be as important in mathematics as the number π, and occurs all over the mathematical world. This is why logarithms to the base e are called 'natural' logarithms.

The table that Gauss had made at the age of fifteen led him to the following guess. For the numbers from 1 to N, roughly 1 out of every $\log(N)$ numbers will be prime (where $\log(N)$ denotes the logarithm of N to the base e). He could then estimate the number of primes from 1 to N as roughly $N/\log(N)$. Gauss was not claiming that this magically gave him an exact formula for the number of primes up to N – it just seemed to provide a very good ballpark estimate.

It was a similar philosophy that he would later apply in his rediscovery of Ceres. His astronomical method made a good prediction for a small region of space to look at, given the data that had been recorded. Gauss had taken the same approach for the primes. Generations had become obsessed with trying to predict the precise location of the next prime, with producing formulas that would generate prime numbers. By not getting hung up on the minutiae of which number was prime or not, Gauss had hit on some sort of pattern. By stepping back and asking the broader question of how many primes there are up to a million rather than precisely which numbers are prime, a strong regularity seemed to emerge.

Gauss had made an important psychological shift in looking at the primes. It was as if previous generations had listened to the music of the primes note by note, unable to hear the whole composition. By concentrating instead on counting how many primes there were as one counted higher, Gauss found a new way to hear the dominant theme.

Following Gauss, it has become customary to denote the number of primes we find in the numbers from 1 to N by the symbol $\pi(N)$ (this is just a name for this count and has nothing to do with the number π). It is perhaps unfortunate that Gauss used a symbol that makes one think of circles and the number 3.1415 . . . Think of this instead as a new button on your calculator. You input a number N and press the $\pi(N)$ button, and the calculator outputs the number of primes up to N. So, for example, $\pi(100) = 25$, the number of primes up to 100, and $\pi(1{,}000) = 168$.

Notice that you can still use this new 'count primes' button to identify precisely when you get a prime. If you input the number 100 and press this button, counting primes between 1 and 100, you get the answer 25. If you now input the number 101, the answer will go up by 1 to 26 primes, and this means that 101 is a new prime number. So whenever there is a difference between $\pi(N)$ and $\pi(N+1)$, you know that $N+1$ must be a new prime.

To reveal quite how stunning Gauss's pattern was, we can look at a graph of the function $\pi(N)$ counting the number of primes from 1 to N. Here's a graph of $\pi(N)$ for numbers N from 1 to 100:

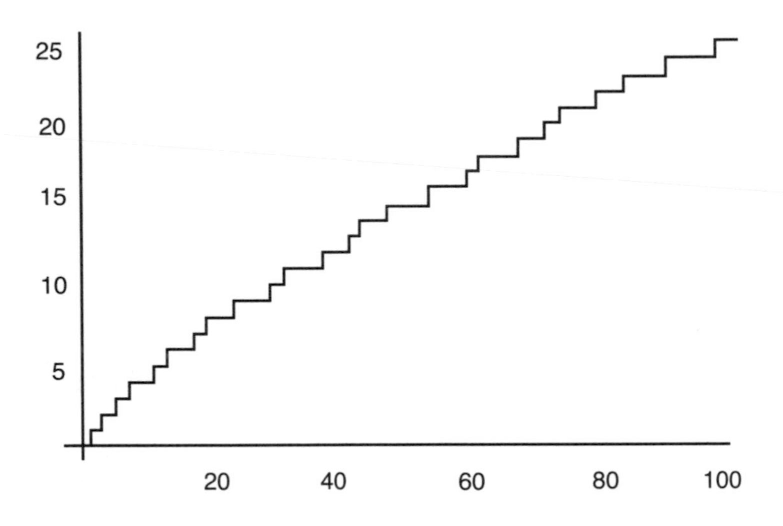

The prime number staircase – the graph counts the cumulative number of primes up to 100.

On this small scale the result is a jumpy staircase, where it is hard to predict how long one has to wait for the next step to appear. We are still seeing at this scale the minutiae of the primes, the individual notes.

Now step back and look at the same function where N is taken over a much wider range of numbers, say counting the primes up to 100,000 (see opposite).

The individual steps themselves become insignificant and we end up seeing the overarching trend of this function creeping up. This was the big theme that Gauss had heard and could imitate using the logarithm function.

The revelation that the graph appears to climb so smoothly, even though the primes themselves are so unpredictable, is one of the most miraculous in mathematics and represents one of the high points in the story of the primes. On the back page of his book of logarithms, Gauss

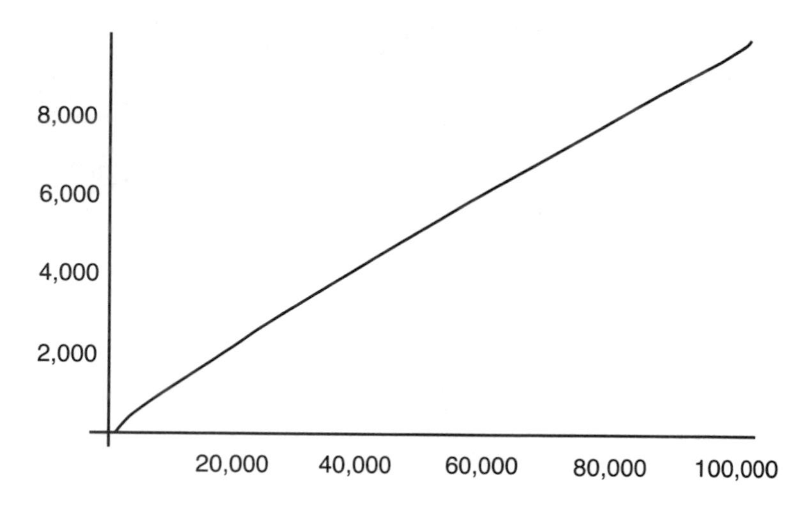

The prime number staircase counting primes up to 100,000.

recorded the discovery of his formula for the number of primes up to N in terms of the logarithm function. Yet despite the importance of the discovery, Gauss told no one what he had found. The most the world heard of his revelation were the cryptic words, 'You have no idea how much poetry there is in a table of logarithms.'

Why Gauss was so reticent about something so momentous is a mystery. It is true that he had only found early evidence of some connection between primes and the logarithm function. He knew that he had absolutely no explanation or proof of why these two things should have anything to do with each other. It wasn't clear that this pattern might not suddenly disappear as you counted higher. Gauss's reluctance to announce unproved results marked a turning point in mathematical history. Although the Greeks had introduced the idea of proof as an important component of the mathematical process, mathematicians before Gauss's time were much more interested in scientific speculation about mathematics. If the mathematics worked, they weren't too concerned about a rigorous justification of why it worked. Mathematics was still the tool of the other sciences.

Gauss broke from the past by stressing the value of proof. For him, presenting proofs was the primary goal of the mathematician, an ethos which has remained fundamental to this day. Without a proof of the connection between logarithms and primes, Gauss's discovery was worthless to him. The freedom that the patronage of the Duke of Brunswick permitted him meant he could be quite choosy, almost indulgent about the

Gauss honoured by the Deutsche Bundesbank on a 10-mark note.

mathematics he produced. His prime motivation was not fame and recognition but a personal understanding of the subject he loved. His seal bore the motto *Pauca sed matura* ('Few but ripe'). Unless a result had reached full maturity it remained an entry in his diary or a doodle at the back of his table of logarithms.

For Gauss, mathematics was a private pursuit. He even encrypted entries in his diary using his own secret language. Some of them are easy to unravel. For example, on July 10, 1796, Gauss wrote Archimedes' famous declaration 'Eureka!' followed by the equation

$$\text{num} = \Delta + \Delta + \Delta$$

which represents his discovery that every number can be written as the sum of three numbers from the list of triangular numbers, 1, 3, 6, 10, 15, 21, 28, . . ., those numbers for which Gauss had produced his schoolroom formula. For example, $50 = 1 + 21 + 28$. But other entries remain a complete mystery. No one has been able to unravel what Gauss meant when he wrote on October 11, 1796, 'Vicimus GEGAN'. Some have blamed Gauss's failure to disseminate his discoveries for holding back the development of mathematics by half a century. If he had bothered to explain half of what he had discovered and not been so cryptic in the explanations he did offer, mathematics might have advanced at a quicker pace.

Some people believe that Gauss kept his results to himself because the Paris Academy had rejected his great treatise on number theory, *Disquisitiones Arithmeticae*, as obscure and dense. Having been stung by rejection, to protect himself from any further humiliation he insisted that every last piece of the mathematical jigsaw be in place before he would consider publishing anything. *Disquisitiones Arithmeticae* did not receive immediate acclaim partly because Gauss continued to be cryptic even in the work he did expose to public view. He always claimed that mathematics was like a piece of architecture. An architect never leaves the scaffolding for people to see how the building was constructed. This was not a philosophy that helped mathematicians to penetrate Gauss's mathematics.

But there were other reasons why Paris was not as receptive to Gauss's ideas as he had hoped. By the end of the eighteenth century, mathematics in Paris was becoming ever more dedicated to serving the demands of an increasingly industrialised state. The Revolution of 1789 had shown Napoleon the need for a more centralised teaching of military engineering, and he responded by founding the École Polytechnique to further his war aims. 'The advancement and perfection of mathematics are intimately connected with the prosperity of the State,' Napoleon declared. French mathematics was dedicated to solving problems of ballistics and hydraulics. But despite this emphasis on the practical needs of the state, Paris still boasted some of the leading pure mathematicians in Europe.

One of the great authorities in Paris was Adrien-Marie Legendre, who was born twenty-five years before Gauss. Portraits of Legendre depict a rather puffed-up gentleman with a round, chubby face. In contrast to Gauss, Legendre came from a wealthy family but he had lost his fortune during the Revolution and been forced to rely instead on his mathematical talents for his livelihood. He too was interested in the primes and number theory, and in 1798, six years after Gauss's childhood calculations, he announced his discovery of the experimental connection between primes and logarithms.

Although it would later be proved that Gauss had indeed beaten Legendre to the discovery, Legendre did nonetheless improve on the estimate for the number of primes up to N. Gauss had guessed that there were roughly $N/\log(N)$ primes up to N. Although this was close, it was found to gradually drift away from the true number of primes as N got larger and larger. Here is a comparison of Gauss's childhood guess, shown as the lower plot, with the true number of primes, the upper plot:

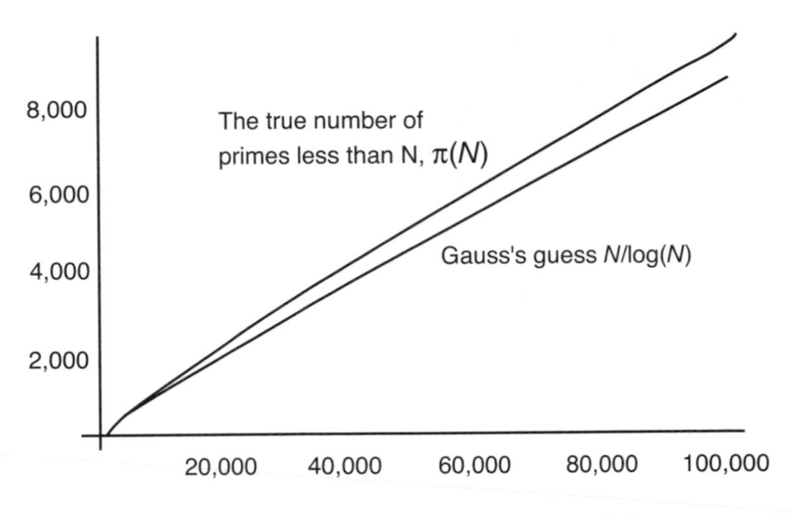

A comparison between Gauss's guess and the true number of primes.

This graph reveals that although Gauss was on to something, there still seemed to be room for improvement.

Legendre's improvement was to replace the approximation $N/\log(N)$ by

$$\frac{N}{\log(N) - 1.80366}$$

thus introducing a small correction which had the effect of shifting Gauss's curve up towards the true number of primes. As far as the values of these functions that were then within computational reach, it was impossible to distinguish the two graphs of $\pi(N)$ and Legendre's estimate. Legendre, steeped in the prevailing preoccupation with the practical application of mathematics, was much less reluctant to stick his neck out and make some prediction about the connection between primes and logarithms. He was not a man scared to circulate unproven ideas, or even proofs with gaps in them. In 1808 he published his guess at the number of primes in a book about number theory entitled *Théorie des Nombres*.

The controversy over who first discovered the connection between primes and logarithms led to a bitter dispute between Legendre and Gauss. It was not limited to the argument about primes – Legendre even claimed that he had been the first to discover Gauss's method for establishing the motion of Ceres. Time and again, Legendre's assertion that he had uncovered some mathematical truth would be countered by an announcement by Gauss that he had already plundered that particular treasure. Gauss commented in a letter of July 30, 1806, written to an astronomical colleague named Schumacher, 'It seems to be my fate to concur in nearly all my theoretical works with Legendre.'

In his lifetime, Gauss was too proud to get involved in open battles of priority. When Gauss's papers and correspondence were examined after his death, it became clear that due credit invariably went to Gauss. It wasn't until 1849 that the world learnt that Gauss had beaten Legendre to the connection between primes and logarithms, which Gauss disclosed in a letter to a fellow mathematician and astronomer, Johann Encke, written on Christmas Eve of that year.

Given the data available at the start of the nineteenth century, Legendre's function was much better than Gauss's formula as an approximation to the number of primes up to some number N. But the appearance of the rather ugly correction term 1.083 66 made mathematicians believe that something better and more natural must exist to capture the behaviour of the prime numbers.

Such ugly numbers may be commonplace in other sciences, but it is remarkable how often the mathematical world favours the most aesthetic possible construction. As we shall see, Riemann's Hypothesis can be interpreted as an example of a general philosophy among mathematicians that, given a choice between an ugly world and an aesthetic one, Nature always chooses the latter. It is a constant source of amazement for most mathematicians that mathematics should be like this, and explains why they so often get wound up about the beauty of their subject.

It is perhaps not surprising that in later life Gauss further refined his guess and arrived at an even more accurate function, one which was also much more beautiful. In the same Christmas Eve letter that Gauss wrote to Encke, he explains how he subsequently discovered how to go one better than Legendre's improvement. What Gauss did was to go back to his very first investigations as a child. He had calculated that amongst the first 100 numbers, 1 in 4 are prime. When he considered the first 1,000 numbers the chance that a number is prime went down to 1 in 6. Gauss realised that the higher you count, the smaller the chance that a number will be prime.

So Gauss formed a picture in his mind of how Nature might have decided which numbers were going to be prime and which were not. Since their distribution looked so random, might tossing a coin not be a good model for choosing primes? Did Nature toss a coin – heads it's prime, tails it's not? Now, thought Gauss, the coin could be weighted so that instead of landing heads half the time, it lands heads with probability $1/\log(N)$. So the probability that the number 1,000,000 is prime should be interpreted as $1/\log(1{,}000{,}000)$, which is about $1/15$. The chances that each number N is a prime gets smaller as N gets bigger because the probability $1/\log(N)$ of coming up heads is getting smaller.

This is just a heuristic argument because 1,000,000 or any other particular number is either prime or it isn't. No toss of a coin can alter that. Although Gauss's mental model was useless at predicting whether a number is prime, he found it very powerful at making predictions about the less specific question of how many primes one might expect to encounter as one counted higher. He used it to estimate the number of primes you should get after tossing the prime number coin N times. With a normal coin which lands heads with probability $\frac{1}{2}$, the number of heads should be $\frac{1}{2}N$. But the probability with the prime number coin is getting smaller with each toss. In Gauss's model the number of primes is predicted to be

$$\frac{1}{\log(2)} + \frac{1}{\log(3)} + \ldots + \frac{1}{\log(N)}$$

Gauss actually went one step further to produce a function which he called the logarithmic integral, denoted by $\mathrm{Li}(N)$. The construction of this new function was based on a slight variation of the above sum of probabilities, and it turned out to be stunningly accurate.

By the time Gauss, in his seventies, wrote to Encke, he had constructed tables of primes up to 3,000,000. 'I very often used an idle quarter of an hour to count through another chiliad [an interval of 1,000 numbers] here and there' in his search for prime numbers. His estimate for primes less than 3,000,000 using his logarithmic integral is a mere seven hundredths of 1 per cent off the mark. Legendre had managed to massage his ugly formula to match $\pi(N)$ for small N, so with the data available at the time it looked as if Legendre's formula was superior. As more extensive tables began to be drawn up, they revealed that Legendre's estimate grew far less accurate for primes beyond 10,000,000. A professor at the University of Prague, Jakub Kulik, spent twenty years single-handedly constructing tables of primes for numbers up to 100,000,000. The eight volumes of this gargantuan effort, completed in 1863, were never published but were deposited in the archives of the Academy of Sciences in Vienna. Although the second volume has gone astray, the tables already revealed that Gauss's method, based on his $\mathrm{Li}(N)$ function, was once again outstripping Legendre's. Modern tables show just how much better Gauss's intuition was. For example, his estimate for the number of primes up to 10^{16} (i.e. 10,000,000,000,000,000) deviates from the correct value by just one ten-millionth of 1 per cent, whilst Legendre's is now off by one-tenth of 1 per cent. Gauss's theoretical analysis had triumphed over Legendre's attempts to manipulate his formula to match the available data.

Gauss noticed a curious feature about his method. Based on what he knew about the primes up to 3,000,000, he could see that his formula Li(N) always appeared to overestimate the number of primes. He conjectured that this would always be the case. And who wouldn't back Gauss's hunch, now that modern numerical evidence confirms Gauss's conjecture up to 10^{16}? Certainly any experiment that produced the same result 10^{16} times would be regarded as pretty convincing evidence in most laboratories – but not in the mathematician's. For once, one of Gauss's intuitive guesses turned out to be wrong. But although mathematicians have now proved that eventually $\pi(N)$ must sometimes overtake Li(N), no one has ever seen it happen because we can't count far enough yet.

A comparison of the graphs of $\pi(N)$ and Li(N) shows such a good match that over a large range it is barely possible to distinguish the two. I should stress that a magnifying glass applied to any portion of such a picture will show that the functions are different. The graph of $\pi(N)$ looks like a staircase, whilst Li(N) is a smooth graph with no sharp jumps.

Gauss had uncovered evidence of the coin that Nature had tossed to choose the primes. The coin was weighted so that a number N has a chance of 1 in log(N) of being prime. But he was still missing a method of predicting precisely the tosses of the coin. That would take the insight of a new generation.

By shifting his perspective, Gauss had perceived a pattern in the primes. His guess became known as the Prime Number Conjecture. To claim Gauss's prize, mathematicians had to prove that the percentage error between Gauss's logarithmic integral and the real number of primes gets smaller and smaller the further you count. Gauss had seen this far-off mountain peak, but it was left to future generations to provide a proof, to reveal the pathway to the summit, or to unmask the connection as an illusion.

Many blame the appearance of Ceres for distracting Gauss from proving the Prime Number Conjecture himself. The overnight fame he received at the age of twenty-four steered him towards astronomy, and mathematics no longer had pride of place. When his patron, Duke Ferdinand, was killed by Napoleon in 1806, Gauss was forced to find other employment to support his family. Despite overtures from the Academy in St Petersburg, which was seeking a successor to Euler, he chose instead to accept a position as director of the Observatory in Göttingen, a small university town in Lower Saxony. He spent his time tracking more asteroids through the night sky and completing surveys of the land for the Hanoverian and Danish governments. But he was always thinking about mathematics. Whilst charting the mountains of Hanover he would ponder Euclid's axiom of parallel lines, and back in the observatory he would continue to expand his table of primes. Gauss had heard the first big theme in the music of the primes. But it was one of his few students, Riemann, who would truly unleash the full force of the hidden harmonies that lay behind the cacophony of the primes.

CHAPTER THREE

Riemann's Imaginary Mathematical Looking-Glass

Do you not feel and hear it? Do I alone hear this melody so wondrously and gently sounding . . . Richard Wagner, *Tristan und Isolde* (Act III, Scene iii)

In 1809 Wilhelm von Humboldt became the education minister for the north German state of Prussia. In a letter to Goethe in 1816 he wrote, 'I have busied myself here with science a great deal, but I have deeply felt the power antiquity has always wielded over me. The new disgusts me . . .' Humboldt favoured a movement away from science as a means to an end, and a return to a more classical tradition of the pursuit of knowledge for its own sake. Previous education schemes had been geared to providing civil servants for the greater glory of Prussia. From now on, more emphasis was to be placed on education serving the needs of the individual rather than the state.

In his role as a thinker and civil servant, Humboldt enacted a revolution with far-reaching effects. New schools, called Gymnasiums, were created across Prussia and the neighbouring state of Hanover. Eventually the teachers in these schools were not to be members of the clergy, as in the old education system, but graduates of the new universities and polytechnics that were built during this period.

The jewel in the crown was Berlin University, founded in 1810 during the French occupation. Humboldt called it the 'mother of all modern universities'. Housed in what had once been the palace of Prince Heinrich of Prussia on the grand boulevard Unter den Linden, the university would for the first time promote research alongside teaching. 'University teaching not only enables an understanding of the unity of science but also its furtherance,' Humboldt declared. Despite his passion for the Ancient World, under his guidance the university pioneered the introduction of new disciplines to sit beside the classical faculties of Law, Medicine, Philosophy and Theology.

For the first time, the study of mathematics was to form a major part of the curriculum in the new Gymnasiums and universities. Students were encouraged to study mathematics for its own sake and not simply as a servant of the other sciences. This contrasted starkly with Napoleon's educational reforms, which saw mathematics harnessed to further French military aims. Carl Jacobi, one of the professors in Berlin, wrote to

Legendre in Paris in 1830 about the French mathematician Joseph Fourier, who had reproached the German school of thought for ignoring more practical problems:

> It is true that Fourier was of the opinion that the principal object of mathematics is public use and the explanation of natural phenomena; but a philosopher like him ought to have known that the sole object of the science is the honour of the human spirit, and that on this view a problem in the theory of numbers is worth as much as a problem of the system of the world.

For Napoleon, it was education that would finally destroy the arcane rules of the *ancien régime*. His recognition that education was the backbone for building his new France had led to the establishment in Paris of some of the institutes which are still famous today. Not only were the colleges meritocratic, allowing students from all backgrounds to attend, but also the educational philosophy put a greater emphasis on education and science serving society. One of the French Revolutionary regional officers wrote to a professor of mathematics in 1794, commending him on teaching a course in 'Republican arithmetic': 'Citizen. The Revolution not only improves our morals and paves the way for our happiness and that of future generations, it even unlooses the shackles that hold back scientific progress.'

Humboldt's approach to mathematics was very different from this utilitarian philosophy that prevailed across the border. The liberating effect of Germany's educational revolution was to have a great impact on mathematicians' understanding of many aspects of their field. It would allow them to establish a new, more abstract language of mathematics. In particular, it would revolutionise the study of prime numbers.

One town that benefited from Humboldt's initiatives was Lüneberg, in Hanover. Lüneberg, once a thriving commercial centre, was now a town in decline. Its narrow streets paved with cobblestones were no longer buzzing with the business it had seen in previous centuries. But in 1829 a new building was erected amidst the tall towers of the three Gothic churches in Lüneberg: the Gymnasium Johanneum.

By the early 1840s the new school was flourishing. Its director, Schmalfuss, was a keen proponent of the neo-humanist ideals initiated by Humboldt. His library reflected his enlightened views: it featured not only the classics and the works of modern German writers, but also volumes from farther afield. In particular, Schmalfuss managed to get his hands on books coming out of Paris, the powerhouse of European intellectual activity during the first half of the century.

Schmalfuss had just accepted a new boy at the Gymnasium Johanneum, Bernhard Riemann. Riemann was very shy and found it difficult to make friends. He had been attending the Gymnasium in the town of Hanover, where he had been boarding with his grandmother, but when she died, in 1842, he was forced to move to Lüneberg where he could board with one of the teachers. Joining the school after all his contemporaries had established friendships did not make life easy for Riemann. He was desperately homesick and was teased by the other children. He would rather walk the long distance back to his father's house in Quickborn than play with his contemporaries.

Riemann's father, the pastor in Quickborn, had high expectations for his son. Although Bernhard was unhappy at school, he worked hard and conscientiously, determined not to disappoint his father. But he had to battle with an almost disabling streak of perfectionism. His teachers would often get frustrated at Riemann's inability to submit his work. Unless it was perfect, the boy could not bear to suffer the indignity of anything less than full marks. His teachers began to doubt whether Riemann would ever be able to pass his final examinations.

It was Schmalfuss who saw a way to bring the young boy on and exploit his obsession with perfection. Early on, Schmalfuss had spotted Riemann's special mathematical skills and was keen to stimulate the student's abilities. He allowed Riemann the freedom of his library, with its fine collection of books on mathematics, where the boy could escape the social pressures of his classmates. The library opened up a whole new world for Riemann, a place where he felt at home and in control. Suddenly here he was in a perfect, idealised mathematical world where proof prevented any collapse of this new world around him, and numbers became his friends.

Humboldt's drive from teaching science as a practical tool to the more aesthetic notion of knowledge for its own sake had filtered down to Schmalfuss's classroom. The teacher steered Riemann's reading away from mathematical texts full of formulas and rules that were aimed at feeding the demands of a growing industrial world, and guided him towards the classics of Euclid, Archimedes and Apollonius. With their geometry, the ancient Greeks sought to understand the abstract structure of points and lines, and they were not hung up on the particular formulas behind the geometry. When Schmalfuss did give Riemann a more modern text, Descartes's treatise on analytical geometry – a subject rife with equations and formulas – the teacher could see that the mechanical method developed in the book did not appeal to Riemann's growing taste for conceptual mathematics. As Schmalfuss later recalled in a letter to a friend, 'already at that time he was a mathematician next to whose wealth a teacher felt poor'.

One of the books that was sitting on Schmalfuss's shelf was a contemporary volume the teacher had acquired from France. Published in 1808, *Théorie des Nombres* by Adrien-Marie Legendre was the first text to record the observation that there seemed to be a strange connection between the function that counted the number of prime numbers and the logarithm function. This connection, discovered by Gauss and Legendre, was only based on experimental evidence. It was far from clear whether, as one counted higher, the number of primes would always be approximated by Gauss's or Legendre's function.

Despite the volume's 859 large quarto pages, Riemann gobbled it up, returning the book to his teacher just six days later with the precocious declaration, 'This is a wonderful book; I know it by heart.' His teacher could not believe it, but when he examined Riemann on its contents during his final examinations two years later, the student answered perfectly. That marked the beginning of the career of one of the giants of modern mathematics. Thanks to Legendre, a seed was sown in the young Riemann's mind that in later life was to blossom in spectacular fashion.

His final examinations over, Riemann was eager to join one of the vigorous new universities that were driving the educational revolution in Germany. His father, though, had other ideas. Riemann's family was poor, and his father hoped that Bernhard would join him in the Church. The life of a clergyman would provide him with a regular income with which he could support his sisters. The only university in the Kingdom of Hanover to offer theology was not one of these new establishments but the University of Göttingen, founded over a century before, in 1734. So in 1846, to comply with his father's wishes, Riemann made his way to the dank town of Göttingen.

Göttingen sits quietly in the gentle hills of Lower Saxony. At its heart lies a small medieval town enclosed by ancient ramparts. This is the Göttingen that Riemann would have known, and it still retains much of its original character. The streets wind narrowly between half-timbered houses topped with red-tiled roofs. The Brothers Grimm wrote many of their fairy tales in Göttingen, and one can imagine Hansel and Gretel running through its streets. In the centre stands the medieval town hall, whose

Bernhard Riemann (1826–66).

walls display the motto 'Away from Göttingen there is no life.' For those at the university that was certainly the feeling. The academic life was one of self-sufficiency. Although theology had predominated in the early years of the university, the winds of academic change sweeping across Germany had stimulated Göttingen's scientific curriculum. By the time Gauss was appointed as professor of astronomy and director of the observatory in 1807, it was science rather than theology for which Göttingen was becoming famous.

The fire for mathematics that the teacher Schmalfuss had ignited in the young Riemann was still burning strong. His father's wish that he study theology had brought him to Göttingen, but it was the influence of the great Gauss and Göttingen's scientific tradition that left its mark during that first year. It was only a matter of time before Greek and Latin lectures gave way to the temptation of courses in physics and mathematics. With trepidation, Riemann wrote to his father suggesting that he would like to switch from theology to mathematics. His father's approval meant everything to Riemann. With a sense of relief he received his father's blessing, and immediately immersed himself in the scientific life of the university.

To a young man of such talent, Göttingen soon began to feel small.

Within a year Riemann had exhausted the resources available to him. Gauss, by now an old man, had become quite withdrawn from the intellectual life of the university – since 1828 he had spent only one night away from the observatory, where he lived. At the university he only lectured on astronomy, and in particular on the method that had made him famous when he'd rediscovered the 'lost' planet Ceres many years before. Riemann had to look elsewhere for the stimulus he needed to take the next step in his development. He could see Berlin was where the buzz of intellectual activity was the loudest.

The University of Berlin had been greatly influenced by the successful French research institutes, such as the École Polytechnique, that had been founded by Napoleon. It had, after all, been founded during the French occupation. One of the key mathematical ambassadors was a brilliant mathematician by the name of Peter Gustav Lejeune-Dirichlet. Although he was born in Germany in 1805, Dirichlet's family was of French origin. A return to his roots took him to Paris in 1822, where he spent five years soaking up the intellectual activity that was bubbling out of the academies. Wilhelm von Humboldt's brother Alexander, an amateur scientist, met Dirichlet on his travels and was so impressed that he secured him a position back in Germany. Dirichlet was something of a rebel. Perhaps the atmosphere on the streets in Paris had given him a taste for challenging authority. In Berlin, he was quite happy to ignore some of the antiquated traditions demanded by the rather stuffy university authorities, and often flouted their demands to demonstrate his command of the Latin language.

Göttingen and Berlin offered emerging scientists such as Riemann contrasting academic climates. Göttingen revelled in its independence and isolation. Few seminars were presented by anyone from beyond the city walls. It was self-sufficient and generated great science from the fuel burning within. Berlin, on the other hand, thrived on the stimulation coming from outside Germany. The ideas feeding through from France mixed with the forward-looking German approach to natural philosophy to create a heady new cocktail.

The different climates of Göttingen and Berlin suited different mathematicians. Some would never have succeeded without contact with new ideas from external sources. The success of other mathematicians can be traced back to an isolation which forced them to find an inner strength and new languages and modes of thought. Riemann would turn out to be someone whose breakthroughs came from contact with the wealth of new ideas that were in the air, and he could see that Berlin was the place to be.

Riemann made his move in 1847 and remained in Berlin for two years.

While there, he was able to get his hands on papers by Gauss which he had not been able to prise from the reticent master in Göttingen. He attended lectures by Dirichlet, who was later to play a part in Riemann's dramatic discoveries about prime numbers. By all accounts, Dirichlet was an inspiring lecturer. One mathematician who attended his lectures put it thus:

> Dirichlet cannot be surpassed for richness of material and clear insight . . . he sits at the high desk facing us, puts his spectacles up on his forehead, leans his head on both hands, and . . . inside his hands he sees an imaginary calculation and reads it out to us – that we understand it as well as if we too saw it. I like that kind of lecturing.

Riemann made friends with several young researchers in Dirichlet's seminars who were equally fired up by their passion for mathematics.

Other forces were also bubbling away in Berlin. The revolution of 1848 that swept away the French monarchy spread from the streets of Paris throughout much of Europe. It found its way to the streets of Berlin while Riemann was studying there. According to accounts of his contemporaries, it had a profound impact on him. On one of the few occasions in his life on which he joined with those around him on anything other than an intellectual level, he enlisted in the student corps defending the king in his Berlin palace. It is reputed that he did a continuous sixteen-hour stint on the barricades.

Riemann's response to the mathematical revolution spreading from the Paris academies was not that of a reactionary. Berlin was importing not only political propaganda from Paris, but also many of the prestigious journals and publications coming out of the academies. Riemann received the latest volumes of the influential French journal *Comptes Rendus* and holed himself up in his room to pore over papers by the mathematical revolutionary Augustin-Louis Cauchy.

Cauchy was a child of the Revolution, born a few weeks after the fall of the Bastille in 1789. Undernourished by the little food available during those years, the feeble young Cauchy preferred to exercise his mind rather than his body. In time-honoured fashion, the mathematical world provided a refuge for him. A mathematical friend of Cauchy's father, Lagrange, recognised the young boy's precocious talent and commented to a contemporary, 'You see that little young man? Well! He will supplant all of us in so far as we are mathematicians.' But he had interesting advice for Cauchy's father. 'Don't let him touch a mathematical book till he is seventeen.' Instead, he suggested stimulating the boy's literary skills so that when eventually he returned to mathematics he would be able to write with his own mathematical voice and not one he had picked up from the books of the day.

It proved to be sound advice. Cauchy developed a new voice that was irrepressible once the floodgates protecting Cauchy from the outside world had been reopened. Cauchy's output grew to be so immense that the journal *Comptes Rendus* had to impose a page limit on articles it printed that is strictly adhered to even today. Cauchy's new mathematical language was too much for some of his contemporaries. The Norwegian mathematician Niels Henrik Abel wrote in 1826, 'Cauchy is mad . . . what he does is excellent but very muddled. At first I understood practically none of it; now I see some of it more clearly.' Abel goes on to observe that of all the mathematicians in Paris, Cauchy was the only one doing 'pure mathematics' whilst others 'busy themselves exclusively with magnetism and other physical subjects . . . he is the only one who knows how mathematics should be done'.

Cauchy was to land himself in trouble with the authorities in Paris for steering students away from practical applications of mathematics. The director of the École Polytechnique, where Cauchy was lecturing, wrote to him criticising him for his obsession with abstract mathematics: 'It is the opinion of many persons that instruction in pure mathematics is being carried too far at the École and that such an uncalled for extravagance is prejudicial to the other branches.' So it was perhaps no wonder that Cauchy's work would be appreciated by the young Riemann.

So exciting were these new ideas that Riemann almost became a recluse. His contemporaries were to see nothing of him while he waded through Cauchy's output. Several weeks later Riemann resurfaced, declaring that 'this is a new mathematics'. What had captured Cauchy and Riemann's imagination was the emerging power of imaginary numbers.

Imaginary numbers – a new mathematical vista

The square root of minus one, the building block of imaginary numbers, seems to be a contradiction in terms. Some say that admitting the possibility of such a number is what separates the mathematicians from the rest. A creative leap is required to gain access to this bit of the mathematical world. At first sight it looks as if it has nothing to do with the physical world. The physical world seems to be built on numbers whose square is always a positive number. Imaginary numbers, however, are more than just an abstract game. They hold the key to the twentieth-century world of subatomic particles. On a larger scale, aeroplanes would not have taken to the skies without engineers taking a journey through the world of imaginary numbers. This new world provides a flexibility denied to those who stick to ordinary numbers.

The story of how these new numbers were discovered begins with the need to solve simple equations. As the ancient Babylonians and Egyptians recognised, if seven fish were to be divided between three people, for example, fractional numbers – $\frac{1}{2}$, $\frac{1}{3}$, $\frac{2}{3}$, $\frac{1}{4}$, and so on – would have to come into the equation. By the sixth century BC, the Greeks had discovered while exploring the geometry of triangles that these fractions were sometimes incapable of expressing the lengths of the sides of a triangle. Pythagoras' theorem forced them to invent new numbers that couldn't be written as simple fractions. For example, Pythagoras could take a right-angled triangle whose two shortest sides are one unit long. His famous theorem then told him that the longest side had length x, where x is a solution of the equation $x^2 = 1^2 + 1^2 = 2$. In other words, the length is the square root of 2.

Fractions are the numbers whose decimal expansions have a repeating pattern. For example, $\frac{1}{7} = 0.142\,857\,142\,857\ldots$ or $\frac{1}{4} = 0.250\,000\,000\ldots$ In contrast, the Greeks could prove that the square root of 2 is not equal to a fraction. However far you calculate the decimal expansion of the square root of 2, it will never settle down into such a repeating pattern. The square root of 2 starts off 1.414 213 562 . . . Riemann used to idle away the hours calculating more and more of these decimal places during his years in Göttingen. His record was thirty-eight places, no mean feat without a computer but perhaps more a reflection on the dull Göttingen nightlife and Riemann's shy persona that this was his evening entertainment. Nonetheless, however far Riemann calculated, he knew that he could never write down the complete number or discover a repeating pattern.

To capture the impossibility of expressing such numbers in any way other than as solutions to equations such as $x^2 = 2$, mathematicians called them *irrational numbers*. The name reflected mathematicians' sense of unease at their inability to write down precisely what these numbers were. Nevertheless, there was still a sense of the reality of these numbers since they could be *seen* as points marked on a ruler, or on what mathematicians call the number line. The square root of 2, for example, is a point somewhere between 1.4 and 1.5. If one could make a perfect Pythagorean right-angled triangle with the two short sides one unit long, then the location of this irrational number could be determined by laying the long side against the ruler and marking off the length.

The negative numbers were discovered similarly out of attempts to solve simple equations such as $x + 3 = 1$. Hindu mathematicians proposed these new numbers in the seventh century AD. Negative numbers were created in response to the growing world of finance, as they were useful for describing debt. It took European mathematicians another millennium before they were happy to admit the existence of such 'fictitious numbers', as they were called. Negative numbers took their place on the number line stretching out to the left of zero.

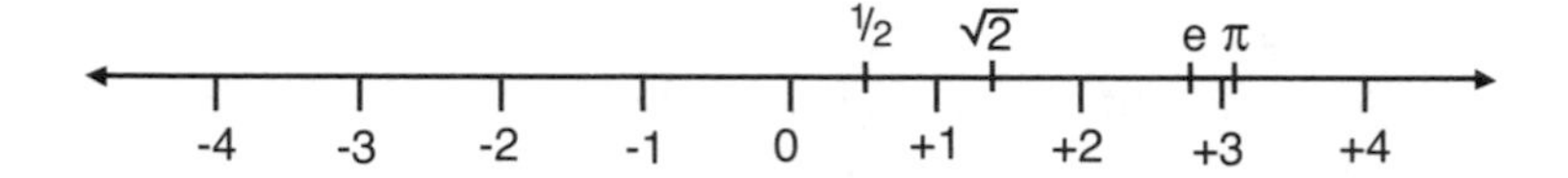

The real numbers – every fraction, negative number or irrational number is represented by a point on the number line.

Irrational numbers and negative numbers allow us to solve many different equations. Fermat's equation $x^3 + y^3 = z^3$ has interesting solutions if you don't insist, as Fermat had, that x, y and z should be whole numbers. For example, we can choose $x = 1$ and $y = 1$, and put z equal to the cube root of 2 – and the equation is solved. But there were still other equations which couldn't be solved in terms of any of the numbers on the number line.

There seemed to be no number which was a solution to the equation $x^2 = -1$. After all, if you square a number, positive or negative, the answer is always positive. So any number that satisfies this equation is not going to be an ordinary number. But if the Greeks could imagine a number like the square root of 2, without being able to write it down as a fraction, mathematicians began to see that they could make a similar imaginative leap and create a new number to solve the equation $x^2 = -1$. Such a creative jump marks one of the conceptual challenges for anyone learning mathematics. This new number, the square root of minus one, was called an *imaginary number* and given the symbol i. In contrast, mathematicians began to refer to the numbers that could be found on the number line as *real numbers*.

To create an answer to this equation, seemingly out of thin air, seems like cheating. Why not accept that the equation has no solution? That is one way forward, but mathematicians like to be more optimistic. Once we accept the idea that there is a new number that *does* solve this equation, the advantages of this creative step far outweigh any initial unease. Once named, its existence seems inevitable. It no longer feels like an artificially created number but a number that had been there all along, unobserved until we'd asked the right question. Eighteenth-century mathematicians had been loath to admit there could be any such numbers. Nineteenth-century mathematicians were brave enough to believe in new modes of thought which challenged the accepted ideas of what constituted the mathematical canon.

Frankly, the square root of -1 is as abstract a concept as the square root of 2. Both are defined as solutions to equations. But would mathematicians have to start creating new numbers for every new equation that came along? What if we want solutions to an equation like $x^4 = -1$? Are we going to have to use more and more letters in our attempts to name all these new solutions? It was with some relief that Gauss finally proved, in his doctoral thesis of 1799, that no new numbers were needed. Using this new number i, mathematicians could finally solve any equation they might come across. Every equation had a solution that consisted of some combination of ordinary real numbers (the fractions and irrational numbers) and this new number, i.

The key to Gauss's proof was to extend the picture we already had of ordinary numbers as lying on a number line: a line running east–west on which each point represents a number. These were the real numbers familiar to mathematicians since the Greeks. But there was no room on the line for this new imaginary number, the square root of -1. So Gauss wondered what would happen if you created a new direction. What if one unit north of the number line were used to represent i? All the new numbers that were needed to solve equations were combinations of i and ordinary numbers, for example $1 + 2i$. Gauss realised that on this two-dimensional map there was a point corresponding to every possible number. The imaginary numbers then simply became coordinates on a map. The number $1 + 2i$ was represented by the point reached by travelling one unit east and two units north.

Gauss would interpret these numbers as sets of directions in his map of the imaginary world. To add two imaginary numbers $A + Bi$ and $C + Di$ just meant following two sets of directions, one after the other. For example, adding together $6 + 3i$ and $1 + 2i$ gets you to the location $7 + 5i$ (see overleaf).

Although this was a very potent picture, Gauss was to keep his map of the imaginary world hidden from public view. Once he had built his proof, he removed the graphic scaffolding so that no trace of his vision remained. He was aware that pictures in mathematics were regarded with some suspicion during this period. The dominance of the French mathematical tradition during Gauss's youth meant that the preferred pathway to the

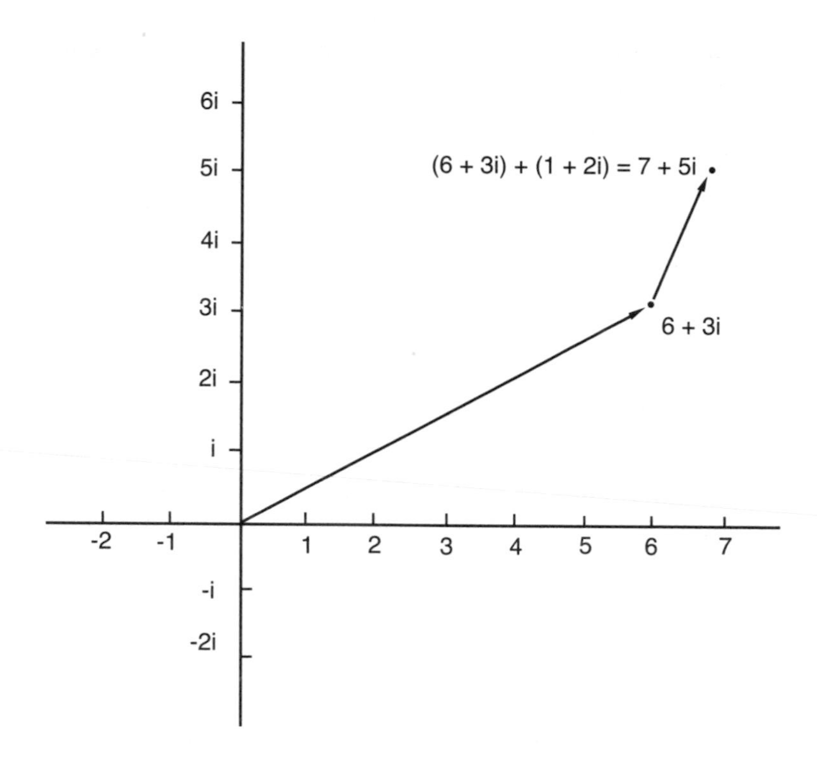

Following directions – how to add two imaginary numbers.

mathematical world was the language of formulas and equations, a language that went hand in hand with the utilitarian approach to the subject. There were also other reasons for this aversion to images.

For several hundred years, mathematicians had believed that pictures had the power to mislead. After all, the language of mathematics had been introduced to tame the physical world. In the seventeenth century, Descartes had attempted to turn the study of geometry into pure statements about numbers and equations. 'Sense perceptions are sense deceptions' was his motto. Riemann had come to dislike this denial of the physical picture when he'd been reading Descartes in the comfort of Schmalfuss's library.

Mathematicians around the turn of the nineteenth century had been burnt by an erroneous pictorial proof of a formula describing the relationship between the number of corners, edges and faces of geometric solids. Euler had conjectured that if a solid has C corners, E edges and F faces, then the numbers C, E and F must satisfy the relationship $C - E + F = 2$. For example, a cube has 8 corners, 12 edges and 6 faces. The young Cauchy had himself constructed a 'proof' in 1811 based on pictorial intuition, but was rather shocked to be shown a solid which didn't satisfy this formula – a cube with a hole at its centre.

The 'proof' had missed the possibility that a solid might contain such a hole. It was necessary to introduce an extra ingredient into the formula which kept track of the number of holes in the solid. Having been tricked by the power of pictures to hide perspectives that weren't initially apparent, Cauchy sought refuge in the security that formulas seemed to provide. One of the revolutions he effected was to create a new mathematical language which allowed mathematicians to discuss the concept of symmetry in a rigorous way without the need for pictures.

Gauss knew that his secret map of imaginary numbers would be anathema to mathematicians at the end of the eighteenth century, so he omitted it from his proof. Numbers were things you added and multiplied, not drew pictures of. Gauss eventually came clean some forty years later about the scaffolding he had used in his doctorate.

Looking-glass world

Even without Gauss's map, Cauchy and other mathematicians had begun to explore what happens if you extend the idea of functions to this new world of imaginary numbers rather than sticking to real numbers. To their surprise, these imaginary numbers opened up new connections between seemingly unrelated parts of the mathematical world.

A function is like a computer program in which you input one number, a calculation is made, and another number is output. The function might be defined by some simple equation like $x^2 + 1$. When you input a number, for example 2, the function calculates $2^2 + 1$ and outputs 5. Other functions are more complicated. Gauss was interested in the function that counted the number of primes. You input a number x, and the function tells you how many primes there are up to x. Gauss had denoted this function by $\pi(x)$. The graph of this function is a climbing staircase, as shown on p. 50. Every time the input encounters a prime number, the output jumps another step. As x goes from 4.9 to 5.1, the number of primes increases from two to three to register the new prime, 5.

Mathematicians soon realised that some of these functions, such as the one built from the equation $x^2 + 1$, could be fed with imaginary numbers as well as ordinary numbers. For example, input $x = 2i$ into the function, and the output is calculated as $(2i)^2 + 1 = -4 + 1 = -3$. The feeding of functions with imaginary numbers had begun in Euler's generation. As early as 1748, Euler had stumbled across strange connections between unrelated bits of mathematics by taking a trip through this looking-glass world. Euler knew that when you fed the exponential function 2^x with ordinary numbers x, you got a graph that climbed rapidly. But when he fed the function with imaginary numbers he got a rather unexpected answer. Instead of the exponentially climbing graph, he started to see undulating waves of the type we now associate with, for example, sound waves. The function that produces these undulating waves is called the *sine function*. The image of the sine function is a familiar repeating curve where every 360 degrees we see the same shape appearing. The sine function is now used in a host of everyday calculations. For example, it can be used to measure the height of a building from ground level by measuring angles. It was Euler's generation who discovered that these sine waves were also the key to reproducing musical sounds. A pure note like the A produced by a tuning fork used to tune a piano can be represented by such a wave.

Euler fed imaginary numbers into the function 2^x. To his surprise, out came waves which corresponded to a particular musical note. Euler showed that the character of each note depended on the coordinates of the corresponding imaginary number. The farther north one is, the higher the pitch. The farther east, the louder the volume. Euler's discovery was the first inkling that these imaginary numbers might open up unexpected new paths through the mathematical landscape. Following Euler, mathematicians began travelling out to this new-found land of imaginary numbers. The search for new connections would become infectious.

Riemann returned to Göttingen in 1849 in order to complete his doctoral thesis for Gauss's consideration. That was the year in which Gauss wrote to his friend Encke of his childhood discovery of the connection between primes and logarithms. Although Gauss probably discussed his discovery with members of the faculty in Göttingen, prime numbers were not yet on Riemann's mind. He was buzzing with the new mathematics from Paris, keen to explore the strange emerging world of functions fed with imaginary numbers.

Cauchy had begun the task of making a rigorous subject out of Euler's first tentative steps into this new territory. Whilst the French were masters of equations and formulaic manipulation, Riemann was ready to capitalise on the German education system's return to a more conceptual view of the world. By November 1851, his ideas had crystallised and he submitted his dissertation to the faculty in Göttingen. His ideas obviously struck a chord with Gauss. He greeted Riemann's doctorate as evidence 'of a creative, active, truly mathematical mind, and of a gloriously fertile originality'.

The Göttingen university library around 1854.

Riemann wrote to his father, keen to tell him about the progress he was making: 'I believe I have improved my prospects with my dissertation. I hope also to learn to write more quickly and more fluently in time, especially if I mingle in society.' But academic life in Göttingen did not at first live up to the thrills of Berlin. It was a somewhat stuffy, insular university, and Riemann lacked the confidence to engage with the old intellectual hierarchy. There were fewer students in Göttingen with whom he could relate. He was mistrustful of other people and never really at ease in a social environment. 'He has done the strangest things here only because he believes that nobody can bear him,' wrote his contemporary Richard Dedekind. Riemann was a hypochondriac and susceptible to bouts of depression. He hid his face behind the security of an increasingly large black beard. He was extremely anxious about his finances, surviving on the uncertainty of half a dozen voluntary students' fees. The workload he undertook combined with the pressures of poverty led to a temporary breakdown in 1854. However, his mood would lighten whenever Dirichlet, the star of the Berlin mathematical tradition, visited Göttingen.

One professor in Göttingen with whom Riemann did manage to strike up a friendship was the eminent physicist Wilhelm Weber. Weber had collaborated on numerous projects with Gauss during their time together in Göttingen. They became a scientific Sherlock Holmes and Dr Watson, Gauss providing the theoretical underpinning which Weber then put into practice. One of their most famous inventions was to realise the potential of electromagnetism to communicate over a distance. They successfully rigged up a telegraph line between Gauss's observatory and Weber's laboratory over which they exchanged messages.

Although Gauss thought the invention a curiosity, Weber saw clearly what their discovery would unleash. 'When the globe is covered with a net of railroads and telegraph wires,' he wrote, 'this net will render services comparable to those of the nervous system in the human body, partly as a means of transport, partly as a means for the propagation of ideas and sensations with the speed of lightning.' The rapid spread of the telegraph, together with the later implementation of Gauss's invention of the clock calculator in computer security, make Gauss and Weber the grandfathers of e-business and the Internet. Their collaboration is immortalised in a statue of the pair in the city of Göttingen.

One visitor to Weber in Göttingen paints a typical picture of a slightly mad inventor: 'a curious little fellow who speaks in a shrill, unpleasant and hesitating voice. He speaks and stutters unceasingly; one has nothing to do but to listen. Sometimes he laughs for no earthly reason, and one feels sorry at not being able to join him.' Weber had a little more of the rebel in him than his collaborator Gauss. He had been one of the 'Göttingen Seven' temporarily dismissed from the faculty for protesting at the arbitrary rule of the Hanoverian king in 1837. For some time after completing his thesis, Riemann worked as Weber's assistant. During this apprenticeship Riemann developed a soft spot for Weber's daughter, but his advances were not reciprocated.

In 1854, Riemann wrote to his father that 'Gauss is seriously ill and the physicians fear that his death is imminent.' Riemann was worried that Gauss might die before examining his habilitation, the degree required to become a professor at a German university. Fortunately Gauss lived long enough to hear Riemann's ideas on geometry and its relationship to physics that had germinated during his work with Weber. Riemann was convinced that the fundamental questions of physics could all be answered using mathematics alone. The developments in physics over the ensuing years would eventually confirm his faith in mathematics. Riemann's theory of geometry is regarded by many as one of his most significant contributions to science, and it would be one of the planks in the platform from which Einstein launched his scientific revolution at the beginning of the twentieth century.

A year later, Gauss died. Although the man had passed, his ideas were to keep mathematicians busy for generations to come. He had left behind his conjectured connection between primes and the logarithm function for the next generation to chew over. Astronomers immortalised the great man in the heavens by naming an asteroid Gaussia. And in the University of Göttingen's anatomical collection one can even find Gauss's brain, pickled for eternity, which was reported to be more richly convoluted than any brain previously dissected.

Dirichlet, whose lectures Riemann had attended in Berlin, was appointed to Gauss's vacant chair. Dirichlet was to bring to Göttingen some of the intellectual excitement that Riemann had enjoyed when he was in Berlin. An English mathematician recorded the impression that a visit to Dirichlet in Göttingen made on him at this time: 'He is rather tall, lanky-looking man with moustache and beard about to turn grey . . . with a somewhat harsh voice and rather deaf: it was early, he was unwashed and unshaved and with his *schlafrock* [dressing gown], slippers, cup of coffee and cigar.' Despite this Bohemian exterior, there burned inside him a desire for rigour and proof that was unequalled at the time. His contemporary in Berlin, Carl Jacobi, wrote to Dirichlet's first patron Alexander von Humboldt that 'Only Dirichlet, not I, nor Cauchy, not Gauss, knows what a perfectly rigorous proof is, but we learn it only from him. When Gauss says he has proved something, I think it is very likely; when Cauchy says it, it is a fifty-fifty bet; when Dirichlet says it, it is certain.'

The arrival of Dirichlet in Göttingen began to shake the social fabric of the town. Dirichlet's wife, Rebecka, was the sister of the composer Felix Mendelssohn. Rebecka loathed the dull Göttingen social scene and threw numerous parties trying to reproduce the Berlin salon atmosphere she had been forced to leave behind.

Dirichlet's less formal approach to the educational hierarchy meant that Riemann was able to discuss mathematics openly with the new professor. Riemann had become rather isolated on his return from Berlin to Göttingen. The combination of Gauss's austere personality in later life and Riemann's shyness meant that Riemann had discussed little with the great master. By contrast, Dirichlet's relaxed manner was perfect for Riemann who, in an atmosphere more conducive to discussion, began to open up. Riemann wrote to his father about his new mentor: 'Next morning Dirichlet was with me for two hours. He read over my dissertation and was very friendly – which I could hardly have expected considering the great distance in rank between us.'

In turn, Dirichlet appreciated Riemann's modesty and also recognised the originality of his work. On occasions Dirichlet even managed to drag Riemann away from the library to join him on walks in the countryside around Göttingen. Almost apologetically, Riemann wrote to his father explaining that these escapes from mathematics did him more good scientifically than if he had stayed at home poring over his books. It was during one of his discussions with Riemann whilst walking through the woods of Lower Saxony that Dirichlet inspired Riemann's next move. It would open up a whole new perspective on the primes.

The zeta function – the dialogue between music and mathematics

During his years in Paris in the 1820s, Dirichlet had become fascinated by Gauss's great youthful treatise *Disquisitiones Arithmeticae*. Although Gauss's book marked a beginning of number theory as an independent discipline, the book was difficult and many failed to penetrate the concise style Gauss preferred. Dirichlet, though, was more than happy to battle with one tough paragraph after another. At night he would place the book under his pillow in the hope that the next morning's reading would suddenly make sense. Gauss's treatise has been described as a 'book of seven seals', but thanks to the labours and dreams of Dirichlet, those seals were broken and the treasures within gained the wide distribution they deserved.

Dirichlet was especially interested in Gauss's clock calculator. In particular, he was intrigued by a conjecture that went back to a pattern spotted by Fermat. If you took a clock calculator with N hours on it and you fed in the primes, then, Fermat conjectured, infinitely often the clock would hit one o'clock. So, for example, if you take a clock with 4 hours there are infinitely many primes which Fermat predicted would leave remainder 1 on division by 4. The list begins 5, 13, 17, 29, . . .

In 1838, at the age of thirty-three, Dirichlet had made his mark in the theory of numbers by proving that Fermat's hunch was indeed correct. He did this by mixing ideas from several areas of mathematics that didn't look as if they had anything to do with one another. Instead of an elementary argument like Euclid's cunning proof that there are infinitely many primes, Dirichlet used a sophisticated function that had first appeared on the mathematical circuit in Euler's day. It was called the *zeta function*, and was denoted by the Greek letter ζ. The following equation provided Dirichlet with the rule for calculating the value of the zeta function when fed with a number x:

$$\zeta(x) = \frac{1}{1^x} + \frac{1}{2^x} + \frac{1}{3^x} + \ldots + \frac{1}{n^x} + \ldots$$

To calculate the output at x, Dirichlet needed to carry out three mathematical steps. First, calculate the exponential numbers 1^x, 2^x, 3^x, . . .,

n^x, . . . Then take the reciprocals of all the numbers produced in the first step. (The reciprocal of 2^x is $1/2^x$.) Finally, add together all the answers from the second step.

It is a complicated recipe. The fact that each number 1, 2, 3, . . . makes a contribution to the definition of the zeta function hints at its usefulness to the number theorist. The downside comes in having to deal with an infinite sum of numbers. Few mathematicians could have predicted what a powerful tool this function would become as the best way to study the primes. It was almost stumbled upon by accident.

The origins of mathematicians' interest in this infinite sum came from music and went back to a discovery made by the Greeks. Pythagoras was the first to discover the fundamental connection between mathematics and music. He filled an urn with water and banged it with a hammer to produce a note. If he removed half the water and banged the urn again, the note had gone up an octave. Each time he removed more water to leave the urn one-third full, then one-quarter full, the notes produced would sound to his ear in harmony with the first note he'd played. Any other notes which were created by removing some other amount of water sounded in dissonance with that original note. There was some audible beauty associated with these fractions. The harmony that Pythagoras had discovered in the numbers $1, \frac{1}{2}, \frac{1}{3}, \frac{1}{4}, \ldots$ made him believe that the whole universe was controlled by music, which is why he coined the expression 'the music of the spheres'.

Ever since Pythagoras' discovery of an arithmetic connection between mathematics and music, people have compared both the aesthetic and the physical traits shared by the two disciplines. The French Baroque composer Jean-Philippe Rameau wrote in 1722 that 'Not withstanding all the experience I may have acquired in music from being associated with it for so long, I must confess that only with the aid of mathematics did my ideas become clear.' Euler sought to make music theory 'part of mathematics and deduce in an orderly manner, from correct principles, everything which can make a fitting together and mingling of tones pleasing'. Euler believed that it was the primes that lay behind the beauty of certain combinations of notes.

Many mathematicians have a natural affinity with music. Euler would relax after a hard day's calculating by playing his clavier. Mathematics departments invariably have little trouble assembling an orchestra from the ranks of their members. There is an obvious numerical connection between the two given that counting underpins both. As Leibniz described it, 'Music is the pleasure the human mind experiences from counting without being aware that it is counting.' But the resonance between the subjects goes much deeper than this.

Mathematics is an aesthetic discipline where talk of beautiful proofs and elegant solutions is commonplace. Only those with a special aesthetic sensibility are equipped to make mathematical discoveries. The flash of illumination that mathematicians crave often feels like bashing notes on a piano until suddenly a combination is found which contains an inner harmony marking it out as different.

G.H. Hardy wrote that he was 'interested in mathematics only as a creative art'. Even for the French mathematicians in Napoleon's academies, the buzz of doing mathematics came not from its practical application but from its inner beauty. The aesthetic experiences of doing mathematics or listening to music have much in common. Just as you might listen to a piece of music over and over and find new resonances previously missed, mathematicians often take pleasure in re-reading proofs in which the subtle nuances that make it hang together so effortlessly gradually reveal themselves. Hardy believed that the true test of a good mathematical proof was that 'the ideas must fit together in a harmonious way. Beauty is the first test: there is no permanent place in the world for ugly mathematics.' For Hardy, 'A mathematical proof should resemble a simple and clear-cut constellation, not a scattered Milky Way.'

Both mathematics and music have a technical language of symbols which allow us to articulate the patterns we are creating or discovering. Music is much more than just the minims and crochets which dance across the musical stave. Similarly, mathematical symbols come alive only when the mathematics is played with in the mind.

As Pythagoras discovered, it is not just in the aesthetic realm that mathematics and music overlap. The very physics of music has at its root the basics of mathematics. If you blow across the top of a bottle you hear a note. By blowing harder, and with a little skill, you can start to hear higher notes – the extra harmonics, the overtones. When a musician plays a note on an instrument they are producing an infinity of additional harmonics, just as you do when you blow across the top of the bottle. These additional harmonics help to give each instrument its own distinctive sound. The physical characteristics of each instrument mean that we hear different combinations of harmonics. In addition to the fundamental note, the clarinet plays only those harmonics produced by odd fractions: $\frac{1}{3}$, $\frac{1}{5}$, $\frac{1}{7}$, . . . The string of a violin, on the other hand, vibrates to create all the harmonics that Pythagoras produced with his urn – those corresponding to the fractions $\frac{1}{2}$, $\frac{1}{3}$, $\frac{1}{4}$, . . .

Since the sound of a vibrating violin string is the infinite sum of the fundamental note and all the possible harmonics, mathematicians became intrigued by the mathematical analogue. The infinite sum $1 + \frac{1}{2} + \frac{1}{3} + \frac{1}{4} + \ldots$ became known as the *harmonic series*. This infinite sum is also the answer Euler got when he fed the zeta function with the number $x = 1$. Although this sum grew only very slowly as he added more terms, mathematicians had known since the fourteenth century that eventually it must spiral off to infinity.

So the zeta function must output the answer infinity when fed the number $x = 1$. If, however, instead of taking $x = 1$, Euler fed the zeta function with a number bigger than 1, the answer no longer spiralled off to infinity. For example, taking $x = 2$ means adding together all the squares in the harmonic series:

$$\frac{1}{1^2}+\frac{1}{2^2}+\frac{1}{3^2}+\frac{1}{4^2}+\ldots=1+\frac{1}{4}+\frac{1}{9}+\frac{1}{16}+\ldots$$

This is a smaller number as it does not include all possible fractions found when $x = 1$. We are now adding only some of the fractions, and Euler knew that this time the smaller sum wouldn't spiral off to infinity but would home in on some particular number. It had become quite a challenge by Euler's day to identify a precise value for this infinite sum when $x = 2$. The best estimate was somewhere around $\frac{8}{5}$. In 1735, Euler wrote that 'So much work has been done on the series that it seems hardly likely that anything new about them may still turn up . . . I, too, in spite of repeated effort, could achieve nothing more than approximate values for their sums.'

Nevertheless, Euler, emboldened by his previous discoveries, began to play around with this infinite sum. Twisting it this way and that like the sides of a Rubik's cube, he suddenly found the series transformed. Like the colours on the cube, these numbers slowly came together to form a completely different pattern from the one he had started with. As he went on to describe, 'Now, however, quite unexpectedly, I have found an elegant formula depending upon the quadrature of the circle' – in modern parlance, a formula depending on the number $\pi = 3.1415\ldots$

By some pretty reckless analysis, Euler had discovered that this infinite sum was homing in on the square of π divided by 6:

$$1 + \tfrac{1}{4} + \tfrac{1}{9} + \tfrac{1}{16} + \ldots = \tfrac{1}{6}\pi^2$$

The decimal expansion of $\frac{1}{6}\pi^2$, like that of π, is completely chaotic and unpredictable. To this day, Euler's discovery of this order lurking in the number $\frac{1}{6}\pi^2$ ranks as one of the most intriguing calculations in all of mathematics, and it took the scientific community of Euler's time by storm. No one had predicted a link between the innocent sum $1 + \frac{1}{4} + \frac{1}{9} + \frac{1}{16} + \ldots$ and the chaotic number π.

This success inspired Euler to investigate the power of the zeta function further. He knew that if he fed the zeta function with any number bigger than 1, the result would be some finite number. After a few years of solitary study he managed to identify the output of the zeta function for every even number. But there was something rather unsatisfactory about the zeta function. Whenever Euler fed the formula for the zeta function with any number less than 1, it would always output infinity. For example, for $x = -1$ it yields the infinite sum $1 + 2 + 3 + 4 + \ldots$ The function behaved well only for numbers bigger than 1.

Euler's discovery of his expression for $\frac{1}{6}\pi^2$ in terms of simple fractions was the first sign that the zeta function might reveal unexpected links between seemingly disparate parts of the mathematical canon. The second strange connection that Euler discovered was with an even more unpredictable sequence of numbers.

Rewriting the Greek story of the primes

Prime numbers suddenly enter Euler's story as he tried to put his rickety analysis of the expression for $\frac{1}{6}\pi^2$ on a sound mathematical footing. As he played with the infinite sums he recalled the Greek discovery that every number can be built from multiplying prime numbers together, and realised that there was an alternative way to write the zeta function. He spotted that every term in the harmonic series, for example $\frac{1}{60}$, could be dissected using the knowledge that every number is built from its prime building blocks. So he wrote

$$\frac{1}{60} = \frac{1}{2} \times \frac{1}{2} \times \frac{1}{3} \times \frac{1}{5} = \left(\frac{1}{2}\right)^2 \times \frac{1}{3} \times \frac{1}{5}$$

Instead of writing the harmonic series as an infinite addition of all the fractions, Euler could take just fractions built from single primes, like $\frac{1}{2}$, $\frac{1}{3}$, $\frac{1}{5}$, $\frac{1}{7}$, . . ., and multiply them together. His expression, known today as *Euler's product*, connected the worlds of addition and multiplication. The zeta function appeared on one side of the new equation and the primes on the other. In one equation was encapsulated the fact that every number can be built by multiplying together prime numbers:

$$\zeta(x) = \frac{1}{1^x} + \frac{1}{2^x} + \frac{1}{3^x} + \ldots + \frac{1}{n^x} + \ldots$$
$$= \left(1 + \frac{1}{2^x} + \frac{1}{4^x} + \ldots\right) \times \left(1 + \frac{1}{3^x} + \frac{1}{9^x} + \ldots\right) \times \ldots \times \left(1 + \frac{1}{p^x} + \frac{1}{(p^2)^x} + \ldots\right) \times \ldots$$

At first sight Euler's product doesn't look as if it will help us in our quest to understand prime numbers. After all, it's just another way of expressing something the Greeks knew more than two thousand years ago. Indeed, Euler himself would not grasp the full significance of his rewriting of this property of the primes.

The significance of Euler's product took another hundred years, and the insight of Dirichlet and Riemann, to be recognised. By turning this Greek gem and staring at it from a nineteenth-century perspective, there emerged a new mathematical horizon that the Greeks could never have imagined. In Berlin, Dirichlet was intrigued by the way Euler had used the zeta function to express an important property of prime numbers – one that the Greeks had proved two thousand years before. When Euler input the number 1 into the zeta function, the output $1 + \frac{1}{2} + \frac{1}{3} + \frac{1}{4} + \ldots$ spiralled off to infinity. Euler saw that the output could spiral off to infinity only if there were infinitely many prime numbers. The key to this realisation was Euler's product, which connected the zeta function and the primes. Although the Greeks had proved centuries before that there were infinitely many primes, Euler's novel proof incorporated concepts completely different to those used by Euclid.

Sometimes it helps to express familiar things in a new language. Euler's reformulation inspired Dirichlet to use the zeta function to prove Fermat's prediction that infinitely many primes would strike 1 on a clock calculator. Euclid's ideas had been no help in confirming Fermat's hunch. Euler's proof, on the other hand, provided Dirichlet with the flexibility to count only the primes that would hit one o'clock. It worked. Dirichlet was the first to use Euler's ideas specifically to find out something new about the primes. It was a big step in the understanding of these unique numbers, but still a long way from the Holy Grail.

With Dirichlet's move to Göttingen it was only a matter of time before his interest in the zeta function would rub off on Riemann. Dirichlet had probably talked to Riemann about the power of these infinite sums. But Riemann's head was still filled with Cauchy's weird world of imaginary numbers. For him, the zeta function simply represented another interesting function that one could feed with imaginary numbers, as opposed to the ordinary numbers his contemporaries had been working with.

A strange new vista began to open up before Riemann's eyes. The more he scribbled away on the pages that covered his desk, the more excited he became. He found himself sucked into a wormhole that took him from the abstract world of imaginary functions into the world of prime numbers. Suddenly he could see a method that might explain why Gauss's guess at the number of primes would remain as accurate as Gauss had predicted. Using the zeta function, it seemed as if the key to Gauss's Prime Number Conjecture might be within Riemann's grasp. It would translate Gauss's hunch into the certain proof that Gauss himself had craved. Mathematicians would finally be certain that the percentage difference between Gauss's logarithmic integral and the true number of primes did indeed get smaller the further one counted. Riemann's discoveries also went well beyond this single idea. He found himself looking at the primes from a completely new perspective. The zeta function was suddenly playing a music that had the potential to reveal the secrets of the primes.

The disabling streak of perfectionism that had afflicted him at school almost prevented Riemann from recording any of his discoveries. He had been influenced by Gauss's insistence that one should publish only perfect proofs with no gaps. Even so, he felt compelled to explain and interpret something of this new music he was hearing. He had just been elected to the Academy in Berlin, and new members were expected to report on their recent research. This provided him with a deadline to produce a paper on these new ideas. It would be a fitting way to show his gratitude to the Academy for Dirichlet's influence and guidance and for the two years he had spent at the university as a doctoral student. After all, Berlin was where he had first learnt of the power of imaginary numbers to open up new vistas.

In November 1859, Riemann published his discoveries in a paper in the monthly notices of the Berlin Academy. These ten pages of dense mathematics were the only ones Riemann was ever to publish on the subject of prime numbers, yet the paper would have a fundamental effect on the way they were perceived. The zeta function provided Riemann with a looking-glass in which the primes appeared transformed. As in *Alice in Wonderland*, Riemann's paper sucked mathematicians from their familiar world of numbers through a rabbit hole into a new and often counterintuitive mathematical land. As mathematicians gradually came to grips with this new perspective over the ensuing decades, they could see the inevitability and brilliance of Riemann's ideas.

Despite its visionary qualities, this ten-page paper was deeply frustrating. Riemann, like Gauss, often covered his tracks when writing. Many tantalising claims are made of results Riemann says he could prove but were in his view not quite ready for publication. In some ways it was a miracle that he actually wrote up his paper on the primes, considering the gaps it contained. Had he continued to procrastinate, we might have been deprived of one conjecture in particular that Riemann admitted even he could not prove. Hidden away almost unnoticed in his ten-page document is a statement of the problem whose solution carries today a price tag of a million dollars: the Riemann Hypothesis.

Unlike many of the claims he made in his paper, Riemann is quite up front about his own limitations when it came to his Hypothesis: 'One would of course like to have a rigorous proof of this, but I have put aside the search for such a proof after some fleeting vain attempts because it is not necessary for the immediate objective of my investigation.' The principal goal of his Berlin paper had been to confirm that Gauss's function would provide a better and better approximation to the number of primes as one counted higher. Although he had discovered the tools that would eventually establish Gauss's Prime Number Conjecture, even this was out of reach. Riemann may not have provided all the answers, but his paper pointed to a completely new approach to the subject which would set the course of number theory to this day.

Dirichlet, who would have been greatly excited by Riemann's discovery, was to die on May 5, 1859, a few months before the publication of the paper. Riemann's reward for his work was the university chair that Gauss had once occupied and Dirichlet's death had now made vacant.

CHAPTER FOUR

The Riemann Hypothesis: From Random Primes to Orderly Zeros

The Riemann Hypothesis is a mathematical statement that you can decompose the primes into music. That the primes have music in them is a poetic way of describing this mathematical theorem. However, it's highly post-modern music. Michael Berry, University of Bristol

Riemann had found a passageway from the familiar world of numbers into a mathematics which would have seemed utterly alien to the Greeks who had studied prime numbers two thousand years before. He had innocently mixed imaginary numbers with his zeta function and discovered, like some mathematical alchemist, the mathematical treasure emerging from this admixture of elements that generations had been searching for. He had crammed his ideas into a ten-page paper, but was fully aware that his ideas would open up radically new vistas on the primes.

Riemann's ability to unleash the full power of the zeta function stems from critical discoveries he made during his Berlin years and in his later doctoral studies in Göttingen. What had so impressed Gauss while he was examining Riemann's thesis was the strong geometric intuition that the young mathematician showed when he was feeding functions with imaginary numbers. After all, Gauss had capitalised on his own private mental picture to map out these imaginary numbers before he dismantled the conceptual scaffolding. The starting point for Riemann's theory of these imaginary functions had been Cauchy's work, and for Cauchy a function was defined by an equation. Riemann had now added the idea that even if the equation was the starting point, it was the geometry of the graph defined by the equation that really mattered.

The problem is that the complete graph of a function fed with imaginary numbers is not something that is possible to draw. To illustrate his graph, Riemann needed to work in four dimensions. What do mathematicians mean by a fourth dimension? Those who have read cosmologists such as Stephen Hawking might well reply 'time'. The truth is that we use dimensions to keep track of anything we might be interested in. In physics there are three dimensions for space and a fourth dimension for time. Economists who wish to investigate the relationship between interest rates, inflation, unemployment and the national debt can interpret the economy as a landscape in four dimensions. As they trek uphill in the direction interest rates, they will be exploring what happens to the economy in the other directions. Although we can't actually draw a picture of this four-dimensional model of the economy, it is still a landscape whose hills and troughs we can analyse.

For Riemann, the zeta function was similarly described by a landscape that existed in four dimensions. There were two dimensions to keep track of the coordinates of the imaginary numbers being fed into the zeta function. The third and fourth dimensions could then be used to record the two coordinates describing the imaginary number output by the function.

The trouble is that we humans exist in three spatial dimensions and so cannot rely on our visual world for a perception of this new 'imaginary graph'. Mathematicians have used the language of mathematics to train their mind's eye to help them 'see' such structures. But if you lack such mathematical lenses, there are still ways to help you to grasp these higher-dimensional worlds. Looking at shadows is one of the best ways to understand them. Our shadow is a two-dimensional picture of our three-dimensional body. From some perspectives the shadow provides little information, but from side-on, for example, a silhouette can give us enough information about the person in three dimensions for us to recognise their face. In a similar way, we can construct a three-dimensional shadow of the four-dimensional landscape that Riemann built using the zeta function which retains enough information for us to understand Riemann's ideas.

Gauss's two-dimensional map of imaginary numbers charts the numbers that we shall feed into the zeta function. The north–south axis keeps track of how many steps we take in the imaginary direction, whilst the east–west axis charts the real numbers. We can lay this map out flat on a table. What we want to do is to create a physical landscape situated in the space above this map. The shadow of the zeta function will then turn into a physical object whose peaks and valleys we can explore.

The height above each imaginary number on the map should record the result of feeding that number into the zeta function. Some information is inevitably lost in the plotting of such a landscape, just as a shadow shows very limited detail of a three-dimensional object. By turning this object we get different shadows which reveal different aspects of the object. Similarly, we have a number of choices for what to record as the height of the landscape above each imaginary number in the map on the table top. There is, however, one choice of shadow which retains enough information to allow us to understand Riemann's revelation. It is a perspective that helped Riemann in his journey through this looking-glass world. So what

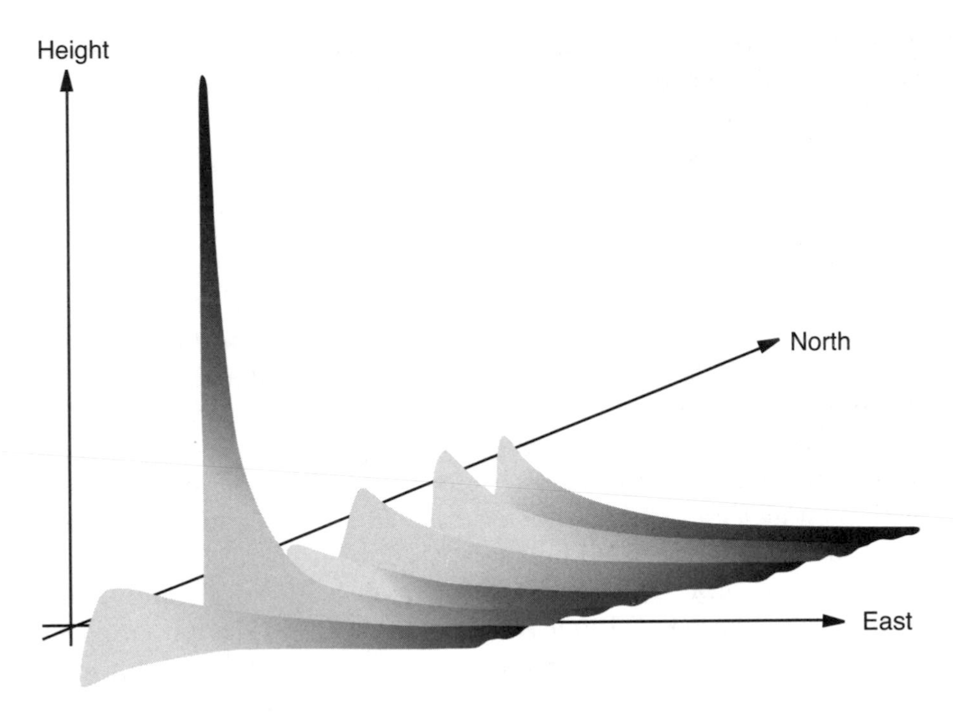

The zeta landscape – Riemann discovered how to continue this picture into a new land to the west.

does this particular three-dimensional shadow of the zeta function look like?

As Riemann began to explore this landscape, he came across several key features. Standing in the landscape and looking towards the east, the zeta landscape levelled out to a smooth plane 1 unit high above sea level. If Riemann turned round and started walking west, he saw a ridge of undulating hills running from north to south. The peaks of these hills were all located above the line that crossed the east–west axis through the number 1. Above this intersection at the number 1 there was a towering peak which climbed into the heavens. It was, in fact, infinitely high. As Euler had learned, feeding the number 1 into the zeta function gives an output which spirals off to infinity. Heading north or south from this infinite peak, Riemann encountered other peaks. None of these peaks, however, were infinitely high. The first peak occurred at just under 10 steps north at the imaginary number 1 + (9.986 . . .)i and was only about 1.4 units high.

If Riemann turned the landscape around and charted the cross-section of the hills running along this north–south divide through the number 1, it would look something like this:

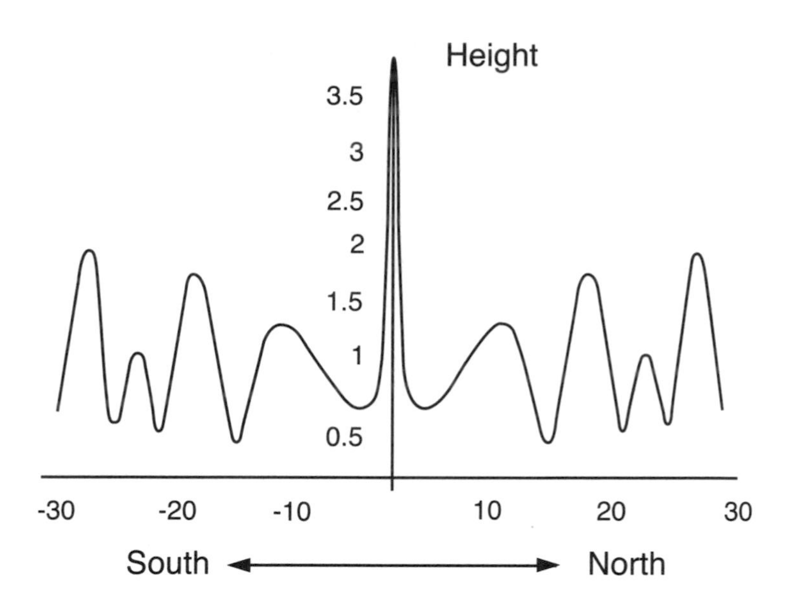

A cross-section through the ridge of mountains along the critical line with east–west coordinate fixed at 1 unit east.

One crucial aspect of the landscape did not fail to attract Riemann's attention. There appeared to be no way to use the formula for the zeta function to build the landscape out to the west beyond the range of mountains. Riemann had the same problem that Euler had experienced when feeding the zeta function with ordinary numbers. Whenever the input was a number west of 1, the formula for the zeta function spiralled off to infinity. Yet in this imaginary landscape, despite the infinite peak over the number 1, the other mountains along this north–south ridge looked passable.

Why did they not just continue undulating, irrespective of the zeta function's results? Surely the landscape didn't end there, at this north–south line. Was there really nothing to the west of this boundary? If you trusted only in the equations, you might believe that the landscape east of 1 was all that could be constructed. The equations didn't make any sense when you put in numbers west of 1. Could Riemann complete the landscape, and if so, how?

Fortunately, Riemann wasn't defeated by the seeming intractability of the zeta function. His education had armed him with a perspective that the French mathematicians lacked. He believed that the equation underlying an imaginary landscape should be regarded as a secondary feature. Of primary importance was the actual four-dimensional topography of the landscape. Although the equations may not make sense, the geometry of the landscape indicated otherwise. Riemann succeeded in finding another formula that could be used to build the missing landscape to the west. His new landscape could then be seamlessly joined onto the original landscape. An imaginary explorer could now pass smoothly from the region defined by Euler's formula to the new landscape constructed by Riemann's new formula without ever knowing there was a border crossing.

Riemann had a complete landscape which covered the entire map of imaginary numbers. He was now ready to make his next move. During his doctorate, Riemann had discovered two crucial and rather counter-intuitive facts about these imaginary landscapes. First, he had learnt that they had an extraordinarily rigid geometry. There was only one way that the landscape could be expanded. The geometry of Euler's landscape to the east completely determined what was possible out to the west. Riemann couldn't massage his new landscape to create hills wherever he fancied. Any changes would cause the seam between the two landscapes to tear.

The inflexibility of these imaginary landscapes was a striking discovery. Once an imaginary cartographer has charted any small region of the imaginary landscape, that would be sufficient to reconstruct the rest of the landscape. Riemann had discovered that the hills and valleys in one region contain information about the topography of the complete landscape. This is certainly counter-intuitive. We would not expect a real-world cartographer, having charted the environs of Oxford, to be able to deduce the complete landscape of the British Isles.

But Riemann had made a second crucial discovery about this strange new brand of mathematics. He had uncovered what could be considered the DNA of these imaginary landscapes. Any mathematical cartographer who knew how to plot on the two-dimensional imaginary map the points where the landscape fell to sea level could reconstruct everything about the entire landscape. The map marking these points was the treasure map of any imaginary landscape. It was a stunning discovery. A cartographer living in our real world wouldn't be able to reconstruct the Alps if you told him all the coordinates of points at sea level around the world. But in these imaginary landscapes the location of all the imaginary numbers where the function outputs zero told you everything. These places are called the *zeros* of the zeta function.

Astronomers are quite used to being able to deduce the chemical make-up of far-away planets without ever visiting them. The light they emit can be analysed by spectroscopy and contains enough information in it to reveal the chemistry of the planet. These zeros behave like the spectrum emitted by a chemical compound. Riemann knew that all he needed to do was to mark out all the points in the map where the height of the complete zeta landscape was zero. The coordinates of all these points at sea level would give him enough information to reconstruct all the hills and valleys above sea level.

Riemann did not forget where all his exploring had started. The big bang that had created this zeta landscape had been Euler's formula for the zeta function, which could be built out of the primes by using Euler's product. If the two things – prime numbers and the zeros – built the same landscape, Riemann knew there had to be some connection between them. One object built in two ways. It was Riemann's genius that revealed how these two things were two sides of the same equation.

Primes and zeros

The connection that Riemann managed to find between prime numbers and the points at sea level in the zeta landscape was about as direct as one could hope for. Gauss had tried to estimate how many primes there were in the numbers from 1 through to any number N. Riemann, though, was able to produce an *exact* formula for the number of primes up to N by using the coordinates of these zeros. The formula that Riemann concocted had two key ingredients. The first was a new function $R(N)$ for estimating the number of primes less than N which substantially improved on Gauss's first guess. Riemann's new function was, like Gauss's, still producing errors, but Riemann's calculations revealed that his formula gave significantly smaller errors. For example, Gauss's logarithmic integral predicted 754 more primes than there were up to 100 million. Riemann's refinement predicted only 97 more – an error of roughly one-thousandth of 1 per cent.

The table overleaf shows how much more accurate Riemann's new function is at guessing the number of primes up to N for values of N from 10^2 to 10^{16}.

Riemann's new function had improved on Gauss but it was still producing errors. Nevertheless, his trip into the imaginary world gave him access to something that Gauss could never have dreamed of – a way to undo these errors. Riemann realised that by using the points in the map of imaginary numbers that marked the places where the zeta landscape was at sea level, he could get rid of these errors and produce an exact formula counting the number of primes. This would be the second key ingredient in Riemann's formula.

N	Number of primes $\pi(N)$ from 1 up to N	Overestimate of Riemann's function $R(N)$	Overestimate of Gauss's function $\mathrm{Li}(N)$
10^2	25	1	5
10^3	168	0	10
10^4	1,229	-2	17
10^5	9,592	-5	38
10^6	78,498	29	130
10^7	664,579	88	339
10^8	5,761,455	97	754
10^9	50,847,534	-79	1,701
10^{10}	455,052,511	-1,828	3,104
10^{11}	4,118,054,813	-2,318	11,588
10^{12}	37,607,912,018	-1,476	38,263
10^{13}	346,065,536,839	-5,773	108,971
10^{14}	3,204,941,750,802	-19,200	314,890
10^{15}	29,844,570,422,669	73,218	1,052,619
10^{16}	279,238,341,033,925	327,052	3,214,632

Euler had made the surprising discovery that feeding an imaginary number into the exponential function produced a sine wave. The rapidly climbing graph usually associated with the exponential function had been transformed by the introduction of these imaginary numbers into an undulating graph of the type more often associated with sound waves. His discovery sparked a rush to explore the strange connections thrown up by these imaginary numbers. Riemann could see that there was a way to extend Euler's find by using his map marking the zeros in the imaginary landscape. In this looking-glass world, Riemann saw how each of these points could be transformed using the zeta function into its own special wave. Each wave would look like a variation on the graph of an undulating sine function.

The character of each wave was determined by the location of the zero responsible for the wave. The farther north the point at sea level, the faster the wave corresponding to this zero would oscillate. If we think of this wave as a sound wave, the note corresponding to a zero sounds higher the farther north the zero is located in the zeta landscape.

Why were these waves or notes helpful for counting primes? Riemann made the stunning discovery that encoded in the varying heights of these waves was the way to correct the errors in his guess for the number of primes. His function $R(N)$ gave a reasonably good count of the number of primes up to N. But by adding to this guess the height of each wave above the number N, he found he could get the exact number of primes. The error had been eliminated completely. Riemann had unearthed the Holy Grail that Gauss had sought: an exact formula for the number of primes up to N.

The equation expressing this discovery can be summed up in words simply as 'primes = zeros = waves'. Riemann's formula for the number of primes in terms of zeros is as dramatic to a mathematician as Einstein's equation $E = mc^2$ that revealed the direct relationship between mass and energy. Here was a formula of connections and transformations. Step by step, Riemann witnessed the primes metamorphose. The primes create the zeta landscape, and the points at sea level in this landscape are the key to unlocking its secrets. A new connection then emerges whereby each of these points at sea level produces a wave, like a musical note. And finally, Riemann came full circle to show how these waves could be used to count primes. Riemann must have been amazed as he completed the circle in such a dramatic fashion.

Riemann knew that, just as there are infinitely many primes, there are infinitely many points at sea level in the zeta landscape. So there are correspondingly infinitely many waves controlling the errors. There is a very graphic way to see how the addition of each extra wave improves Riemann's formula for the number of primes. Before adding the waves corresponding to zeros, the graph of Riemann's function $R(N)$ (depicted overleaf on the top) doesn't look anything like the staircase counting the number of primes (on the bottom). One is smooth, the other jagged.

Even just adding the errors predicted by the first thirty waves created by the first thirty zeros we encounter when we head north in the landscape has a dramatic effect. Riemann's graph has already transformed itself from the smooth graph corresponding to $R(N)$ and is beginning to look much more like the staircase graph describing the number of primes (see page 93).

Each new wave contorts the smooth graph that little bit more. Riemann realised that by the time you added on the infinite number of waves, one for each point at sea level he encountered as he headed north across the zeta landscape, the resulting graph would be an exact match for the prime number staircase.

A generation before, Gauss had discovered what he believed was the coin that Nature had tossed to choose the primes. The waves that Riemann had discovered were the actual results of Nature's tosses. The heights of each of these waves at the number N would predict at each toss whether the prime number coin landed heads or tails. Whilst Gauss's discovery of the connection between primes and logarithms had predicted

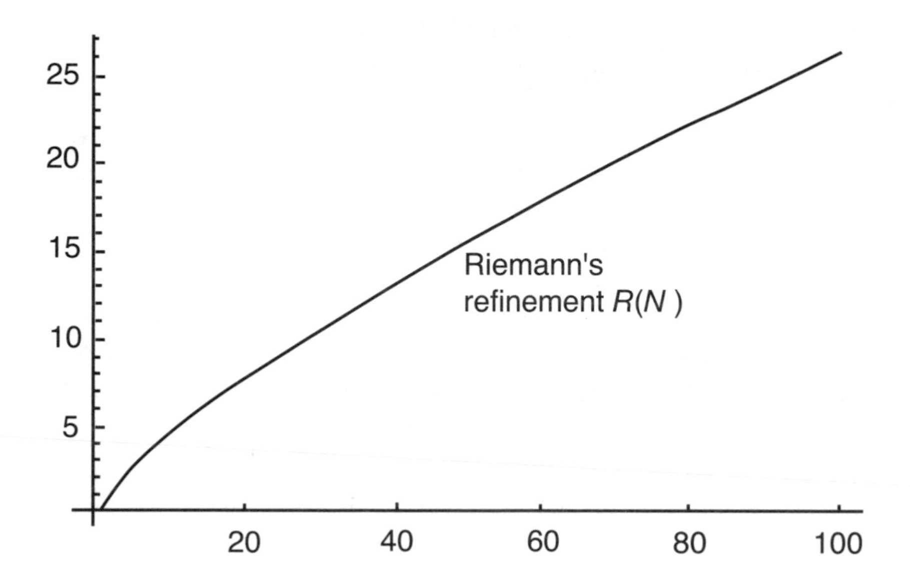

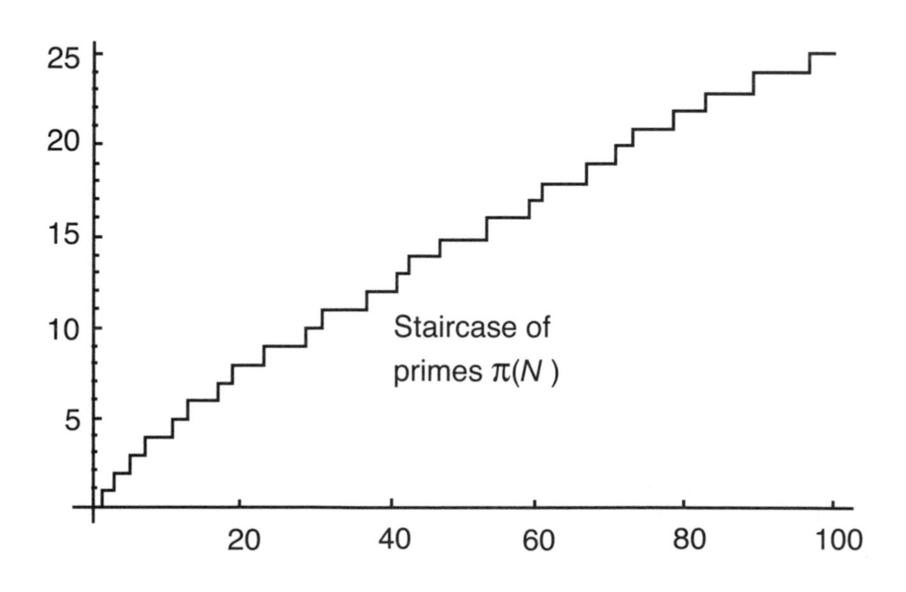

The challenge: to get from Riemann's smooth graph on the top to the jagged graph counting primes on the bottom.

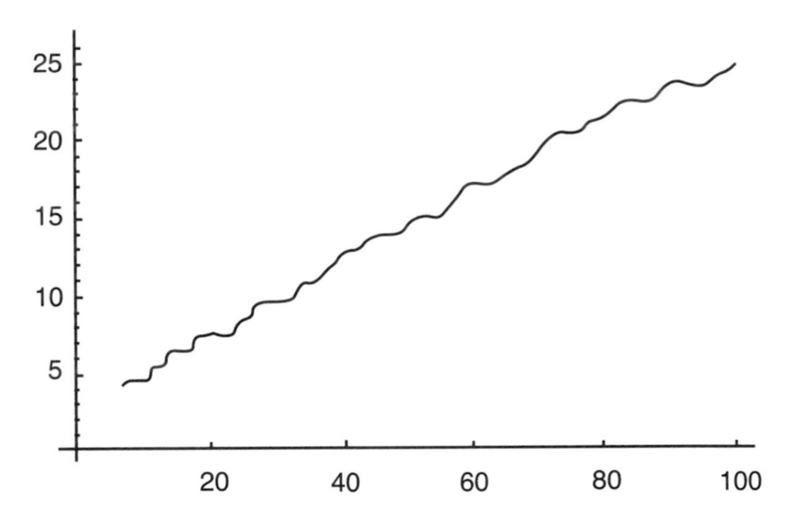

The effect of adding the first thirty waves to Riemann's smooth graph.

how the primes behave on average, Riemann's had found what controlled the minute details of the primes. Riemann had uncovered a record of the winning prime number lottery tickets.

The music of the primes

For centuries, mathematicians had been listening to the primes and hearing only disorganised noise. These numbers were like random notes wildly dotted on a mathematical stave, with no discernible tune. Now Riemann had found new ears with which to listen to these mysterious tones. The sine-like waves that Riemann had created from the zeros in his zeta landscape revealed some hidden harmonic structure.

Pythagoras had unveiled the musical harmony hidden in a sequence of fractions by banging his urn. Mersenne and Euler, both masters of the primes, were responsible for the mathematical theory of harmonics. But none of them had any inkling that there were direct connections between music and the primes. It was a music that required nineteenth-century mathematical ears to hear it. Riemann's imaginary world had thrown up simple waves that together could reproduce the subtle harmonies of the primes.

One mathematician above all others would have been able to see how Riemann's formula captured the hidden music of the primes: Joseph Fourier. An orphan, Fourier had been educated in a military school run by Benedictine monks. A wild child until the age of thirteen, he became entranced by mathematics. Fourier had been destined for a life as a monk, but the events of 1789 released him from the expectations that pre-Revolutionary life had imposed on him. Now he could indulge his passion for mathematics and the military.

Fourier was a great enthusiast of the Revolution, and soon came to the notice of Napoleon. The emperor was setting up the academies that were to turn out the teachers and engineers who would bring about his cultural and military revolution. Napoleon, recognising Fourier's exceptional abilities not only as a mathematician but also as a teacher, appointed him to take charge of mathematical education at the École Polytechnique.

Napoleon was so impressed by what his protégé achieved that he made Fourier part of the Legion of Culture set up to 'civilise' Egypt following the French invasion of 1798. The expedition was driven by Napoleon's desire to disrupt Britain's growing colonial supremacy, but the opportunity to study the ancient world was also on his agenda. His army of intellectuals were put to work as soon as they had boarded Napoleon's flagship *L'Orient* on its way to North Africa. Every morning Napoleon would announce the subject he expected his academic ambassadors to entertain him with that evening. As sailors laboured over rigging and sails, below the decks Fourier and his colleagues wrestled with a variety of Napoleon's pet interests, ranging from the age of the Earth to the question of whether the other planets are inhabited.

On arrival in Egypt, not everything went to plan. Having taken Cairo by force at the Battle of the Pyramids in July 1798, Napoleon was disappointed to find that the Egyptians did not seem to appreciate the cultural force-feeding served up by the likes of Fourier. When three hundred of his men had their throats cut during an evening brawl, Napoleon decided to cut his losses and return to the turmoil that was brewing back in Paris. He set off without telling any of his intellectual army that he was abandoning them. Fourier, left stranded in Cairo, didn't have sufficient rank to take to his heels without risking being shot for desertion and was forced to stay on in the desert. He managed to return to France in 1801, after the French had decided to leave the 'civilising' of Egypt in the hands of the British.

While he was in Egypt, Fourier developed an addiction to the searing heat of the desert. Back in Paris he kept his rooms so hot that friends compared them to the furnaces of hell. He believed that extreme heat helped keep the body healthy and even cured some diseases. His friends would find him wrapped up like an Egyptian mummy, sweating in a room that was as hot as the Sahara.

Fourier's predilection for heat extended to his academic work. He earned his place in the history of mathematics for his analysis of the propagation of heat, work described by the British physicist Lord Kelvin as 'a great mathematical poem'. Fourier had been spurred on in his efforts after the Academy in Paris announced that it would offer its Grand Prix des Mathématiques in 1812 to whoever could unravel the mysteries of how heat moved through matter. Fourier was awarded the prize in recognition of the novelty and importance of his ideas. But he also had to swallow some criticism of his work from, among others, Legendre. The judges of the Grand Prix pointed out that much of his treatise contained mistakes and his mathematical explanation was far from rigorous. Fourier deeply resented the Academy's critique but recognised that there was still work to be done.

As he set out to correct the errors in his analysis, Fourier tried to understand the nature of graphs representing physical phenomena – for example, the graph that showed how temperature evolved over time, or the graph representing a sound wave. He knew that sound could be represented by a graph where the horizontal axis charts time and the vertical axis controls the volume and pitch of the sound at each instant.

Fourier started with a graph of the simplest sound. If you set a tuning fork vibrating, you find when you plot the resulting sound wave that it is a pure, perfect sine curve. Fourier began to explore how more complicated sounds could be produced by taking combinations of these pure sine waves. If a violin plays the same note as the tuning fork, the sound is very different. As we have seen (see p. 78), the violin string doesn't just vibrate at the fundamental frequency, determined by the length of the string. There are additional notes, the harmonics, which correspond to simple fractions of the length of the string. The graphs of each of these additional notes are still sine waves, but of higher frequencies. It is the combination of all these pure notes, dominated by the lowest, fundamental note, that creates the sound of a violin, whose graph looks like the teeth on a saw.

Why does a clarinet sound so characteristically different to the sound of a violin playing the same note? The graph of the sound wave created by the clarinet looks like a square wave function, like the crenellation on the top of a castle wall, instead of the spiky graph of the violin. The reason for the difference is that the clarinet is open at one end, whereas the string in the violin is fixed at both ends. This means that the harmonics produced by the clarinet vary from those of the violin, so the graph depicting the sound of the clarinet is built from sine waves oscillating at different frequencies.

Fourier realised that even the complicated graph depicting the sound of an entire orchestra could be broken down into simple sine curves of the fundamental and harmonics for each and every instrument. Since each of the pure sound waves can be reproduced by a tuning fork, Fourier had proved that by playing a huge number of tuning forks simultaneously you could create the sound of a whole orchestra. Someone with a blindfold on would not be able to tell whether it was a real orchestra or thousands of tuning forks. This principle is at the heart of how sound is encoded on a CD: the CD instructs your speakers how to vibrate to create all the sine waves that make up the sound of the music. This combination of sine waves gives you the miraculous sensation of having an orchestra or a band performing live in your living room.

It wasn't only the sound of musical instruments that could be reproduced by adding together pure sine waves with different frequencies. For example, the static white noise created by an untuned radio or a running tap can be represented as an infinite sum of sine waves. In contrast to the distinct frequencies required to reproduce the sound of the orchestra, white noise is built from a continuous range of frequencies.

Fourier's revolutionary insights went beyond reproducing sound alone. He began to understand how to use sine waves to plot graphs that depicted other physical or mathematical phenomena. Many of Fourier's contemporaries doubted that such a simple graph as the sine wave could be used as a building block to construct complicated graphs of the sound of an orchestra or a running tap. In fact, a number of senior mathematicians in France voiced their vigorous opposition to Fourier's ideas. Emboldened, however, by his prestigious association with Napoleon, Fourier did not fight shy of challenging the authorities. He showed how an appropriate selection of sine waves oscillating at different frequencies could be used to create a whole range of complicated graphs. By adding the heights of the sine waves you could reproduce the shapes of these graphs, in just the same way that a CD combines the pure tones of tuning forks to reproduce complex musical sounds.

This is precisely what Riemann succeeded in doing in that ten-page paper. He reproduced the staircase graph that counted the number of primes in exactly the same fashion by adding together the heights of the wave functions he derived from the zeros in the zeta landscape. This is why Fourier would have recognised Riemann's formula for the number of primes as the discovery of the basic tones that make up the sound of the primes. This complicated sound is represented by the staircase graph. Riemann's waves that he created from the zeros, the points at sea level in the landscape, were like the sounds of tuning forks, single clear notes with no harmonics. When played simultaneously these basic waves reproduced the sound of the primes. So what does Riemann's prime number music sound like? Is it the sound of an orchestra, or does it resemble the white noise of a running tap? If the frequencies of Riemann's notes are in a continuous range, then the primes make white noise. But if the frequencies are isolated notes, then the sound of the primes resembles the music of an orchestra.

Given the randomness of the primes, we might well expect the combination of notes played by the zeros in Riemann's landscape to be nothing more than noise. The north–south coordinate of each zero determines the pitch of its note. If the sound of the primes were indeed white noise, there would have to be a concentration of zeros in the zeta landscape. And Riemann knew, from his dissertation for Gauss, that such a concentration of points at sea level would actually force the *whole* of the landscape to be at sea level. This clearly wasn't the case. The sound of the primes was not white noise at all. The points at sea level had to be isolated points, so they had to produce a collection of isolated notes. Nature had hidden in the primes the music of some mathematical orchestra.

The Riemann Hypothesis – order out of chaos

What Riemann had done was to take each of the points on the map of the imaginary world that sat at sea level. Out of each point he had created a wave, like a note from some mathematical instrument. By combining all these waves, he had an orchestra that played the music of the primes. The north–south coordinate of each point at sea level controlled the frequency of the wave – how high the corresponding note sounded. In contrast, the east–west coordinate controlled, as Euler had learnt, how loud each note would be played. The louder the note, the larger the fluctuations of its undulating graph.

Riemann was intrigued to know whether any of the zeros were playing significantly louder than the others. Such a zero would produce a wave whose graph was fluctuating higher than the other waves and hence would play a bigger part in counting the primes. After all, it is the heights of these waves that control the difference between Gauss's guess and the true number of primes. Was there some instrument in this prime number orchestra that was playing a solo above the sound of the others? The farther east a point at sea level was located, the louder the note would be. To determine the balance of the orchestra, Riemann had to go back and look at the coordinates of each of the zeros in his imaginary map.

Remarkably, so far his analysis had worked without him having to know the locations of any of the points at sea level. He knew that there were some zeros that stretched out to the west which were easy to locate, but they contributed nothing interesting to the sound of the primes because they had no pitch. In their dismissive manner, mathematicians would later call these the trivial zeros. It was the location of the remaining zeros that Riemann was after.

As he began to explore the precise location of these points, he got a big surprise. Instead of being randomly dotted around the map, making some notes louder than others, the zeros he calculated seemed to be miraculously arranged in a straight line running north–south through the landscape. It appeared as if every point at sea level had the same east–west coordinate, equal to $\frac{1}{2}$. If true, it meant that the corresponding waves were perfectly balanced, none performing louder than any other.

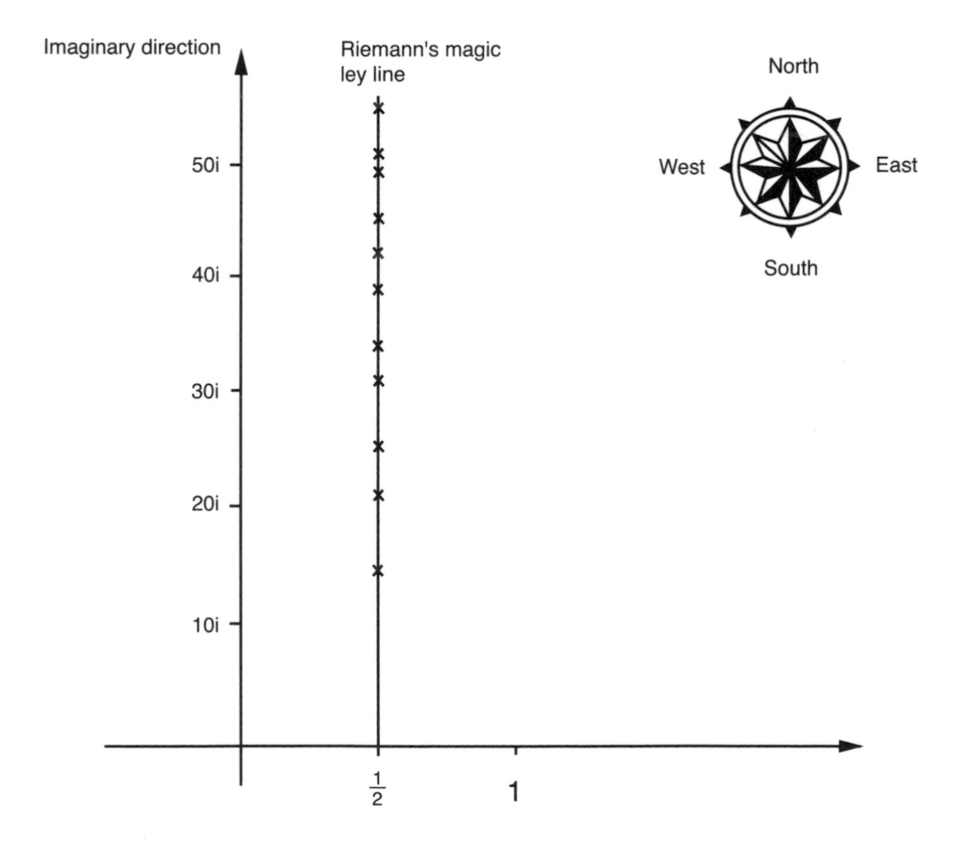

Riemann's treasure map of the primes – the crosses mark the locations of the points at sea level in the zeta landscape.

The first zero that Riemann calculated had coordinates ($\frac{1}{2}$, 14.134 725 . . .): you head east $\frac{1}{2}$ step, and north about 14.134 725 steps. The next zero had coordinates ($\frac{1}{2}$, 21.022 040 . . .). (How he managed to calculate the location of any of these zeros remained a mystery for years.) He calculated a third zero which was located at ($\frac{1}{2}$, 25.010 856 . . .). These zeros did not appear to be scattered at random. Riemann's calculations indicated that they were lining up as if along some mystical ley line running through the landscape. Riemann speculated that the regimented behaviour of the few zeros he was able to calculate was no coincidence. His belief that every point at sea level in his landscape would be found on this straight line is what has become known as the *Riemann Hypothesis*.

Riemann looked at the image of the primes in the mirror that separated the world of numbers from his zeta landscape. As he watched, he saw the chaotic arrangement of the prime numbers on one side of the mirror transform into the strict regimented order of the zeros on the other side of the mirror. Riemann had finally found the mysterious pattern that centuries of mathematicians had been yearning to see as they stared at the primes.

The discovery of this pattern was totally unexpected. Riemann was lucky to be the right person in the right place at the right time: he could not have anticipated what was awaiting him on the other side of the looking-glass. But what lay there completely transformed the task of understanding the mysteries of the primes. Now mathematicians had a new landscape to explore. If they could navigate the land of the zeta function and chart the places at sea level, the primes might yet yield their secrets. What Riemann had also discovered was evidence of some ley line running through this landscape whose significance runs to the very heart of mathematics. The importance of Riemann's ley line can be judged by the name by which mathematicians now refer to it – *the critical line*. Suddenly, the puzzle of the randomness of the primes in the real world has been replaced by the quest to understand the harmony of this imaginary looking-glass landscape.

Since there are an infinite number of zeros, the few scraps of evidence that Riemann had discovered seem quite precarious facts on which to base a theory. All the same, Riemann knew that this ley line had an important significance. He already knew that the east–west axis represented a line of symmetry for the zeta landscape. Anything that happened to the north would be reflected by the same behaviour to the south. Riemann had further discovered the much more significant fact that this line running north–south through the point at $\frac{1}{2}$ was also an important line of symmetry.

This might well have given Riemann cause to believe that Nature would also have used this line of symmetry to order the zeros.

What is so extraordinary about the events surrounding Riemann's major discovery is that his calculations of the locations of the first few zeros did not appear anywhere in the dense paper about prime numbers he wrote for the Berlin Academy. In fact, one is hard pushed to find any statement of this discovery in his published paper. He writes that many of the zeros appear on this straight line, and that it is 'very likely' that all of them do. But Riemann admits in this paper that he didn't try too hard to prove his Hypothesis.

After all, Riemann was after the much more immediate goal of proving Gauss's Prime Number Conjecture: to show why Gauss's guess for the primes became more accurate the more primes one counted. Although this proof also remained elusive, Riemann realised that if his hunch about this ley line were true then it would imply that Gauss had indeed been right. As Riemann had discovered, the errors in Gauss's formula could be described by the location of each of the zeros. The farther east a zero, the louder the wave. The louder the wave, the bigger the error. This is why Riemann's prediction about the location of the zeros was so important to mathematics. If he was right, and the zeros did all lie on his magic ley line, it would imply that Gauss's guess would always be incredibly accurate.

The publication of his ten-page paper marked a brief period of happiness for Riemann. He was honoured with the chair that his mentors Gauss and Dirichlet had both occupied. His sisters had joined him in Göttingen after his brother, who had been supporting them, died in 1857. It boosted Riemann's morale to have his family close by, and he was less prone to the bouts of depression that he had suffered in previous years. With the salary of a professor, he no longer had to endure the poverty of his student years. And at last he could afford proper lodgings and even a housekeeper so that he could dedicate his time to pursuing the ideas that were dancing around in his head.

Yet he was never to return to the subject of prime numbers. He went on to follow his geometric intuition and developed a notion of the geometry of space which was to become one of the cornerstones of Einstein's Theory of Relativity. This period of good fortune culminated in 1862 in his marriage to Elise Koch, a friend of his sister's. But barely a month later he fell ill with pleurisy. From then on, ill health was to plague him constantly, and on many occasions he sought refuge in the Italian countryside. He became particularly attached to Pisa, which was where in August 1863 his only child, Ida, was born. Riemann enjoyed his trips to Italy not just for the clement weather but also for the intellectual climate he encountered; in his lifetime it was the Italian mathematical community that was most receptive to his revolutionary ideas.

His last visit to Italy was an escape not from the dank atmosphere of Göttingen but from an invading army. In 1866 the armies of Hanover and Prussia clashed in Göttingen. Riemann found himself stranded in his lodgings – Gauss's old observatory, outside the city walls. Judging by the state in which he left the place, Riemann fled in a hurry to Italy. The shock proved too much for his frail constitution. Seven years after the publication of his paper on the primes, Riemann died from consumption, aged only thirty-nine.

Faced with the mess Riemann left behind, his housekeeper destroyed many of his unpublished scribblings before she was stopped by members of the faculty in Göttingen. The remaining papers were handed over to Riemann's widow and disappeared for years. It is intriguing to speculate what might have been found if his housekeeper had not been so keen to clear his study. One statement which Riemann made in his ten-page paper indicates that he believed he could prove that most zeros are on his ley line. His perfectionism stopped him from elaborating, and he wrote simply that his proof was not yet ready for publication. The proof was never found amongst his unpublished papers, and to this day mathematicians haven't been able to reproduce one. Those missing pages of Riemann's are as provocative as Fermat's claim to have had a proof of his Last Theorem.

Some unpublished notes that did survive the housekeeper's fire resurfaced after fifty years. Frustratingly, they indicate that Riemann had indeed proved much more than he published. Sadly, though, many of the papers detailing results that he tantalisingly hinted he had some understanding of have probably been lost for eternity in the kitchen fire of his overzealous housekeeper.

CHAPTER FIVE

The Mathematical Relay Race: Realising Riemann's Revolution

A problem in number theory is as timeless as a true work of art. David Hilbert, Introduction to Legh Wilber Reid, *The Elements of the Theory of Algebraic Numbers*

Euclid in Alexandria. Euler in St Petersburg. The Göttingen trio – Gauss, Dirichlet, Riemann. The problem of prime numbers had been passed on like a baton from one generation to another. The new perspectives of each generation provided impetus for a fresh surge along the track. Each wave of mathematicians left its characteristic mark on the primes, a reflection of their era's particular cultural outlook on the mathematical world. However, Riemann's contribution took him so far ahead of the field that it would be nearly thirty years before anyone was in a position to capitalise on his rush of new ideas.

Then in 1885, out of nowhere, it looked as if the game was up. Although word got around more slowly than would Bombieri's April Fool email over a century later, news began to spread that a little-known figure had not only picked up Riemann's baton but had crossed the finishing line. A Dutch mathematician, Thomas Stieltjes, was boasting that he had a proof of Riemann's Hypothesis – a proof confirming that all the zeros were on Riemann's magic ley line, passing through $\frac{1}{2}$.

Stieltjes was an unlikely victor. As a student he had failed his university examinations three times, much to the despair of his father, a member of the Dutch parliament and an eminent engineer responsible for building the docks of Rotterdam. But Thomas's failure was not due to laziness. He had simply been distracted by the pleasure of reading real mathematics in Leiden's library rather than mugging up on the technical exercises required for his examinations.

Gauss had been a favourite author of Stieltjes, and the Dutchman was eager to follow in the master's footsteps. He took up a post at the observatory in Leiden, just as Gauss had worked at the observatory in Göttingen. The job had magically become available thanks to a word to the director of the observatory from Stieltjes's influential father, but he was never aware of this helping hand. As he trained his telescope on the skies, it was the mathematics of celestial motion that captured his imagination rather than measuring the positions of new stars. As his ideas blossomed, he decided to write to one of the eminent mathematicians at the famous French academies, Charles Hermite.

Hermite was born in 1822, four years before Riemann. Now in his sixties, he had become one of the standard-bearers for Cauchy and Riemann's work on functions of imaginary numbers. Cauchy's influence on Hermite went beyond mathematics. As a young man Hermite had been an agnostic, but Cauchy, a devout Roman Catholic, had caught Hermite in a weak moment during a severe illness and had converted him to Catholicism. The result was a strange mix of mathematical mysticism, akin to the cult of the Pythagoreans. Hermite believed that mathematical existence was some supernatural state which mortal mathematicians were only occasionally allowed to glimpse.

Perhaps that's why Hermite responded so enthusiastically to the letter sent by an obscure assistant at the Leiden Observatory, persuaded that this stargazer was blessed with a heightened mathematical vision. Soon they were conducting a furious mathematical correspondence which saw 432 letters exchanged over a period of twelve years. Hermite was impressed with the young Dutchman's ideas, and despite Stieltjes being without a degree, gave him his support and ensured that Stieltjes was rewarded with a professorship at the University of Toulouse. In one letter to Stieltjes about his work, Hermite wrote, 'Vous avez toujours raison et j'ai toujours tort.' ('You are always right and I'm always wrong.')

It was during this correspondence that Stieltjes made his extraordinary claim to have proved the Riemann Hypothesis. Hermite's faith in his young protégé gave him no cause to doubt that Stieltjes had indeed come up with a proof. After all, he had already made great contributions to other branches of mathematics.

Since Riemann's conjecture had not had time to mature into the seemingly intractable challenge it currently represents, Stieltjes's announcement was greeted with less excitement than it would have been today. Riemann had not trumpeted his hunch about the zeros, but had buried it deep in his ten-page paper with little evidence to support it. It would take a new generation to appreciate the importance of the Riemann Hypothesis. Stieltjes's announcement was exciting nonetheless, because a proof of the Riemann Hypothesis would prove Gauss's Prime Number Conjecture, which at the time was the Holy Grail of Number Theory. For numbers up to 1,000,000, Gauss's guess at the number of primes was 0.17 per cent off the mark. By 1,000,000,000 the percentage error had fallen to 0.003 per cent. Gauss believed that, as the numbers got bigger and bigger, the percentage error would get smaller and smaller. By the late nineteenth century, Gauss's conjecture had been around for long enough to earn its prospective conqueror great kudos. The evidence in support of Gauss's hunch was certainly compelling.

By the time Stieltjes wrote to Hermite about his proof, the best progress towards cracking Gauss's conjecture had been made in the 1850s in Euler's old stomping ground of St Petersburg. The Russian mathematician Pafnuty Chebyshev couldn't actually prove that the percentage error between Gauss's guess and the true number of primes became smaller and smaller, but he did manage to show that the error for the number of primes up to *N* would never be more than 11 per cent, however big you chose *N*. This may sound a far cry from the 0.003% that Gauss had achieved for the number of primes up to a billion, but the significance of Chebyshev's result was that he could guarantee that however far one counted primes, the error would not suddenly become overwhelmingly large. Before Chebyshev's result, Gauss's conjecture had been based solely on a small amount of experimental evidence. Chebyshev's theoretical analysis provided the first real support for some connection between logarithms and primes. However, there was still a long way to go to prove that the connection would remain as tight as Gauss was conjecturing.

Chebyshev managed to achieve this control on the error by purely elementary means. Riemann, labouring away in Göttingen on his sophisticated imaginary landscape, knew of Chebyshev's work. There is evidence of a letter that he had prepared to send to Chebyshev outlining his own progress; the surviving pages of Riemann's notes include several drafts in which he tries out different spellings of the Russian's name. It's not clear whether Riemann did eventually send his letter to Chebyshev. With or without the letter, Chebyshev was never to improve his estimate of the error in counting the primes.

This is why Stieltjes's announcement was still so exciting for mathematicians at the time. Nobody yet suspected how difficult the Riemann Hypothesis would be to prove, but proving Gauss's conjecture was a recognised achievement. Hermite was keen to see the details of Stieltjes's proof but the young man was rather reticent. The proof was not quite ready yet. Despite much prodding over the next five years, Stieltjes came up with nothing to support his claim. To try to counter his mounting frustration at Stieltjes's reluctance to explain his ideas, Hermite came up with what he believed to be a cunning ruse to smoke out the proof. Hermite proposed that the Grand Prix des Sciences Mathématiques of the Paris Academy for 1890 be dedicated to a proof of Gauss's Prime Number Conjecture. Hermite sat back, confident that the prize would be going to his friend Stieltjes.

This was Hermite's plan: to win the prize it wasn't necessary for Stieltjes to claim anything as impressive as having cracked the Riemann Hypothesis. Rather, he only needed to chart a small section of the imaginary landscape – the border between Euler's landscape and Riemann's extension. It was enough to prove there were no zeros on this line, running north–south through the number 1. Riemann's landscape could be used to judge the errors in Gauss's formula, which were determined by how far east each of Riemann's zeros would be found in the landscape. The more easterly the zero, the bigger the error. The error would be very small if the Riemann Hypothesis were correct, but Gauss's conjecture would still be true even if the Riemann Hypothesis were false – provided all the zeros lay strictly to the west of the north–south border through the number 1.

The deadline for the prize passed, and Stieltjes remained silent. But Hermite was not to be completely disappointed. Unexpectedly, his student Jacques Hadamard submitted an entry. Although Hadamard's paper fell short of a complete proof, his ideas were enough to award him the prize. Spurred on by the award, by 1896 he had managed to fill in the gaps in his previous ideas. He couldn't show that all the zeros were on Riemann's critical line through $\frac{1}{2}$, but he could prove that there were no zeros as far east as the border through 1.

Finally, a century after Gauss's discovery of a connection between primes and the logarithm function, mathematics had a proof of Gauss's Prime Number Conjecture. No longer a conjecture, thenceforth it was known as the Prime Number Theorem. The proof was the most significant result about the primes since the Greeks confirmed there were infinitely many such numbers. Even though we will never be able to count to the farthest reaches of the universe of numbers, Hadamard had proved that no surprises awaited any intrepid traveller. The early experimental evidence discovered by Gauss was not some misleading trick being played by Nature.

Hadamard could never have achieved what he did without Riemann's head start. His ideas for the proof were steeped in Riemann's analysis of the zeta landscape, yet he was still far from proving Riemann's conjecture. In his paper explaining the proof, Hadamard acknowledges that his work didn't match the achievement of Stieltjes, who until his death in 1894 was still claiming to have a proof of the Riemann Hypothesis. Stieltjes was the first in a long line of reputable mathematicians who have announced proofs but failed to deliver the goods.

Hadamard soon learnt that he would have to share the glory of proving the Prime Number Theorem. Simultaneously, a Belgian mathematician,

Charles de la Vallée-Poussin, had also come up with a proof. Hadamard and de la Vallée-Poussin's great achievement marked the start of a journey that would continue into the twentieth century, with mathematicians now eager to push on in their exploration of Riemann's landscape. Hadamard and de la Vallée-Poussin had established the base camp in preparation for the major ascent towards Riemann's critical line. It was during this period that the problem began to assume its position as the Everest of mathematical exploration even though, ironically, its proof depended on navigating the lowest points in the zeta landscape. With Gauss's Prime Number Theorem finally claimed, it was time for Riemann's great problem to emerge from the hidden depths of his dense Berlin paper.

It was another Göttingen resident, David Hilbert, who brought Riemann's remarkable insight to the world's attention. This charismatic mathematician helped more than anyone to launch the twentieth-century drive to claim the ultimate prize of the Riemann Hypothesis.

Hilbert, the mathematical Pied Piper

The town of Königsberg in Prussia had achieved some mathematical notoriety during the eighteenth century, thanks to the riddle about its bridges that Euler had solved in 1735. During the late nineteenth century the town made its way back onto the mathematical map as the birthplace of David Hilbert, one of the giants of twentieth-century mathematics.

Although Hilbert was very fond of his home town, he could see that it was within Göttingen's city walls that the mathematical fire was burning brightest. Thanks to the legacy of Gauss, Dirichlet, Dedekind and, most of all, Riemann, Göttingen had become a mathematical Mecca. Perhaps more than anyone else at that time, Hilbert appreciated what a mathematical sea change Riemann had brought about. Riemann recognised that seeking to understand the structures and patterns underpinning the mathematical world was more fruitful than focusing on formulas and tedious calculations. Mathematicians began to listen to the mathematical orchestra in a new way. No longer obsessed with individual notes, they were starting to hear the underlying music of the objects they studied. Riemann had begun a renaissance in mathematical thinking that took hold in Hilbert's generation. As Hilbert wrote in 1897, he wanted to implement 'Riemann's principle according to which proofs should be impelled by thought alone and not by computation'.

Hilbert made his mark in the academic circles of Germany by doing just that. He had learnt as a child that the Greeks had proved that there

David Hilbert (1862–1943).

were infinitely many prime numbers that were required to build all possible numbers. He had read as a student that things appeared differently if you considered equations rather than numbers. It had become a challenge at the end of the nineteenth century to show that, in contrast to the primes, there were a finite number of equations that could be used to generate certain infinite sets of equations. Mathematicians of Hilbert's day were trying to prove this by laboriously constructing the equations. Hilbert stunned his contemporaries with a proof that this finite set of building blocks must exist even though he couldn't construct the set. Just as Gauss's schoolteacher had looked on incredulously as the student slickly added the numbers from 1 to 100, Hilbert's superiors seriously doubted that anything other than hard graft could explain the theory of equations.

It was quite a challenge to the mathematical orthodoxy of the day. If you couldn't see the finite list, it was hard to accept its existence, even though the proof confirmed that it was there. To be told that something could not be seen but was unmistakably there was disconcerting for those still wedded to the French tradition of equations and explicit formulas. Paul Gordan, one of the experts in this field, declared of Hilbert's work, 'This is not mathematics. This is theology.' Hilbert nonetheless stuck to his guns even though he was only in his twenties. It was eventually accepted that Hilbert was right, and even Gordan conceded the argument: 'I have convinced myself that theology has its merits.' Hilbert then turned to the study of numbers, something he described as 'a building of rare beauty and harmony'.

In 1893 he was asked by the German Mathematical Society to write an account of the state of number theory at the close of the century. It was a daunting task for someone in his early thirties. A hundred years before, the subject had barely existed as a coherent entity. Gauss's *Disquisitiones Arithmeticae*, published in 1801, had uncovered such fertile ground that by the end of the century number theory had blossomed so much that it had become overgrown. To help tame the subject, Hilbert was joined by an old friend, Hermann Minkowski. They had known each other ever since they were students together in Königsberg. Minkowski had made his mark in number theory by winning the Grand Prix des Sciences Mathématiques at the age of eighteen. He was only too happy to work on a project that would bring to life what he called 'the insinuating melodies of this powerful music'. Their collaboration forged Hilbert's passion for the primes, which Minkowski claimed would 'wiggeln und waggeln' under their spotlight.

Hilbert's 'theology' earned him much respect amongst a number of influential mathematicians in Europe. In 1895 a letter arrived from Göttingen from one of the professors, Felix Klein, offering him a position in the hallowed university. Hilbert did not hesitate and accepted straight away. During the meeting to discuss his appointment, the faculty questioned Klein's support and speculated that he was appointing a lackey who would not be able to hold his own. Klein assured them that, on the contrary, 'I have asked the most difficult person of all.' That autumn Hilbert made his way to the town where Riemann, his inspiration, had been a professor, hoping to further the mathematical revolution.

Before long the faculty became aware that Hilbert was not content simply to challenge mathematical orthodoxies. The professors' wives were appalled by the behaviour of this newcomer to the faculty. As one of them wrote, 'he is upsetting the whole situation here. I learned that the other night he was seen in some restaurants playing billiards in the backroom with the students.' As time went by, Hilbert began to win the hearts of the

Göttingen ladies, and got himself a reputation as a womaniser. At his fiftieth birthday party, his students performed a song with a verse for each letter of the alphabet detailing one of Hilbert's conquests.

The bohemian professor acquired a bicycle to which he became deeply attached. He was often to be seen cycling through the streets of Göttingen carrying flowers from his garden for one of his flames. He would give lectures in his shirtsleeves, which was unheard of at the time. In draughty restaurants he would borrow feather boas from women diners. It wasn't clear whether Hilbert was deliberately courting controversy or simply seeking the most obvious solution to every problem. Clearly, though, his mind was focused more on mathematical questions than on the niceties of social etiquette.

Hilbert set up a twenty-foot blackboard in his garden. On it, between tending his flower beds and performing stunts on his bicycle, he would chalk his mathematics. He loved parties and would always play his music loudly by choosing the largest available needle for his phonograph. When he eventually heard Caruso singing live he was quite disappointed: 'Caruso sings on the small needle.' But Hilbert's mathematics far outstripped his personal eccentricities. In 1898, he shifted his attention away from number theory and turned instead to the challenge of geometry. He had become intrigued with new types of geometry proposed by several mathematicians during the nineteenth century which claimed to violate one of the fundamental axioms of geometry as proposed by the Greeks. Because of his intense belief in the abstract power of mathematics, the physical reality of objects was irrelevant to him, and he began to study the connections and abstract structures underlying these new geometries. It was the relationship between the objects that was important. Hilbert once famously declared that the theory of geometry would still make sense if points, lines and planes were replaced by tables, chairs and beer mugs.

A century before, Gauss had considered the challenge posed by these new models of geometry, but had held back from voicing such heretical thoughts. Surely it was impossible that the Greeks had got it wrong. Nevertheless, he had begun to question one of Euclid's fundamental axioms of geometry, about the existence of parallel lines. Euclid had considered this question: if you draw a line and then a point off the line, how many lines are there parallel to the first line which run through the point? It seemed obvious to Euclid that the answer was that there was one and only one such parallel line.

At the age of sixteen, Gauss had already begun to speculate that there might be equally consistent and valid geometries in which there were no such things as parallel lines. In addition to Euclid's geometry and this new geometry with no parallel lines, there might even be a third class of geometries in which there was more than one parallel line. If that were so, there would be geometries in which the angles of a triangle no longer added up to 180 degrees, which the Greeks had believed impossible. If there were several possible geometries, Gauss wondered, which of them best described the physical world? The Greeks had certainly believed that their model provided a mathematical description of physical reality. But Gauss was not at all convinced that the Greeks were right.

In later life, whilst surveying the state of Hanover, Gauss used some of the measurements he'd made in the neighbourhood of Göttingen to see whether a triangle of light beams shone between three hilltops might not contradict Euclid by having angles that did not sum to 180 degrees. Gauss thought that the line followed by a path of light might bend in space. Perhaps three-dimensional space was curved in the same manner as the two-dimensional surface of the globe. He had in mind the so-called great circles, such as lines of longitude, along which the shortest path between two points on the surface of the Earth is measured. In this two-dimensional geometry there are no parallel lines of longitude since they all meet at the poles. No one had contemplated the idea that three-dimensional space might also bend.

We realise now that Gauss was working on too small a scale to observe any significant bending of space to counter the view of a Euclidean world. Arthur Eddington's confirmation of the bending of light from stars during the solar eclipse of 1919 supported Gauss's hunch. Gauss never went public with his ideas, perhaps because his new geometries seemed to be at variance with the task of mathematics, which was to represent physical reality. The friends he did mention his idea to, Gauss pledged to secrecy.

The idea of these new geometries was eventually floated publicly in the 1830s by the Russian Nikolai Ivanovic Lobachevsky and the Hungarian János Bolyai. The discovery of these non-Euclidean geometries, as Gauss called them, did not rock the mathematical boat as much as Gauss had feared they might, but was simply dismissed as too abstract. As a result they were ignored for many years. Nonetheless, by Hilbert's time they were beginning to emerge as a perfect expression of his own more abstract approach to the mathematical world.

Some mathematicians claimed that any geometry that did not satisfy Euclid's assumption about parallel lines must contain some hidden contradiction that would cause it to collapse. As Hilbert began to explore this possibility, he saw that there was a strong logical bond between non-Euclidean and Euclidean geometry. He discovered the only way these non-Euclidean geometries could contain contradictions was if Euclid's geometry also contained contradictions. This seemed like some kind of progress. Mathematicians at the time believed that Euclid's geometry was logically sound. Hilbert's discovery meant that these non-Euclidean models would have the same logical foundations. If one geometry collapsed, it would bring down all other geometries with it. But then Hilbert had a rather unsettling realisation. No one had actually *proved* that Euclid's geometry had no hidden contradictions.

Hilbert began to think about how to go about proving that Euclid's geometry did not contain contradictions. Although none had been found during the two thousand years since Euclid, that was not to say that they weren't there. Hilbert decided that the first thing to do was to recast geometry in terms of formulas and equations. This practice had been initiated by Descartes (hence the name Cartesian geometry) and adopted by the French mathematicians of the eighteenth century. Geometry could be reduced to arithmetic via the use of equations which described lines and points, and a point could be changed into numbers describing its co-ordinates in space. Mathematicians believed that the theory of numbers contained no contradictions, so Hilbert hoped that by replacing geometry by numbers the question of whether Euclidean geometry contained contradictions could be settled.

However, instead of an answer to the problem, Hilbert found something even more unsettling: no one had actually proved that the theory of numbers itself did not contain contradictions. Suddenly Hilbert was reeling. The fact that mathematics had worked both theoretically and practically for centuries and had not produced contradictions had given mathematicians confidence in what they were doing. 'Allez en avant, et la foi vous viendra' ('Go forward, and faith will come to you') was the answer given by the eighteenth-century French mathematician Jean Le Rond d'Alembert to those who questioned the foundations of the subject. To mathematicians, the existence of the numbers they were studying was as real as the organisms that biologists were classifying. Mathematicians had been happily plying their trade, making deductions from the assumptions they believed were self-evident truths about numbers. No one had contemplated the possibility that these assumptions might lead to contradictions.

Hilbert had been pushed further and further back and was now having to question the very basis on which mathematics had been constructed. Now that the question had been asked, it was impossible to ignore these foundational problems. Hilbert himself believed that no contradictions would ever be discovered and that mathematicians were equipped to dispel any such doubts, to prove that the subject was built upon a solid bedrock. His question marked the coming of age of mathematics. The nineteenth century had seen the transition from mathematics as practical handmaiden to science to the theoretical pursuit of fundamental truths more akin to the philosophy of a former resident of Königsberg, Immanuel Kant. Hilbert's deliberations on the very foundations of the subject gave him the platform from which to launch this new practice of abstract mathematics. His new approach would characterise mathematics during the twentieth century.

At the close of the year 1899, Hilbert was given the perfect opportunity to draw together the dramatic changes that his new ideas were bringing about in geometry, number theory and the logical foundations of mathematics. He received an invitation to deliver one of the main lectures to the International Congress of Mathematicians to be held in Paris the following year. It was a great honour for a mathematician who was still under forty.

Hilbert felt daunted by the task of addressing his community at the beginning of a new century. Surely it called for a truly momentous lecture, one that would live up to the occasion. Hilbert began consulting friends about his idea to use the lecture to speculate on the future of mathematics. This was highly unconventional and went against the unspoken rule that only complete, fully formed ideas were to be made public. It would take some nerve to forgo the security offered by presenting proofs of established theorems and, instead, to speculate on the uncertainties of the future. But Hilbert was never one to shy away from controversy. In the end he decided to challenge the international community with what hadn't been proved rather than what had.

He still had his doubts. Was it wise to use the occasion to try something so new? Perhaps he should follow convention and talk about what he had achieved rather than what he couldn't solve. Because of his procrastination he missed the deadline for submitting the title of his lecture and wasn't listed as a speaker at the congress. By the summer of 1900 his friends were worried that he might miss out completely on this wonderful opportunity to present his ideas, but one day they all found on their desks the text of Hilbert's lecture. It was entitled simply 'Mathematical problems'.

Hilbert believed that problems are the lifeblood of mathematics but also that they should be chosen with care. 'A mathematical problem should be difficult in order to entice us,' he wrote, 'yet not completely inaccessible, lest it mock at our efforts. It should be to us a guide post on the mazy paths to hidden truths, and ultimately a reminder of our pleasure in the successful solution.' The twenty-three problems he had chosen to present were perfectly selected to meet his stringent criterion. In the sultry August heat in the Sorbonne in Paris, Hilbert rose to deliver his lecture and challenge the mathematical explorers of the new century.

In the late nineteenth century, many areas of study had been influenced by the distinguished physiologist Emil du Bois-Reymond's philosophical movement which held that there are limits to our ability to understand nature. The catch phrase in philosophical circles had been 'Ignoramus et ignorabimus' – we are ignorant and we shall remain ignorant. But Hilbert's dream for the new century was to sweep aside such pessimism. He ended his introduction to the twenty-three problems with a rousing battle cry: 'This conviction of the solvability of every mathematical problem is a powerful incentive to the worker. We hear within us the perpetual call: There is the problem. Seek its solution. You can find it by pure reason, for in mathematics there is no ignorabimus.'

The problems which Hilbert set the mathematicians of the new century captured the revolutionary spirit of Bernhard Riemann. The first two problems on Hilbert's list concerned the foundational questions that had begun to obsess him, but the others ranged far and wide over the mathematical landscape. Some of them were open-ended projects rather than questions that seemed as if they should have clear answers. They included one related to Riemann's dream that the fundamental questions of physics would turn out to be answerable using mathematics alone.

The fifth problem arose out of Riemann's notion that the different fields of mathematics, such as algebra, analysis and geometry, were intimately related and could not be understood in isolation from one another. Riemann had demonstrated how algebraic properties of equations could be deduced from the geometry of the graphs defined by those equations. It had taken some courage to oppose the dogma that algebra and analysis must be kept away from the potentially misleading power of geometry. This is why the likes of Euler and Cauchy were so against the graphical depiction of imaginary numbers. For them, imaginary numbers were solutions to equations such as $x^2 = -1$ and should not be confused with pictures. But to Riemann it was obvious that the subjects were connected.

Hilbert mentioned Fermat's Last Theorem in the build-up to the announcement of his twenty-three problems. Curiously, despite the public perception, even in Hilbert's day, of this problem as one of the great unresolved questions in mathematics, it never featured as one of Hilbert's choices. In Hilbert's view it was 'a striking example of the inspiring effect which such a very special and apparently unimportant problem may have on science'. Gauss had expressed the same sentiment when he declared that one could choose a host of other equations and ask whether these had solutions or not. There was nothing special about Fermat's choice.

Hilbert took Gauss's critique of Fermat's Last Theorem as the inspiration for his tenth problem: is there an algorithm (a mathematical procedure that works something like computer software) that can decide in a finite amount of time whether any equation has solutions? Hilbert hoped that his question would move mathematicians' attention away from the particular and persuade them to focus on the abstract. For example, Hilbert had always appreciated how Gauss and Riemann had inspired a new perspective on the primes. No longer were mathematicians obsessed with whether a particular number was prime – instead, they were seeking to understand the music that flowed through all the primes. Hilbert hoped that his question about equations might have a similar effect.

Although one reporter at the meeting described the resulting discussion as 'desultory', this was more to do with the oppressive August weather than the intellectual appeal of Hilbert's lecture. As Hilbert's closest friend Minkowski commented, 'through this lecture, which indeed every mathematician in the world without exception will be sure to read, your attractiveness for young mathematicians will increase'. The risk Hilbert took in presenting such an unconventional lecture cemented his reputation in the twentieth century as a pioneer of new mathematical thinking. Minkowski believed that these twenty-three problems would prove to be hugely influential, and told Hilbert that 'you really have taken a total lease on mathematics for the twentieth century'. His words turned out to be prophetic.

Within his list of broad open-ended problems there was one, the eighth, that was very specific: to prove the Riemann Hypothesis. In an interview, Hilbert explained that he believed the Riemann Hypothesis to be the most important problem 'not only in mathematics but absolutely the most important'. In the same interview he was asked what he thought the most important technological achievement would be: 'To catch a fly on the Moon. Because the auxiliary problems which would have to be solved for such a result to be achieved imply the solution of almost all the material difficulties of mankind.' An insightful analysis, given the way the twentieth century panned out.

He believed that a proof of the Riemann Hypothesis would do for mathematics what Lunar fly-catching would do for technology. After posing the Hypothesis as his eighth problem, he went on to explain to the International Congress that a full understanding of Riemann's prime number formula might put us in a position to understand many of the other mysteries of the primes. He mentioned both Goldbach's Conjecture and the problem of the existence of infinitely many twin primes. The appeal of proving the Riemann Hypothesis was twofold: as well as closing a chapter in the history of mathematics, it would open many new doors.

Hilbert didn't think the Riemann Hypothesis would remain unproved for so long. In a lecture he gave in 1919, he declared he was optimistic that he would live to see it cracked, and perhaps the youngest member of the audience would live to see Fermat's Last Theorem proved. But he boldly predicted that no one then present would be alive to witness the solution of the seventh problem on his list – whether 2 to the power of the square root of 2 was the solution to an equation. Hilbert may have had great mathematical insight, but it wasn't matched by his powers of prediction. Within ten years the seventh problem had fallen. It is also just possible that some young graduate at Hilbert's 1919 lecture lived to witness Andrew Wiles's proof of Fermat's Last Theorem in 1994. Despite exciting progress over the last few decades, the Riemann Hypothesis might indeed still be unresolved when Hilbert wakes up, like Barbarossa, in five hundred years' time.

There was one occasion when Hilbert thought that he wouldn't have to wait so long. One day he received a paper from a student which purported to prove the Riemann Hypothesis. Before too long Hilbert found a gap in the proof, but the method impressed him. Tragically, the student died a year later and Hilbert was asked to give an address beside the grave. He praised the boy's ideas and hoped that one day they might stimulate a proof of the great conjecture. Then with the words, 'Consider if you will a function defined on imaginary numbers . . .' in a wholly inappropriate departure that illustrates perfectly the stereotype of mathematicians disconnected from social reality, Hilbert launched into the details of the incorrect proof. Whether the story is true or not, it is believable. Mathematicians sometimes develop tunnel vision.

Hilbert's lecture quickly pushed the Riemann Hypothesis into the limelight. It was now seen as one of mathematics' greatest unsolved problems. Although his obsession with the Hypothesis produced no direct contribution to its solution, the new programme that Hilbert was proposing for twentieth-century mathematics would be deeply influential. Even his questions on physics and foundational questions about the axioms of mathematics would by the end of the century have played their part in improving our understanding of the primes. Meanwhile, though, Hilbert was responsible for bringing to Göttingen a mathematician who would be next in line to carry the baton passed from Gauss to Dirichlet to Riemann.

Landau, most difficult of men

A position in Göttingen had become vacant following the tragically early death of Hilbert's closest friend, Minkowski. Aged only forty-five, Minkowski suffered devastating appendicitis. Hilbert had just succeeded in solving Waring's Problem, which had to do with writing numbers as sums of cubes, fourth powers and beyond. He knew that Minkowski would appreciate the result because it extended the work for which Minkowski had been awarded the Grand Prix des Sciences Mathématiques by the French Academy when he was only eighteen. 'Even on the hospital bed, lying mortally afflicted, he was concerned with the fact that at the next meeting of the seminar, when I would talk on my solution of Waring's Problem, he would not be able to be present.'

Minkowski's death affected Hilbert deeply. As one student at Göttingen related, 'I was in class when Hilbert told us about Minkowski's death, and Hilbert wept. Because of the great position of a professor in those days and the distance between him and the students, it was almost more of a shock for us to see Hilbert weep than to hear that Minkowski was dead.' Hilbert was keen to find a successor whose passion for number theory was the equal of Minkowski's.

By all accounts, Hilbert's choice, Edmund Landau, was not an easy man. It seems that it was a toss-up between appointing him or someone else. Hilbert asked his colleagues, 'Who of the two is the most difficult?' When the reply came that without doubt Landau was, Hilbert said that Göttingen must have Landau. Theirs was not to be a department of yes-men. Hilbert wanted colleagues who would challenge both social and mathematical conventions.

Landau was tough on his students and lived up to his billing as a prickly member of the department. His students used to dread weekend invitations to his house, where they had to humour his passion for mathematical games. A newly-wed student of Landau's was just about to depart for his honeymoon. The train had almost left the station at Göttingen when Landau stormed down the platform, pushed the manuscript of his latest book through the carriage window and demanded, 'I want it proof-read by the time you return!'

Landau soon assumed the mantle of successor to the tradition of

Edmund Landau (1877–1938).

Riemann and Gauss and was the central figure in Europe in developing the work of de la Vallée-Poussin and Hadamard. His temperament was perfectly suited to striking out from the base camp they had established and heading up the slopes of Mount Riemann. To prove Gauss's Prime Number Theorem, Hadamard and de la Vallée-Poussin had shown that there were no zeros on the north–south border through the number 1. The challenge now was to prove there were no zeros before one reached Riemann's critical line through $\frac{1}{2}$.

Landau was joined in his expedition by Harald Bohr. Bohr was based in Copenhagen but was one of the regular pilgrims who made their way across Europe to Göttingen. Bohr's brother, Niels, was ultimately to become world-famous as one of the creators of the theory of quantum physics. Harald had already made a name for himself as a key player in the Danish football team that secured silver at the 1908 Olympic Games.

Together, Landau and Bohr made the first successful push to navigate the points at sea level in Riemann's landscape. They were able to show that most of the zeros like to be bunched up against Riemann's ley line. They considered the number of zeros from 0.5 to 0.51 and compared it to the number of zeros outside this thin strip of land. They were able to prove that the zeros in this strip at least accounted for a large proportion of the zeros. Riemann had predicted that all the zeros were on the line through $\frac{1}{2}$. Landau and Bohr couldn't prove anything as definite as that, but they had made a start.

To make their argument work, the strip didn't necessarily need to be of width 0.01. However narrow it was, even of width $1/10^{30}$, say, Landau and Bohr could still prove that most zeros were inside this vertical band of land. Yet, frustratingly, neither of them could show that this meant that most must actually be on Riemann's line through $\frac{1}{2}$, something which Riemann claimed he had proved but had never published. This may seem counter-intuitive. If all the zeros lie in a vanishingly small band, then why can't we conclude that most of them must actually lie on the critical line? Such are the mysteries of mathematics. Suppose, for example, that for every number N, there are 10^N zeros in the narrow band between $\frac{1}{2} + 1/10^{N+1}$ and $\frac{1}{2} + 1/10^N$. Such a hypothetical set-up would satisfy the result established by Bohr and Landau without there requiring any of the zeros to lie on the critical line through $\frac{1}{2}$.

Göttingen by this time was beginning to live up to the motto, blazoned across the town hall, declaring that outside its medieval walls there was no life. More than anything else, Hilbert's influence turned Göttingen from Riemann's quiet university town into the mathematical powerhouse it had become in the early twentieth century. In Riemann's time it was Berlin that was buzzing with intellectual energy, but when Hilbert was later offered a position at Berlin University he turned it down. The small medieval town steeped in Gauss's heritage was the perfect environment for mathematical activity.

Hilbert was able to bring the world's best mathematicians to Göttingen thanks to money bestowed by a mathematics professor, Paul Wolfskehl, who died in 1908. In his will Wolfskehl had left 100,000 marks as a prize for the first person to come up with a proof of Fermat's Last Theorem. This was the prize that Wiles had read about as a child and sparked his interest in trying to prove Fermat's riddle. (The prize money that Wiles eventually received for his proof was significantly devalued by the hyper-inflation in Germany that followed the two world wars.) Wolfskehl's will stipulated that for every year that the theorem remained unproved, the interest accrued by the prize money should be used to fund visitors to Göttingen.

Landau took charge of checking the solutions sent to the faculty in Göttingen. Eventually the task became so overwhelming that Landau resorted to passing manuscripts to his students together with a standard letter of rejection for them to complete. It read, 'Thank you for your solution of Fermat's Last Theorem. The first mistake occurs on page . . . line . . .' Hilbert assumed the much more pleasant job of spending the interest generated by the unclaimed prize money. It gave him the flexibility to bring many mathematicians to Göttingen, so much so that he hoped Fermat's Last Theorem would remain unproved. 'Why should I kill the goose that lays the golden eggs?' he asked.

It was generally accepted that any young mathematician who wanted to make his way in the world first made his way to Göttingen. One student compared Hilbert's influence on mathematics to listening to 'the sweet flute of the Pied Piper . . . seducing so many rats to follow him into the deep river of mathematics'. Not unexpectedly, many of these mathematical rats came from the academies of Continental Europe that had blossomed in the political and intellectual revolutions that had swept through Europe in the nineteenth century.

In contrast, Great Britain was suffering from her traditional inability to absorb good ideas coming from the Continent. Just as the shores of England had remained remarkably resistant to the political turmoil of the French Revolution, mathematics in England had missed Riemann's revolution. Imaginary numbers were still regarded as a dangerous Continental notion. Indeed, mathematics in England had not flourished significantly since the dispute in the seventeenth century between Newton and Leibniz over who should get the credit for discovering calculus. Even if Newton had been the first, his country's mathematical development would for many years be hamstrung by its refusal to recognise the superiority of Leibniz's development of the subject. However, things were about to change.

Hardy, the mathematical aesthete

By 1914 Landau and Bohr had completed their work showing that most zeros were, at least, bunched up against Riemann's critical line. But how far had mathematicians got in charting the zeros that were on the line? Out of the infinite number of points at sea level that mathematicians knew were there, only seventy-one had so far been identified as lining up along Riemann's critical line.

Then came an important psychological breakthrough. After two centuries in the wilderness of disinterest in ideas from the Continent, an English mathematician, G.H. Hardy, seized Riemann's baton and managed to prove that infinitely many of the zeros were indeed lining up on the north–south line running through $\frac{1}{2}$. Hilbert was very impressed with Hardy's contribution. Indeed, when Hilbert discovered that Hardy was having trouble with the authorities in Trinity College, Cambridge about his accommodation, he wrote a letter to the Master. Hardy, said Hilbert, was not only the best mathematician in Trinity, he was the best in England and he should therefore have the best rooms in the college.

Hardy's fame outside mathematical circles owes much to his eloquent memoir *A Mathematician's Apology*, but he earned his mathematical laurels for his contributions to the theory of prime numbers and the Riemann Hypothesis. If Hardy had proved that infinitely many zeros were on the line, then was the game up? Had Hardy proved the Riemann Hypothesis? After all, if there are infinitely many zeros and Hardy had proved that infinitely many of them were on Riemann's line, aren't we home and dry?

The infinite, unfortunately, is a slippery character. Hilbert liked to illustrate the mysteries of the infinite by using the idea of a hotel with an infinite number of rooms. You might check each odd-numbered room and find them all occupied, but even though you've checked an infinite number of rooms there are still all the even-numbered rooms to account for. In Hardy's case, checking rooms to see whether they are occupied was replaced by checking zeros to see whether they were on the critical line. Unfortunately, Hardy had not even managed to prove that at least half the zeros were on the line. He'd accounted for an infinite number of rooms, but as a proportion of all the rooms left it represented zero per cent. Hardy's achievement was extraordinary but there was still a long way to go. Hardy had taken a bite out of the zeros, but what he was left with was as enormous and intractable as before.

That tantalising first taste was to act like a drug for Hardy. Nothing would obsess him as much as his desire to prove that all the zeros were on Riemann's line – with the possible exception of his passion for cricket and his running battle with God. As it was for Hilbert, the Riemann Hypothesis was at the top of Hardy's wish list, as is clear from the New Year's resolutions he wrote on one of the numerous postcards he sent to friends and colleagues:

(1) Prove the Riemann hypothesis.
(2) Make 211 [the first prime after the double century] not out in the fourth innings of the last test match at the Oval.
(3) Find an argument for the non-existence of God which shall convince the general public.
(4) Be the first man at the top of Mt. Everest.
(5) Be proclaimed the first president of the U.S.S.R., of Great Britain and Germany.
(6) Murder Mussolini.

Prime numbers had fascinated Hardy from an early age. As a child he amused himself in church by breaking down the numbers of the hymns into their prime building blocks. He loved to pore over books containing curiosities about these fundamental numbers, which he declared were 'better than the football reports for light breakfast table reading'. In fact, Hardy believed that anyone who enjoyed reading the football would appreciate the joys of prime numbers. 'It is a peculiarity of the theory of numbers that much of it could be published, and would win new readers for the *Daily Mail*.' He believed that the primes retained enough mystery to intrigue the reader, yet were simple enough for anyone to begin to explore their magic. Hardy, more than any mathematician at the time, worked hard to communicate some of this passion for his subject, and did not believe it should remain the secret pleasure of those in the ivory towers of academe.

As the third of his New Year's resolutions indicates, the church where he first cracked hymn numbers into primes also had a profound effect on Hardy. From an early age he became fiercely opposed to the idea of a God and the trappings of religion. He was to have a running battle with God through the whole of his life, attempting to prove his impossibility. His fight became so personal, he paradoxically conjured up the very character whose existence he vehemently wanted to deny. On trips to watch cricket he would take an anti-God battery to ward off any possibility of rain. Even though the sky was cloudless, he would arrive with four sweaters, an umbrella and a bundle of work under his arm. As he explained to his neighbouring spectators at the ground, he was trying to trick God into thinking that he hoped it was going to rain so that he could catch up on some work. God, his personal enemy, he believed would send sunshine to scupper any such plans Hardy had for doing mathematics.

One summer's day, Hardy was frustrated to see the cricket match he was attending abruptly curtailed when the batsman complained of being unsighted by a flashing light emanating from the stands where he was sitting. His anger turned to joy when an enormous clergyman was asked to remove the huge silver cross from around his neck that was catching the light of the sun. Hardy could not contain himself and spent the lunch break firing off postcards to his friends describing cricket's vanquishing of the clergy.

When the cricket season ended in September, Hardy would often visit Harald Bohr in Copenhagen before the English academic term began. They had a daily ritual of work. Every morning they would place a piece of paper on the table on which Hardy would write their task for the day:

to prove the Riemann Hypothesis. Hardy had been hopeful that ideas developed by Bohr on his visits to Göttingen might provide a pathway to a proof. The rest of the day might be spent walking and talking, or scribbling away. Time after time their efforts failed to yield the breakthrough Hardy so hoped for.

Then, on one occasion, shortly after Hardy had set off on his return to England for the start of the new academic year, Bohr received a postcard. His heart raced as he read Hardy's words: 'Have proof of Riemann Hypothesis. Postcard too short for proof.' Finally, Hardy had broken the impasse. The postcard, though, had a rather familiar ring to it. Fermat's tantalising marginal comments flashed into Bohr's mind. Hardy was too much of a prankster to have missed the irony in the postcard. Bohr decided to delay celebrations and await Hardy's further elaboration. Sure enough, the postcard turned out not to be the breakthrough that Bohr had hoped for – Hardy was playing one of his games with God.

As Hardy boarded the ship to cross the North Sea from Denmark to England, the sea was unusually rough. The ship itself was not especially large, and Hardy began to fear for his life. So he took out his own very individual insurance policy. It was then that he sent Bohr that postcard announcing his fictional discovery. If the first passion in his life was to prove the Riemann Hypothesis, the second was his battle with God. Hardy knew that God would never allow the ship to sink and leave the world with the impression that Hardy and his proof had drowned and were lost for ever. His ploy worked, and he arrived safely back in England.

It is probably fair to say that Hardy's addiction to the Riemann Hypothesis, combined with his colourful and charismatic character, helped to boost the problem to the top of mathematics' most-wanted list. His eloquent writing style, encapsulated in *A Mathematician's Apology*, was instrumental in promoting the importance of number theory and what he regarded as its central problem. It is striking that for all Hardy's talk in the *Apology* of beauty and aesthetics in mathematics, the beauty of the proofs that Hardy was responsible for is often obscured by a mass of technical details that are required to see the proofs through to their conclusion. As often as not, success was a result not so much of a great idea as of hard graft.

The one book that was probably responsible for Hardy's desire to become a mathematician was not a mathematics book at all. It was a story about the delights of a life at high table at Trinity College. The description of drinking port in the Senior Combination Room that he'd read in a novel, *A Fellow of Trinity*, fascinated him. Hardy admitted that he chose mathematics because 'it is the one and only thing I can do at all well . . . until I obtained one, mathematics meant to me primarily a Fellowship at Trinity'.

To get there he was subjected to the gruelling round of examinations that the Cambridge system demanded. Hardy later realised that the emphasis the examination system placed on solving artificial technical problems and mathematical puzzles meant that, even after completing a degree in mathematics, few were aware what it was really all about. One of the Göttingen professors in 1904 parodied the problems that British students were expected to answer: 'On an elastic bridge stands an elephant of negligible mass; on his trunk sits a mosquito of mass *m*. Calculate the vibrations of the bridge when the elephant moves the mosquito round by rotating its trunk.' Students were expected to quote Newton's *Principia* as if it were the Bible. Results were known by line number rather than by what they actually meant. Hardy believed that this system contributed to Britain's time in the mathematical wilderness. British mathematicians were being taught to play their mathematical scales ever faster, but they were completely unaware of the beautiful mathematical music they could play once they had mastered their scales.

Hardy put his own mathematical enlightenment down to the French mathematician Camille Jordan's book *Cours d'Analyse*, which opened his eyes to the mathematics that had been flourishing on the Continent. 'I shall never forget the astonishment with which I read that remarkable work . . . and learnt for the first time as I read it what mathematics really meant.'

Hardy's election to Trinity in 1900 released him from the burden of taking examinations, and set him free to explore the real world of mathematics.

Littlewood, the mathematical bully boy

Hardy was joined at Trinity in 1910 by a mathematician eight years his junior, J.E. Littlewood. Together they would spend the next thirty-seven years like a mathematical Scott and Oates, exploring the new land that had been opened up on the Continent. Their collaboration generated nearly a hundred joint papers. Bohr used to joke that there were three great English mathematicians during this period: Hardy, Littlewood and Hardy–Littlewood.

The two mathematicians each brought their own qualities to the collaboration. Littlewood was the bully boy who went in with all guns

Hardy and Littlewood in Trinity College, Cambridge in 1924.

blazing in his assault on a problem. He revelled in the satisfaction of bringing a difficult problem to its knees. Hardy, in contrast, valued beauty and elegance. Invariably this carried over to the writing of their papers. Hardy would take Littlewood's rough draft and would add what they called the 'gas' to produce the elegant prose that invariably accompanied their proofs.

It's curious that these two mathematicians' styles were mirrored in their physical appearance. Hardy was a beautiful man, one of those whose appearance maintains the stamp of youth well beyond their sell-by date. In his early days as a Fellow of Trinity he was often challenged by staff in the Senior Combination Room, who thought he must be an undergraduate who had lost his way in the labyrinth of Trinity's corridors. Littlewood was rough-hewn – 'a character straight out of Dickens', as one mathematician remarked. He was strong and agile in mind and body. Like Hardy, he loved cricket and was a hard-hitting batsman. His other passion was music, something Hardy never felt an affinity for. As an adult he taught himself to play the piano, and had an intense love for the music of Bach, Beethoven and Mozart. He thought life too short to waste on lesser composers.

The other thing which separated them was sexuality. It was recognised that Hardy was very likely homosexual. Nevertheless, he was very discreet about it even though in Cambridge homosexuality was almost more acceptable than marriage. This was at a time when Oxford and Cambridge dons would have to leave their fellowships if they ever married. Littlewood declared Hardy to be a 'non-practising homosexual'. By all accounts Littlewood was something of a ladies' man. Although not up to Hilbert's standards, he did become very friendly with a local doctor's wife, with whom he spent summer holidays in Cornwall. Many years later, one of her children was looking into a mirror and commented on the striking resemblance to Uncle John. 'That's not surprising,' she replied, 'he's your father.'

As befitting two mathematicians, Hardy and Littlewood's collaboration was based on very clear axiomatic foundations:

Axiom 1: It didn't matter whether what they wrote to each other was right or wrong.

Axiom 2: There was no obligation to reply, or even read, any letter one sent to the other.

Axiom 3: They should try not to think about the same things.

And the most important axiom of them all:

Axiom 4: To avoid any quarrels, all papers would be under their joint name regardless of whether one of them had contributed nothing to the work.

Bohr summed up their relationship thus: 'Never was such an important and harmonious collaboration founded on such apparently negative axioms.' Mathematicians today still talk about 'playing under Hardy–Littlewood rules' when they carry out joint work. Bohr found that Hardy remained true to his second axiom when they collaborated together in Copenhagen. He remembered the voluminous mathematical letters from Littlewood that arrived each day, and Hardy calmly tossing them into the corner of the room with a dismissive 'I suppose I shall have to read them some day.' When Hardy was in Copenhagen, there was only one thing on his mind: the Riemann Hypothesis. Unless Littlewood was sending him a proof of the Hypothesis, the letter flew into the corner.

There is a story recounted by Harold Davenport, a student of Littlewood's, that Hardy and Littlewood almost fell out over the Riemann Hypothesis. Hardy had written a murder mystery in which one mathematician proves the Riemann Hypothesis, only to be killed by a second mathematician who then claims authorship of the proof. Littlewood was most upset. It wasn't that Hardy had violated Axiom 4 and failed to include Littlewood as an author. Littlewood was convinced that the murderer was modelled on him, and he objected to the manuscript ever seeing the light of day. Hardy conceded, and mathematics was deprived of this literary gem.

Littlewood had come up through the ranks of Cambridge mathematical undergraduates, performing all the tricks required by the examination system. He made it to the top of the pile, earning himself the much coveted title of senior wrangler, jointly shared with another student, Mercer. The senior wranglers were celebrities in Cambridge, and photographs of them would be on sale at the end of the academic year. Perhaps his fellow students had already guessed that this was just the beginning of Littlewood's outstanding career. When a friend tried to buy one of the photographs he was told, 'I'm afraid we're sold out of Mr Littlewood but we have plenty of Mr Mercer.'

Littlewood could see that the exams were not what mathematics was really about, but simply some technical game he was required to play and win before he could move on to the next stage. 'The game we were playing came easily to me and I even felt some satisfaction in successful craftsmanship.' Littlewood was eager to practise the craft he had learnt as an undergraduate and put it to more creative use. His introduction to serious mathematical research was to be something of a baptism of fire.

Fresh from the exams, Littlewood was keen to get stuck into research over the long summer vacation. He asked his tutor, Ernest Barnes, for a suitable problem on which he could cut his teeth. Barnes, who went on later to become bishop of Birmingham, thought for a while, and recalled an interesting function which no one had really got to grips with. Perhaps Littlewood could investigate where the function output zero. Barnes wrote out a definition of the function for Littlewood to take away for the summer. 'It's called the zeta function,' said Barnes innocuously. Littlewood left Barnes's rooms with the paper in his hand, oblivious of the fact that Barnes had just suggested he might like to spend the summer proving the Riemann Hypothesis.

Barnes had failed to provide Littlewood with the historical background of the problem, which would have indicated its difficulty. Littlewood's tutor may even have been unaware that there were any connections between the zeros and prime numbers, and thought of it solely as an interesting problem: where does this function output the value zero? As Peter Sarnak, one of the leading lights in modern attempts on the Riemann Hypothesis, explains, 'It was really the only analytic function mathematicians still did not understand as we entered the twentieth century.' As Sir Peter Swinnerton-Dyer, who became one of Littlewood's students, reflected at Littlewood's memorial service, the fact that 'Barnes thought [the Riemann Hypothesis] suitable for even the most brilliant research student, and that Littlewood should have tackled it without demur' illustrated the dire state of British mathematics before Hardy and Littlewood had made an impact.

Littlewood battled all the summer, wrestling with the innocent-looking problem that Barnes had given him. Although he had no luck finding the locations of the zeros, he was very pleased with something else he came across. Just as Riemann had discovered some fifty years before, Littlewood realised that these zeros could tell you something about prime numbers. Although this had been known on the Continent since Riemann's time, in England the connection between the zeta function and the primes was still unappreciated. Littlewood was thrilled by what he thought was a new link and in September 1907 he wrote it up for his dissertation in support of a research fellowship at Trinity. That Littlewood thought his discovery was original is further confirmation of how isolated British mathematics had become.

Hardy, who was one of the few in England aware of the recent progress made by Hadamard and de la Vallée-Poussin, knew that the result was not as original as Littlewood had hoped. Nonetheless, Hardy recognised Littlewood's potential, and although he failed that year to be elected a Fellow of the college, there was a gentleman's agreement to elect him next time round. He joined Hardy at Trinity in October 1910.

Cambridge was beginning to blossom as it opened its doors to the influences of the intellectual tradition across the Channel. Travel between the Continent and England was becoming easier, and Hardy and other academics were making the effort to visit many of the European centres of learning. The new contacts they made encouraged a flow of new journals, books and ideas from abroad. Trinity College in particular became an extraordinarily vibrant community in the early twentieth century. The Senior Combination Room was no longer a gentleman's club, but a place of research. Conversation at high table did not confine itself to port and claret but was infused with the ideas of the day. Also at Trinity, working alongside Hardy and Littlewood, were the two most eminent philosophers active in England: Bertrand Russell and Ludwig Wittgenstein. Both were wrestling with the same foundational problems of mathematics that had so concerned Hilbert. And Cambridge was buzzing with new breakthroughs in physics made by the likes of J.J. Thomson, who was awarded a Nobel prize for his discovery of the electron, and Arthur Eddington, who had confirmed Gauss and Einstein's belief that space was indeed curved and non-Euclidean.

The great collaboration between Hardy and Littlewood was fuelled by the timely arrival from Göttingen of a book by Landau about prime numbers. The publication in 1909 of his two-volume work *Handbuch der Lehre von der Verteilung der Primzahlen* ('Handbook of the Theory of the Distribution of Prime Numbers') proselytised the wonders of the connections between primes and the Riemann zeta function. Before Landau's book, the story of Riemann and the primes was largely unknown in the broader mathematical community. As Hardy acknowledged in his obituary of Landau (jointly written with Hans Heilbronn), 'The book transformed the subject, hitherto the hunting ground of a few adventurous heroes, into one of the most fruitful fields in the last thirty years.' It would be Landau's book that would in 1914 inspire Hardy to prove that infinitely many zeros sat on Riemann's critical line. Fired up by the experiences of wrestling with the zeta function as a student, Littlewood too was encouraged to make the first of his great contributions to the subject.

To prove a theorem that Gauss believed to be true but couldn't prove is generally regarded as a true test of a mathematician's mettle. To disprove such a theorem puts one in a different league altogether. It is not often that Gauss had a hunch which turned out to be false. He had produced a function, the logarithmic integral Li(N), that he had predicted would guess the number of primes up to any number N, with increasing accuracy as N got bigger. Hadamard and de la Vallée-Poussin had carved their names into mathematical history by proving Gauss right. But Gauss had made a second conjecture: that his guess would always *overestimate* the number of primes – it would never predict that there were fewer primes than there really were in the range from 1 to N. This contrasted with Riemann's refinement, which fluctuated between overestimating and underestimating the correct number of primes.

By the time Littlewood started to think about Gauss's second conjecture, it had been confirmed to be true for all numbers up to 10,000,000.

Any experimental scientist would have accepted 10 million pieces of evidence as utterly convincing support for Gauss's hunch. Sciences with less of an addiction to proof and more respect for experimental results would have been perfectly happy to accept Gauss's conjecture as a foundation stone upon which they could start to build new theories. By Littlewood's day, some hundred years later, the mathematical edifice might well have towered way above this foundation. But in 1912 Littlewood discovered that, contrary to expectations, Gauss's hypothesis was a mirage. The foundation stone crumbled into dust under his scrutiny. He proved that as you counted higher you would eventually come to regions of numbers where Gauss's guess would switch from overestimating to underestimating the number of primes.

Littlewood also succeeded in demolishing another idea that was beginning to get a toehold. Many believed that Riemann's refinement of Gauss's guess at the number of primes would always be the more accurate. Littlewood showed that Riemann's refinement might look more accurate as we count through the first million numbers, but in the farther reaches of the universe of numbers Gauss's guess would sometimes give the better prediction.

Littlewood's discoveries were particularly striking because Gauss's guess starts to underestimate the number of primes only in regions of numbers that we will probably never be able to calculate. Littlewood could not even predict how far we would need to go before we could observe any of these phenomena. Indeed, to this day no one has actually counted far enough to arrive at a region of numbers where Gauss's guess underestimates the primes. It is only through Littlewood's theoretical analysis and the power of mathematical proof that we can be sure that somewhere along the line Gauss's original prediction is false.

Some years later, in 1933, a graduate student of Littlewood's named Stanley Skewes estimated that by the time one had counted the primes up to $10^{10^{10^{34}}}$, one will have witnessed Gauss's guess finally underestimate the number of primes. That is a ridiculously large number. Encounters with large numbers often elicit comparisons with the number of atoms in the visible universe, which is according to the best estimates approximately 10^{78}, but the number suggested by Skewes defies even that. It is a number that begins with a 1 and then has so many zeros after it that even if you wrote a 0 on each atom in the universe you still wouldn't have got anywhere near it. Hardy was to declare that the Skewes Number, as it became known, was surely the largest number that had ever been contemplated in a mathematical proof.

The proof of Skewes's estimate was interesting for another reason. It is one of the many thousands of proofs which begin 'suppose the Riemann Hypothesis is true'. Skewes could make his proof work only by assuming that Riemann's conjecture is correct: that all the points at sea level in the zeta landscape are on the line through $\frac{1}{2}$. Without making this assumption, mathematicians in the 1930s were unable to guarantee how far we would need to count before Gauss's guess underestimated the number of primes. In this particular case mathematicians finally found a way to avoid having to cross the summit of Mount Riemann. In 1955 Skewes produced an even larger number that would still work in the event that the Riemann Hypothesis turned out to be false.

It was curious that, in contrast to their reluctance to accept Gauss's second conjecture, mathematicians were beginning to have sufficient faith in the truth of the Riemann Hypothesis that they were prepared to build upon it while it remained unproved. The Riemann Hypothesis was now becoming an essential structural component in the mathematical edifice. But it was probably as much a matter of pragmatism as of faith. More and more mathematicians were finding themselves coming up against the Riemann Hypothesis as an obstacle to their mathematical progress. Only by assuming that it was true could they proceed any further. But as Littlewood illustrated with Gauss's second conjecture, mathematicians have to be prepared for the possible collapse of all that is built on the foundations of the Riemann Hypothesis, should someone discover a zero off the line.

Littlewood's proof had a huge psychological effect on the perception of mathematics and especially on the appreciation of the primes. It sent out a stark warning to anyone impressed by a vast accumulation of numerical evidence. It revealed that prime numbers are masters of disguise. They hide their true colours in the deep recesses of the universe of numbers, so deep that witnessing their true nature may be beyond the computational powers of humankind. Their true behaviour can be seen only through the penetrating eyes of abstract mathematical proof.

Littlewood's proof also provided the perfect ammunition for those who argued that mathematics differs in some essential way from the other sciences. No longer could mathematicians be happy with the experimentalism of the seventeenth- and eighteenth-century brand of mathematics in which theories were advanced after minimal calculations. Empiricism was no longer a suitable vehicle in which to navigate the mathematical world. Millions of pieces of data might be sufficient evidence on which to base theories in the other sciences, but Littlewood had proved that in mathematics that would be treading on thin ice. From now on, proof was everything. Nothing could be trusted without conclusive evidence.

As more mathematicians found themselves forced to assume the truth of the Riemann Hypothesis, it became more imperative than ever to make sure that in some distant part of Riemann's landscape there weren't zeros straying off the critical line. Until that had been done, mathematicians would always live in fear that the Riemann Hypothesis might be disproved.

CHAPTER SIX

Ramanujan, the Mathematical Mystic

An equation means nothing to me unless it expresses a thought of God. Srinivasa Ramanujan

While Hardy and Littlewood were fighting their way across Riemann's strange landscape, some five thousand miles away in the offices of the Madras Port Authority in India, a young clerk named Srinivasa Ramanujan had become obsessed by the mysterious, intoxicating ebb and flow of the primes. Instead of attending to his tedious task of keeping the accounts, which he was employed to do, he spent all his waking hours filling notebooks with observations and calculations as he searched for what it was that made these strange numbers tick. As Ramanujan counted primes, he had no way of knowing about the sophisticated perspective that was being developed in the West. With no formal education, he lacked the respect that Littlewood and Hardy showed the subject of number theory, and in particular the primes, which Hardy referred to as 'the most difficult of *all* branches of pure mathematics'. Unconstrained by any mathematical tradition, Ramanujan dived into the primes with an almost childlike enthusiasm. His naivety, combined with his extraordinary aptitude for mathematics, would turn out to be his great strength.

In Cambridge, Hardy and Littlewood pored over the wonderful story unfolding in Landau's book on prime numbers. In India, Ramanujan's obsession with the primes had been inspired by a far more elementary book, and with consequences just as far-reaching. There are certain turning points in a young scientist's life which can often be identified as key to their future development. For Riemann it was the book by Legendre that he had been given as a schoolboy. That book had planted the seed that was to germinate in his later life. For Hardy and Littlewood, Landau's book was equally influential. Ramanujan, aged fifteen, had been inspired by his discovery in 1903 of a copy of George Carr's *A Synopsis of Elementary Results in Pure and Applied Mathematics*. Save for the connection to Ramanujan, this book and its author are of little importance, but the structure of the book was. It was a list of some 4,400 classical results – just the results, no proofs. Ramanujan had taken up the challenge and spent the following years going through the book and justifying each statement. Unfamiliar with Western-style proofs, Ramanujan was forced to create his own mathematics. Spared the straightjacket of accepted modes of

Srinivasa Ramanujan (1887–1920).

thought, he was left free to roam. And it wasn't long before he was filling his notebooks with ideas and results that went beyond Carr's book.

Euler had cut his teeth on Fermat's many unproved statements. We can see the spirit of Euler in Ramanujan's approach to problems. Ramanujan had a fantastic intuition for how to twist formulas this way and that until new insights emerged. He was very excited when he discovered for himself the connection that imaginary numbers provided between the exponential function and the equations describing sound waves. Joy turned to despair when the young Indian clerk found out a few days later that Euler had beaten him to this great discovery by some hundred and fifty years. Humbled and dispirited, Ramanujan hid his calculations in the roof of his house.

Mathematical creativity is difficult to understand at the best of times, but the way Ramanujan worked was always something of a mystery. He used to claim that his ideas were given to him in his dreams by the goddess Namagiri, the Ramanujans' family goddess and consort of Lord Narasimha, the lion-faced, fourth incantation of Vishnu. Others in

Ramanujan's village believed that the goddess had the power to exorcise devils. For Ramanujan himself she was the explanation for the flashes of insight that sparked his continuous stream of mathematical discoveries.

Ramanujan is not the only mathematician for whom the dream-world was fertile territory for mathematical exploration. Dirichlet had tucked Gauss's *Disquisitiones Arithmeticae* under his pillow at night, hoping for inspiration in his efforts to understand the often cryptic statements the book contained. It's as though the mind is released from the constraints of the real world, free to explore avenues that the waking mind has sealed off. Ramanujan, it seemed, was able to induce this dreamlike state in his waking hours. Such a trance is in fact very close to the state of mind that most mathematicians try to achieve.

Hadamard, who had made his name proving the Prime Number Theorem, was fascinated by what went on in the mind of a creative mathematician. He set down his ideas in a book entitled *The Psychology of Invention in the Mathematical Field*, published in 1945, in which he makes a powerful case for the role of the subconscious. Neurologists are becoming increasingly interested in the workings of the mathematical mind since it could shed light on how the brain works. It is often periods of rest or even dreams that give the mind the freedom to play around with ideas which have been sown in the brain during our conscious periods of work.

In his book, Hadamard divided the act of mathematical discovery into four stages: preparation, incubation, illumination and verification. If Ramanujan had been given the gift of the third of these stages, he was distinctly lacking in the fourth. Illumination itself was sufficient for Ramanujan. He just did not see the point in verification. Perhaps it was not having the responsibility of proof around his neck that allowed Ramanujan the freedom to discover new pathways through the mathematical wilderness. This intuitive style was quite at odds with the scientific traditions of the West. As Littlewood wrote later, 'The clear-cut idea of what is meant by a proof he did not possess at all; if the total mixture of evidence and intuition gave him certainty, he looked no further.'

Indian schools owed much to the ideas of Imperial Britain. However, the English education system that had served Littlewood and Hardy so well singularly failed to nurture the young Ramanujan in India. In 1907, when Littlewood's dissertation was being so warmly received in Cambridge, Ramanujan failed his college exams for the third and final time. He would have passed the exams had they been confined to mathematics, but he was required to study English, history, Sanskrit and even physiology. As an orthodox Brahmin, Ramanujan was a strict vegetarian, and dissecting frogs and rabbits was beyond the pale. But his failure, which meant he could not enter the University of Madras, did not extinguish the mathematical fire raging inside him.

By 1910 Ramanujan was keen to have his ideas recognised somehow. He was particularly excited by his discovery of a formula that appeared to count the primes with extraordinary accuracy. At first he had experienced the frustration that most people feel when they try to tame this wild sequence of numbers. But Ramanujan knew how fundamental the primes were to mathematics and did not give up on his belief that there should be some mathematical formula which would explain them. He still innocently believed that all mathematics and its patterns could be expressed precisely by the power of equations and formulas. As Littlewood later explained, 'How great a mathematician might Ramanujan have been 100 or 150 years ago; what would have happened if he had come into touch with Euler at the right moment? . . . But the great day of formulae seems to be over.' Ramanujan, however, had not been subjected to the nineteenth- and twentieth-century shift induced by Riemann. He was out to find a formula that produces the primes. Having spent hours calculating tables of primes, he saw a pattern emerge. He was keen to explain his tentative findings to someone who might appreciate his ideas.

His impressive-looking notebooks and the power of the Brahmin network had secured Ramanujan a job as an accountant with the Port Authority in Madras. He had begun to publish some of his ideas in the *Journal of the Indian Mathematical Society*, and by now his name had come to the attention of the British authorities. C.L.T. Griffith, who worked at the College of Engineering in Madras, recognised that Ramanujan's work was that of a 'remarkable mathematician' but he felt unable to follow or criticise it. So he decided to get the opinion of one of the professors who had taught him as a student in London.

Without formal training, Ramanujan had evolved a very personal mathematical style. It is perhaps not surprising, then, that when Professor Hill of University College, London received Ramanujan's papers claiming to have proved that

$$1 + 2 + 3 + \ldots + \infty = -\tfrac{1}{12}$$

he dismissed most of them as meaningless. Even to the untrained eye, this formula looks ridiculous. To add up all the whole numbers and get a negative fraction is clearly the work of a madman! 'Mr Ramanujan has fallen into the pitfalls of the very difficult subject of Divergent Series,' he wrote back to Griffith.

Hill, however, was not totally dismissive. Ramanujan was sufficiently encouraged by the comments that were passed on to him to try his own luck, and wrote directly to a number of mathematicians in Cambridge. Two recipients failed to penetrate the message behind Ramanujan's strange arithmetic and declined the Indian's request for help. But then Ramanujan's letter landed on Hardy's desk.

Mathematics seems to bring out the cranks. Maybe Fermat is partly responsible. Landau's standard rejection letter testifies to the number of crank responses he received claiming Wolfskehl's prize for the solution of Fermat's Last Theorem. Mathematicians are quite accustomed to receiving unsolicited letters with mad numerological theories – Hardy used to be deluged with manuscripts which, as his friend C.P. Snow recalled, claimed to have solved the prophetic mysteries of the Great Pyramid or decoded the cryptograms that Francis Bacon had secreted in the plays of Shakespeare.

Ramanujan had recently been given a copy of Hardy's *Orders of Infinity* by Ganapathy Iyer, a Professor of Mathematics in Madras with whom he regularly discussed mathematics on the beach in the evenings. As he read Hardy, Ramanujan must have recognised that here at last was someone who might appreciate his ideas, but later he admitted that he had feared his infinite sums would prompt Hardy 'to point out to me the lunatic asylum as my goal'. Ramanujan was particularly excited by Hardy's statement that 'no definite expression has been found as yet for the number of prime numbers less than any given number'. Ramanujan had discovered an expression which he believed very nearly captured this number. He was very keen to find out what Hardy thought of his formula.

Hardy's first impression on finding in the morning post Ramanujan's envelope covered in Indian stamps was not immediately favourable. It contained a manuscript filled with wild, fantastic theorems about counting primes, alongside well-known results presented as if they were original discoveries. In the covering letter Ramanujan declared that he had 'found a function which exactly represents the number of prime numbers'. Hardy knew that this was a stunning claim, but no formula had been supplied. Worst of all – no proofs of anything! For Hardy, proof was everything. He once told Bertrand Russell across the high table at Trinity, 'If I could prove by logic that you would die in five minutes, I should be sorry you were going to die, but my sorrow would be very much mitigated by pleasure in the proof.'

According to C.P. Snow, Hardy, having quickly looked over Ramanujan's work, 'was not only bored, but irritated. It seemed like a curious kind of fraud.' But by the evening the wild theorems were beginning to work their magic, and Hardy summoned Littlewood for after-dinner discussions. By midnight they had cracked it. Hardy and Littlewood, equipped with the knowledge to decode Ramanujan's unorthodox language, could now see that these were not the outpourings of a crank but the works of a genius – untrained, but brilliant.

They both recognised that Ramanujan's crazy infinite sum was none other than the rediscovery of how to define the missing part of Riemann's zeta landscape. The clue to decoding Ramanujan's formula is to rewrite the number 2 as $1/(2^{-1})$ (2^{-1} is another way of writing $\frac{1}{2}$). Applying the same trick to each number in the infinite sum, Hardy and Littlewood rewrote Ramanujan's formula as

$$1+2+3+\ldots+n+\ldots=1+\frac{1}{2^{-1}}+\frac{1}{3^{-1}}+\ldots+\frac{1}{n^{-1}}+\ldots=-\frac{1}{12}$$

Staring them in the face was Riemann's answer to how to calculate the zeta function when fed with the number -1. With no formal training, Ramanujan had run the whole race on his own and reconstructed Riemann's discovery of the zeta landscape.

Ramanujan's letter could not have arrived at a better time. Thanks to Landau's book, both Littlewood and Hardy had become fascinated by the wonders of Riemann's zeta function and its connection with primes. Now here was Ramanujan claiming to have an incredibly accurate formula for the number of primes within any given range of numbers. That morning Hardy had dismissed the claim, convinced that Ramanujan was some mathematical crank. But the evening's work had put this Indian package in an entirely different light.

Hardy and Littlewood must have been astounded by Ramanujan's assertion that his formula could calculate the number of primes up to 100,000,000 'with generally no error and in some cases with an error of 1 or 2'. The trouble was that no formula was given. In fact the whole letter was deeply frustrating for two mathematicians for whom rigorous proof was absolutely essential. It was full of formulas and assertions with no justification or explanation of where they came from.

Hardy wrote a very positive reply to Ramanujan but begged for proofs and more details of his formulas for primes. Littlewood added a note asking to be sent the formula for the number of primes and 'as much proof as possible quickly'. Both mathematicians were fired up in anticipation of Ramanujan's response and spent many high-table dinners trying to decode more of Ramanujan's first letter. As Bertrand Russell related in a letter to a friend, 'In Hall I found Hardy and Littlewood in a state of wild excitement, because they believe they have discovered a second Newton, a Hindu clerk in Madras on £20 a year.'

A second letter from Ramanujan duly arrived. It contained several formulas for the number of primes but was still short on proof. 'How maddening his letter is in the circumstances,' Littlewood wrote, and went on to speculate that Ramanujan might have been afraid that Hardy was going to steal his discoveries. As Hardy and Littlewood pored over this second letter they found that Ramanujan had come up with another of Riemann's fundamental discoveries. Riemann's refinement of Gauss's formula for counting primes was very accurate, and Riemann had discovered how to use the zeros in the zeta landscape to remove the errors that his formula was still producing. From absolutely nowhere Ramanujan had reconstructed part of the formula that Riemann had discovered fifty years before. Ramanujan's formula included Riemann's refinement of Gauss's guess for the number of primes, but it was missing the corrections that Riemann had built from the zeros in his landscape.

Was Ramanujan saying that the errors coming from the points at sea level cancelled each other out in some miraculous way? Fourier had provided a musical perspective to these errors. Each zero was like a tuning fork, and when the forks are all sounded together they create the noise of the primes. Sometimes sound waves can combine to produce silence if they cancel one another out. An aeroplane cuts down the noise of the engines by creating sound waves within the cabin to counteract it. Was Ramanujan claiming that Riemann's waves from the zeros might be creating silence?

During the Easter vacation, Littlewood took a copy of Ramanujan's letter with him on a trip to Cornwall with his lover and her family. 'Dear Hardy,' he wrote back – they never addressed each other by their first names – 'The stuff about primes is wrong.' Littlewood had managed to prove that there was no way the errors from these waves could cancel one another out to justify Ramanujan's claim that his reconstruction of Riemann's formula was as accurate as he claimed. There would always be some noise, however far one counted.

As it happened, the analysis by Littlewood, stimulated by Ramanujan's letter, did lead to an interesting new insight into Riemann's work. The Riemann Hypothesis had become important for mathematicians because it implied that the difference between Gauss's guess and the true number of primes up to N, when compared with the size of N, would be very small – essentially no more than the square root of N. But if any zero was not on Riemann's ley line the error would be bigger than that. Now here in Ramanujan's letter was the suggestion that one could go one better than Riemann. Perhaps, as one counted more primes, the error would become even smaller than the square root of N. Littlewood's work in Cornwall had smashed that hope. Littlewood could prove that infinitely often the error caused by the zeros would be at least as big as the square root of N. The Riemann Hypothesis was the best-case scenario. Ramanujan was simply wrong, but Hardy was impressed. As he wrote later, 'I am not sure that in some ways, his failure was not more wonderful than any of his triumphs.'

'I have a vague theory as to how his mistakes have come about.' Littlewood speculated in his letter to Hardy that Ramanujan must be under the illusion that the zeta landscape had no points at sea level. Indeed, if this were true then Ramanujan's formulas would have been spot-on. Littlewood was nonetheless excited. 'I can believe that he's at least a Jacobi,' he declared, comparing Ramanujan to one of the mathematical stars of Riemann's generation. Hardy wrote to Ramanujan, 'To have proved what you claimed would have been about the most remarkable mathematical feat in the whole history of mathematics.' It was clear that although Ramanujan was bursting with talent, he desperately needed to be brought up to date by making him familiar with the current state of knowledge. Littlewood wrote to Hardy about his hunch: 'it is not surprising that he would have been caught, unsuspicious as he presumably is of the diabolical malice inherent in the primes'. As Hardy remarked, 'He had been carrying an impossible handicap, a poor and solitary Hindu pitting his brains against the accumulated wisdom of Europe.'

They determined to do whatever it would take to get Ramanujan to Cambridge. They dispatched E.H. Neville, a Fellow of Trinity, to persuade Ramanujan to join them. At first, Ramanujan was reluctant to leave India because as a strict Brahmin he believed that crossing the seas would make him an outcaste. A friend, Narayana Iyer, could see how Ramanujan was longing to go to Cambridge and devised a plan. Iyer was convinced that Ramanujan's devotion to mathematics and to his goddess Namagiri could together produce a revelation which would persuade Ramanujan that he could leave for Cambridge. He took Ramanujan off to Namagiri's temple to seek divine inspiration. After three nights sleeping on the stone floor, Ramanujan woke with a start. He quickly shook his friend awake. 'I have seen in a flash of brilliant light Namagiri commanding me to cross the sea.' Iyer smiled. His plan had worked.

Ramanujan was also worried that his family might not approve. But the family God Namagiri was to intervene again. Ramanujan's mother dreamt that her son was seated in a big hall surrounded by Europeans, and that the goddess Namagiri had commanded her not to stand in her son's way. His final concern was whether in Cambridge he would be subjected to further humiliating examinations. Neville was able to dispel this last fear, and the scene was set for Ramanujan to leave the sprawl of tiny houses that was Madras for the grand halls and libraries of Cambridge, the backdrop to his mother's dream.

Cambridge culture clash

By 1914 Ramanujan was in Cambridge, and so began one of the great collaborations in the history of mathematics. Hardy always spoke passionately about his time working with Ramanujan. They revelled in each other's mathematical ideas, happy to have found a kindred spirit in their love of numbers. Hardy would later reflect on those years with Ramanujan as some of the happiest of his life, and referred to their relationship poignantly as 'the one romantic incident in my life'.

Hardy and Ramanujan's partnership was like a classical interrogation team: there's a good guy and a bad guy. The good guy is the eternal optimist full of mad proposals. The bad guy is the pessimist, doubting everything, spotting the card disappear up the sleeve. Ramanujan needed the critical Hardy to check his wild enthusiasm as they interrogated their mathematical suspect.

Finding common ground, however, wasn't always easy. It was certainly a culture clash. Whilst Hardy and Littlewood insisted on rigorous, Western-style proofs, Ramanujan's theorems simply spilled forth, inspired by his goddess Namagiri. Hardy and Littlewood often hadn't a clue where their new colleague's ideas were coming from. Hardy remarked that, 'It seemed ridiculous to worry him about how he had found this or that known theorem, when he was showing me half a dozen new ones almost every day.'

Ramanujan had to contend with more than just a mathematical culture shock. He was alone in an alien world of mortarboards and black gowns. He could find no vegetarian food and wrote home for food parcels of tamarind and coconut oil. Without his familiar world of mathematics, the transition might have been impossible. Neville, the Fellow who'd gained his trust in India, described those early days. 'He felt the petty miseries of life in a strange civilisation, the vegetables that were unpalatable because they were unfamiliar, the shoes that tormented feet that had been unconfined for twenty-six years. But he was a happy man, revelling in the mathematical society which he was entering.' He could be seen every day waddling across the college quad in slippers, having cast aside his English shoes in despair. But once ensconced in Hardy's room, his notebooks open in front of him, he could escape into his formulas and equations whilst Hardy gazed on, caught up in Ramanujan's web of bewitching theorems. Ramanujan had exchanged the mathematical isolation in India for the cultural loneliness of Cambridge, but had gained a companion with whom to explore his mathematical world.

Hardy found Ramanujan's education a difficult balancing act. He worried that if he insisted too much on forcing Ramanujan to expend energy proving his results, 'I might destroy his confidence or break the spell of his inspiration.' He gave Littlewood the task of familiarising Ramanujan with modern rigorous mathematics. Littlewood found the job was virtually impossible. Whatever Littlewood tried to introduce to Ramanujan, the response was an avalanche of original ideas which stopped Littlewood in his tracks.

Although Ramanujan's attempts to provide exact formulas to count the primes had helped launch his ship to England, it was in related fields that he eventually made his mark. He had been rather put off tackling the primes head-on after reading Hardy and Littlewood's pessimistic comments about how malicious the primes were. One can only speculate what Ramanujan might have discovered had he not been made to feel the West's fear of the primes. He did continue his exploration of related properties of numbers with Hardy. The ideas that he and Hardy developed would contribute to the first progress to be made on Goldbach's Conjecture – that every even number is the sum of two prime numbers. This progress came in a roundabout way, but its starting point was Ramanujan's innocent belief that there should be precise formulas for expressing important sequences like the number of primes. In the same letter in which he had claimed a formula for the primes, he made statements indicating that he believed he understood how to generate another previously untamed sequence: the partition numbers.

How many different ways are there to divide five stones into separate piles? The number of piles can vary between five piles of one stone to one pile of five, with other possibilities in between:

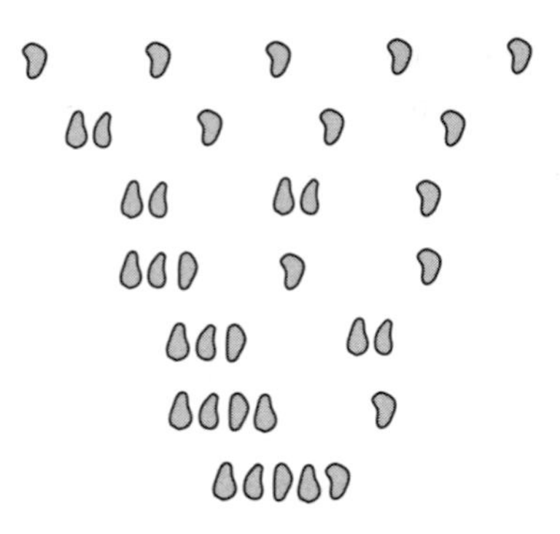

The seven ways to partition five stones.

These are called the *partitions* of the number 5. As this illustration shows, there are seven possible partitions of 5.

Here are the number of partitions of the numbers from 1 to 15:

Number	1	2	3	4	5	6	7	8	9	10	11	12	13	14	15
Partitions	1	2	3	5	7	11	15	22	30	42	56	77	101	135	176

This is one of the sequences of numbers that we encountered in Chapter 2. They crop up in the physical world nearly as often as those of Fibonacci. For example, the density of energy levels in certain simple quantum systems reduces to understanding how the partition numbers grow.

The numbers don't look as randomly distributed as the primes, but Hardy's generation had all but given up on finding an exact formula which would output the numbers in this list. Mathematicians thought that, at best, there might be a formula that produced an estimate which didn't stray too far from the actual number of partitions of N, in much the same way that Gauss's formula for the primes gave a close approximation to the number of primes up to N. But Ramanujan had never learned to fear such sequences. He was out for a formula that would tell you that there were exactly five ways of dividing four stones into different piles, or that there were 3,972,999,029,388 ways of dividing 200 stones into different piles.

Whereas Ramanujan had failed with the primes, he was spectacularly successful with the partition numbers. The combination of Hardy's skill at negotiating complex proofs and Ramanujan's blind insistence that a formula must exist carried them both through to its discovery. Littlewood never understood 'why Ramanujan was so certain there was one'. And when one looks at the formula – which involves the square root of 2, π, differentials, trigonometric functions, imaginary numbers – one has to wonder from where it was conjured:

$$p(n)=\frac{1}{\pi\sqrt{2}}\sum_{1\le k\le N}\sqrt{k}\left(\sum_{h \bmod k}\omega_{h,k}\,e^{-2\pi i\frac{hn}{k}}\right)\frac{d}{dn}\left(\frac{\cosh\left(\frac{\pi\sqrt{n-\frac{1}{24}}}{k}\sqrt{\frac{2}{3}}\right)-1}{\sqrt{n-\frac{1}{24}}}\right)+O\left(n^{-1/4}\right)$$

Littlewood later commented that, 'We owe the theorem to a singularly happy collaboration of two men, of quite unlike gifts, in which each contributed the best, most characteristic, and most fortunate work that was in him.'

There is a curious twist to the story. Instead of giving the exact figure, Hardy and Ramanujan's complicated formula produces an answer that is correct when rounded to the nearest whole number. So, for instance, when the formula is fed with the number 200 it outputs a value to which the nearest whole number is 3,972,999,029,388. So the formula is good enough to indicate the right answer, but in a frustrating way doesn't quite capture these partition numbers. (Later, a variation of their formula would be discovered which gave the answer on the nose.)

Although Ramanujan couldn't pull off the same trick for the primes, his work with Hardy on the partition function did have an impact on Goldbach's Conjecture, one of the great unsolved problems in the theory of prime numbers. Most mathematicians had given up on even trying to crack this problem. No idea for making any inroads had ever been offered. Some years earlier, Landau had declared that it was simply unattackable.

Hardy and Ramanujan's work on the partition function initiated a technique now called the Hardy–Littlewood Circle Method. The name comes from all the little diagrams that accompanied their calculations which depicted circles in the map of imaginary numbers around which Ramanujan and Hardy would try to do integration. The reason Littlewood's name is attached to the method rather than Ramanujan's is because of the use to which he and Hardy put it to make the first substantial contribution towards proving Goldbach's conjecture. They couldn't show that every even number could be expressed as a sum of two prime numbers, yet in 1923 they managed to prove what to mathematicians was the next best thing: that all odd numbers bigger than some fixed enormous number could be written as the sum of three prime numbers. But there was one condition they needed to impose to make their proof work – that the Riemann Hypothesis was true. This was yet another result predicated on Riemann's Hypothesis becoming Riemann's Theorem.

Ramanujan had helped develop this technique but sadly never lived to witness the stunning role it played in mathematics. By 1917 Ramanujan was becoming increasingly depressed. Britain was gripped by the horrors of the First World War. Trinity had just failed to elect Ramanujan to a fellowship. Russell's fellowship had just been terminated because of his anti-war sentiments, and the College was not disposed to tolerate Ramanujan's own pacifist stance. He may have finally learned how to squeeze his feet into western shoes and to sport a mortarboard and gown, but his soul remained in Southern India.

Cambridge was becoming like a prison. Ramanujan was used to the freedom that life in India bestowed. The warm climate allowed people to spend much of their time outside. In Cambridge, he had to huddle inside the thick college walls to protect himself from the chill wind that whipped off the North Sea. The social divide meant he had little contact beyond the formal interactions of academic life. He was also beginning to find that Hardy's insistence on mathematical rigour prevented him from letting his mind run free through the mathematical landscape.

The decline of his mental state was matched by a physical deterioration. Trinity College didn't understand Ramanujan's strict religious dietary requirements. In India he had been used to his wife putting things in his hand to eat as he filled his notebooks. Although the College kitchens offered the same service for Fellows such as Hardy and Littlewood, the food at high table was completely unpalatable to Ramanujan. He simply could not survive on his own and was dreadfully lonely, having left his wife and family behind in India. Malnutrition led to suspected tuberculosis, and he entered a succession of nursing homes.

Ramanujan tried to keep himself going by thinking about mathematics, but without much success. His dreams were filled with delirious mathematical images. The pain in his abdomen he believed was being caused by the infinite spike in Riemann's landscape where the formula for the zeta function surges off to infinity. Was this some awful punishment for breaking the Brahmin law not to cross the seas? Had he misunderstood Namagiri's message? His wife had not written to him once since his arrival in Cambridge. The pressure became too much to bear.

After a partial recovery, Ramanujan, still depressed, tried to commit suicide by hurling himself in front of a London underground train. He failed, thanks to the intervention of a guard who brought the train to a halt just in front of Ramanujan's prostrate body. In 1917, attempted suicide was a criminal offence, but through Hardy's intervention charges were dropped provided Ramanujan entered a sanatorium in Matlock, in Derbyshire, for full-time medical supervision for twelve months.

Now he was stuck miles from anywhere without even the stimulation of his daily meetings with Hardy. 'I have been here a month,' he wrote to Hardy, 'and I have not been allowed fire even for a single day. They promised me fire on those days in which I do some serious mathematical work. That day hasn't come yet and I am left in this dreadfully cold open room.'

Hardy eventually succeeded in getting Ramanujan transferred to a nursing home in Putney, in London. Despite Hardy's admission that Ramanujan had been the one true love of his life, their relationship was still devoid of much emotion apart from the excitement of doing mathematics together. Hardy, on a visit to Ramanujan while he lay ill, was unable to come up with any comforting sentiments. Instead he offered the number of the taxi in which he arrived, 1,729, as an example of rather a dull number. Even on his sickbed, Ramanujan was unstoppable. 'No, Hardy! No, Hardy! It is a very interesting number. It is the smallest number expressible as the sum of two cubes in two different ways.' He was right: $1{,}729 = 1^3 + 12^3 = 10^3 + 9^3$.

Ramanujan's fortunes did pick up with his election to the Royal Society, Britain's most prestigious scientific institution, and eventually to a fellowship at Trinity. Hardy's influence on these elections was the only way he knew to express the love he spoke of. But Ramanujan never recovered his health. When the First World War ended, Hardy suggested that maybe Ramanujan should return to India for a period to recuperate. On April 26, 1920, Ramanujan died in Madras, aged thirty-three, of what is now believed to have been amoebiasis, an infection of the large intestine, which Ramanujan probably contracted before he left for England.

Although ultimately Ramanujan was unsuccessful in mastering the primes, his first letter to Hardy has had a lasting effect on their story. Mathematicians recognise that the answer to this unsolved riddle can appear at any time and from anywhere. One new insight might project some previously unknown name out of the shadows of obscurity into the limelight. As Ramanujan illustrated, sometimes knowledge and expectations can hold back progress. Academics reared in the traditional seats of learning are not necessarily best placed to break the mould. There is always the possibility that another package will arrive on some mathematician's desk which will herald the arrival of an unknown genius ready to realise Ramanujan's dream of cracking the enigma of the primes.

The ideas that Ramanujan left behind were to fuel the work of generations of mathematicians and still do. Indeed, it could be said that it is only within the last few decades that the real worth of Ramanujan's ideas has finally been fully appreciated. Even by the time of Hardy's death, the true significance of Ramanujan's formulas was not apparent. Hardy himself was quite dismissive of one of Ramanujan's conjectures, commenting in a paper that 'we seem to have drifted into one of the back-waters of mathematics'. But years later, the importance of Ramanujan's Tau Conjecture, as it became known, may be judged from the fact that its eventual solution, by Pierre Deligne, was rewarded with a Fields Medal in 1978. One of Ramanujan's champions, Bruce Berndt, has compared him to J.S. Bach, who remained largely ignored for years after his death.

Berndt has dedicated much of his life to poring over Ramanujan's unpublished notebooks. He follows in a line of mathematicians who have been bewitched by the mass of formulas and equations that Ramanujan generated. Whilst exploring these notebooks, Berndt uncovered a curious table detailing the number of primes below 100,000,000. They are correct or very close to correct, and are more accurate than the numbers produced by the formula Ramanujan had first sent to Hardy. But there is no indication of how they were derived.

Did Ramanujan have access to some secret formula for the primes that was as successful as his formula for the partition function? Are there clues still to be found hidden in Ramanujan's notebooks? In 1976, the mathematical community thrilled to the discovery of a lost notebook by Ramanujan filled with new mathematics. Its discovery can only increase speculation that hidden away in the archives of Trinity College or in boxes in Madras there might be more treasures yet to be unearthed that would explain Ramanujan's ability to count primes so accurately.

Ramanujan's death came as a great shock to Hardy, who only two months before had received a letter which 'was quite cheerful and full of mathematics'. Hardy was devastated to be deprived of such a wonderful travelling companion in his hikes across the mathematical terrain. 'His originality has been a constant source of suggestion to me ever since I knew him and his death is one of the worst blows I have ever had.'

As Hardy grew older, he was himself beset by depression. He had always been someone who thought of himself as young. Now the sight of his old face repelled him, and he would always insist on reversing all the mirrors when he entered a room. He hated the effect of age on his abilities to do mathematics, and his volume *A Mathematician's Apology* is a haunting account of a mathematician at the end of his career. To do mathematics, a mathematician 'must not be too old. Mathematics is not a contemplative but a creative subject; no one can draw much consolation from it when he has lost the power or desire to create; and that is apt to happen to mathematicians rather soon.'

Like Ramanujan before him, Hardy too was to attempt to take his own life, by pills rather than jumping in front of a train. But he vomited them up and was left with a black eye. C.P. Snow recalls his visits to the sick Hardy. 'He was self-mocking. He had made a mess of it. Had anyone ever made a bigger mess?' Hardy's one consolation, he wrote in the *Apology*, was Ramanujan. 'I still say to myself when I am depressed and find myself forced to listen to pompous and tiresome people, "Well, I have done one thing you could never have done, and that is to have collaborated with Littlewood and Ramanujan on something like equal terms."'

CHAPTER SEVEN

Mathematical Exodus: From Göttingen to Princeton

Since the mathematical sciences are so vast and varied, it is necessary to localise their cultivation, for all human activity is tied to places and persons. David Hilbert, speaking at a party to celebrate Landau's arrival in Göttingen as a professor in 1913

Landau's father, Leopold, discovered that there was a young mathematical prodigy living on his street in Berlin. He was intrigued and sent him an invitation to come to his house for tea. Although Carl Ludwig Siegel was rather shy, he agreed to go and meet the great Göttingen mathematician's father. In his library, Landau senior took down the two volumes on prime numbers written by his son and handed them to Siegel. It would probably be too difficult for Siegel now, he explained, but perhaps later he might be able to read it. Siegel was to treasure Edmund Landau's book, which would have a lasting impact on his mathematical development.

Siegel's coming of age coincided with the start of the First World War. The idea of serving in the army filled the young, reticent boy with dread. He began to develop a deep-felt loathing of all things military. Despite the interest that Landau's father had taken in his mathematical development, Siegel had initially opted to study astronomy in Berlin, believing that the subject couldn't possibly have any relevance to war. But the astronomy course started late, so to fill the time he attended some mathematics lectures. It wasn't long before he'd caught the bug. Exploring the universe of numbers became Siegel's passion. Soon he was equipped to make sense of the volumes on prime numbers that Landau's father had given him.

By 1917, the war inevitably encroached on Siegel's life, and when he refused to serve in the army he was confined to a mental institution as punishment. Landau's father intervened to get him released. 'If it had not been for Landau I would have died,' Siegel later admitted. In 1919 the young man, still recovering from the ordeal, joined his mathematical idol Landau at Göttingen, where his mathematical talents were to flower.

Siegel found he had to put up with Landau's rather infuriating personality. On one occasion, when Siegel was already a senior mathematician, he visited Landau in Berlin. During dinner the professor spent the whole meal painstakingly explaining an extremely detailed and technical proof, insisting on providing the minutest detail. Siegel listened patiently, but by the time Landau had finished it was so late that he had missed the last bus home and was forced to walk all the way back to his lodgings. During the long walk he thought about Landau's proof, which concerned points at sea level in a landscape similar to the one constructed by Riemann. By the time he had arrived home he had come up with a slick alternative proof to the laborious one that had made him miss his bus. The next day, in a moment of chutzpah, Siegel sent Landau a postcard thanking him for dinner and giving the succinct details of his alternative proof – all of which fitted on the same card.

When Siegel arrived in Göttingen, Germany was weighed down by the costs of war reparations, and he was forced to lodge with one of the professors in the department. Another professor bought him a bicycle so that he could pedal round the town's medieval streets. At first Siegel was rather intimidated by the Göttingen mathematical hierarchy, especially the great Hilbert. So he worked quietly alone, determined to make the breakthrough that would impress the great names he passed in the corridors of the department. He sat in Hilbert's lectures, soaking up the formidable man's ideas. He knew that answering just one of Hilbert's twenty-three problems would be his passport to success.

At first he was far too timid in the presence of giants such as Hilbert to air his ideas. He eventually plucked up the courage when several members of the senior faculty invited him to join them on a swimming trip to the River Leine. Hilbert appeared far less intimidating in a swimming costume, and Siegel felt brave enough to share with Hilbert his thoughts on the Riemann Hypothesis. Hilbert responded very enthusiastically, and his support for the shy mathematician secured Siegel a position at the University of Frankfurt in 1922.

During his lifetime, Siegel successfully contributed to a number of Hilbert's problems, but it was his unconventional breakthrough on the eighth problem, the Riemann Hypothesis, that stamped his name firmly on the mathematical map.

Rethinking Riemann

As Siegel began to apply himself to the solution of Hilbert's eighth problem, he was becoming aware that some mathematicians were growing disillusioned about Riemann's contribution to the subject. Siegel's mentor, Landau, was perhaps the most vocal critic of what Riemann had actually managed to achieve in his ten-page paper published in 1859. Although he acknowledged it to be a 'most brilliant and fruitful paper', he went on to qualify his praise: 'Riemann's formula is far from the most important thing in prime number theory. He just created the tools which when refined made it possible later to prove many other things.'

Meanwhile, in Cambridge, Hardy and Littlewood were becoming equally dismissive. By the late 1920s, Hardy's inability to solve the Riemann Hypothesis was beginning to frustrate him. Littlewood too began to wonder whether the fact that they couldn't prove it meant that it wasn't actually true:

> I believe this to be false. There is no evidence whatever for it. One should not believe things for which there is no evidence. I should also record my feeling that there is no imaginable *reason* why it should be true . . . Nonetheless life would be more comfortable if one could believe firmly that the hypothesis is false.

Riemann had indeed been rather lacking when it came to providing evidence that the zeros were located where his Hypothesis predicted. In his ten-page paper there wasn't a single calculation of one of these points at sea level. Hardy believed that Riemann's hunch about the zeros in his landscape was nothing more than heuristic speculation.

The fact that in his paper Riemann appeared not to have calculated the locations of these zeros contributed to the image of him as the thinking man's mathematician, a man of ideas not interested in getting his hands dirty doing calculations. After all, this was the ethos of the revolution Riemann had initiated. Hilbert had similarly dedicated his life to promoting this new approach to mathematics. As he wrote in one of his papers, 'I have tried to avoid the huge computational apparatus of Kummer [Ernst Kummer, Dirichlet's successor in Berlin], so that here too Riemann's principle should be realised, according to which proofs should be impelled by thought alone and not by computation.' Hilbert's colleague in Göttingen, Felix Klein, was fond of saying that Riemann worked primarily by means of 'great general ideas' and 'often relied on his intuition'.

Hardy, however, was not content to rely on intuition. He and Littlewood managed to develop a method for calculating precisely the locations of some of the early zeros. If the Riemann Hypothesis was false, then, armed with their formula, there was a very small chance that they might quickly locate a zero that was not on Riemann's critical line. The method they developed exploited the symmetry that Riemann had discovered in his landscape between the land to the east and the land to the west of the ley line passing through $\frac{1}{2}$. They used their method in conjunction with an efficient means devised by Euler to approximate the value of infinite sums of numbers. By the end of the 1920s the Cambridge mathematicians had successfully located 138 zeros. As Riemann had predicted, they were indeed all on the line through $\frac{1}{2}$. It was clear, however, that Hardy and Littlewood's formula was running out of steam. It was becoming computationally unfeasible to pinpoint exactly the location of any of the zeros to the north of these 138.

It seemed that these calculations could not be pushed much further. Hardy had proved by theoretical analysis infinitely many zeros had to be on the line. Now there was a growing feeling that any zero that might be off the line wouldn't be seen until one had travelled very far north in the landscape. As Littlewood had demonstrated, prime numbers, more than any other creature in the mathematical zoo, like to hide their true colours in the far distant reaches of the universe of numbers. As a result, mathematicians began to give up on explicitly locating zeros and began to concentrate on other, more theoretical features of the landscape that might reveal the mysteries of Riemann's thinking.

All this was changed by a most unexpected discovery. While Siegel was struggling in Frankfurt with his ideas on the Riemann Hypothesis, he received a letter from the mathematical historian Erich Bessel-Hagen who had been working through Riemann's unpublished notes. Riemann's wife, Elise, had rescued some of them from the zealous housekeeper who had burnt many of his papers. She had given the majority of Riemann's scientific papers to Riemann's contemporary Richard Dedekind, but a few years later she began to regret having handed over anything that might contain personal details. She asked Dedekind to return them. Even if a manuscript was mostly mathematics, if it contained the slightest hint of a shopping list or the name of a family friend, Elise wanted the offending pages returned.

The remaining scientific papers were eventually deposited by Dedekind in the library in Göttingen. Bessel-Hagen had been trying to make sense of the mass of papers contained in the archives, but with little success. As with most mathematicians' private jottings, they were a chaotic jumble of half-formed ideas and formulas. Bessel-Hagen wondered whether Siegel might do better at decoding these hieroglyphics.

Siegel wrote to the librarian in Göttingen to ask whether he could consult Riemann's *Nachlass*, as it is now known. The librarian arranged for the documents to be sent to Siegel's local library in Frankfurt. Siegel was looking forward to the task: it would be a welcome distraction from the frustrations of the lack of progress he was making in his research. The package duly arrived, and he hurried down to the library with a visiting colleague. As he opened the package, out spilled a mass of papers crammed with complicated numerical calculations. These pages would give the lie to the picture of Riemann as a man who for seventy years had been billed as a mathematician of intuition and concepts unable to provide much hard evidence to back his ideas. Pointing to the mass of calculations, Siegel exclaimed ironically, 'Here are Riemann's great general thoughts!'

Several minor mathematicians had previously rummaged through these pages in search of a clue to Riemann's Hypothesis, but none could make sense of the mass of fragmented equations. Most baffling was the huge amount of sheer arithmetic computation that Riemann seemed to have done in his spare time. What were all these sums? It took a mathematician of the stature of Siegel to see what Riemann had been doing.

As Siegel gazed at these pages, he began to see that Riemann had held true to his teacher's dictum. As Gauss had always stressed, an architect removes the scaffolding once he has erected the building. The brittle leaves that Siegel now held in his hands were packed to the edges with calculations. Riemann had been poor, having had to support his sisters in later life, and he could only afford low-quality paper, from which he squeezed every last space. Hilbert's thinker turned out to be a master-calculator, and it was upon those calculations that he had built his conceptual view of the world, finding patterns in the evidence he collected. Some of the calculations were not innovative, for example the square root of 2 calculated to 38 decimal places, but Siegel was intrigued by others the like of which he had not seen anywhere before. As he scoured the pages, the chaotic jumble of random calculations began to make some sense. He realised that Riemann was calculating zeros.

Siegel discovered that Riemann was using an extraordinary formula which enabled him to calculate the heights in his zeta landscape very accurately. The first part of the formula was based on a trick that Hardy and Littlewood had discovered. Riemann had anticipated their contribution by some sixty years. The second piece of the formula was completely new: Riemann had also discovered a way to add up the remaining infinite sum that was much cleverer than the method currently being used. In contrast to Euler's methods that had been used to locate the first 138 zeros, the points at sea level in the zeta landscape, Riemann's formula would maintain a head of steam as he calculated farther north.

Sixty-five years after Riemann's death, the august mathematician was still streets ahead of the competition. Hardy and Landau had been wrong to believe that Riemann's paper was just a remarkable set of heuristic insights. Instead it was based on solid calculation and theoretical ideas that Riemann had chosen not to reveal to the world. Within a few years of Siegel's discovery of Riemann's secret formula, it would be used by Hardy's students in Cambridge to confirm that the first 1,041 zeros were on Riemann's line. The formula, however, would truly come into its own with the dawn of the computer age.

It is rather odd that it took mathematicians so long to realise that Riemann's notes might contain such gems. There are certainly clues in Riemann's ten-page paper, and in letters he wrote to other mathematicians at the time, that he was sitting on something. In the paper he mentions a new formula but goes on to say that he 'has not yet sufficiently simplified it to announce it'. The mathematicians in Göttingen had been poring over this published paper for seventy years and, unbeknownst to them, a few blocks down the road was the magic formula for locating zeros. Klein, Hilbert and Landau were all happy to pass judgement on Riemann, though none of them had so much as glanced at the unpublished *Nachlass*.

In fairness, a glance at Riemann's jottings is enough to indicate the magnitude of the task. As Siegel wrote, 'No part of Riemann's writings related to the zeta function is ready for publication; occasionally one finds disconnected formulas on the same page; frequently just one side of an equation had been written down.' It was like poring over the first draft of an unfinished symphony. The final composition owes much to Siegel's mathematical virtuosity in extracting the formula from the mess of Riemann's notes. It fully deserves the name by which it is now known – the Riemann–Siegel formula.

Thanks to Siegel's perseverance, a new side to Riemann's character had been revealed. Riemann had certainly championed the importance of abstract thinking and general concepts. But he knew it was important not to neglect the power of computation and numerical experiment. Riemann had never forgotten the eighteenth-century tradition from which his mathematics had emerged.

The *Nachlass* housed in the library in Göttingen was only part of what was rescued from Riemann's housekeeper. Elise Riemann wrote to Dedekind on May 1, 1875, describing some of the personal material she wanted back in family hands. This included 'a small black book containing records of Riemann's sojourn in Paris in spring 1860'. Only a few months before, Riemann had published his great ten-page paper on the primes, rushing to get it into print so that it would coincide with his election to the Berlin Academy. Now in Paris, after the flurry of publication, he had time to flesh out his ideas. The weather there was appalling; hail and snow prevented Riemann from exploring the city. Instead, he sat in his room, setting down his thoughts on paper. It is not unreasonable to speculate that, along with his personal impressions of Paris, in that little 'black book' Riemann recorded his thoughts about the points at sea level in his zeta landscape. The book has never been recovered, although there are a few clues to its fate.

Riemann's son-in-law wrote to Heinrich Weber on July 22, 1892, that 'at first mother could not come to terms with the idea that Riemann's papers should no longer remain in private hands; to her, they are something sacred and she doesn't like to think of them being made accessible to any student, who would then also be able to read the marginal notes, some of which are purely personal'. Unlike Fermat, whose nephew was only too keen to publish his uncle's marginalia, Riemann's family was reluctant to make public notes that Riemann never intended to publish. It seems then that at this stage the little black book was still in family hands.

Speculation about the location of this notebook is rife. There is evidence that Bessel-Hagen subsequently acquired some of the remaining unpublished material that had remained in family hands. It is unclear whether he bought it at auction or acquired it through a personal contact. Some of the papers found their way into the archive at the University of Berlin, but it appears that Bessel-Hagen decided to hang on to the rest of his collection. He died of starvation in the winter of 1946, in the chaos that followed the end of the Second World War. His belongings have never been found.

Another story has the little black book finding its way into Landau's hands. It is said that, in view of the uncertainty of the inter-war years, he gave the notebook to his mathematical son-in-law, I.J. Schoenberg, who escaped to America in 1930. Again the trail runs dry. Given that there is now a million dollars at stake, the search for Riemann's little black book has turned into a treasure hunt.

Without Riemann's crib-sheet and Siegel's determination, how long might it have taken us to unearth the magic formula? It is so sophisticated that we might still not have known it today. What other gems have we been deprived of because of the disappearance of that black book? Riemann claimed he could prove that most zeros were on the critical line, yet no one has yet matched that claim with a proof. What might there be still buried in the archives of German libraries? Did the black book find its way to America? Did it survive the housekeeper's fire only to be lost in the fires of the Second World War?

By 1933, mathematicians across Germany were finding it increasingly difficult to concentrate on mathematics. The swastika was flying over the library in Göttingen. The faculty was home to many Jewish and left-wing mathematicians. Street campaigns at the time specifically focused on the mathematics department as a 'fortress of Marxism' and by the mid-thirties most of the faculty lost their jobs in Hitler's purge of the universities. Many of them sought refuge overseas. Landau, although he was Jewish, was allowed to stay because he had been appointed before the outbreak of the First World War. The non-Aryan clause in the civil-service law of April 1933 did not apply to long-serving professors or those who had fought in the war.

Things got worse. By the winter of 1933, Landau's lectures were being picketed by Nazi students, including one of the most brilliant mathematicians of his generation, Oswald Teichmüller. A Jewish professor in Göttingen described Teichmüller as 'a very young, scientifically gifted man, but completely muddled and notoriously crazy'. One day, as Landau arrived at the lecture theatre, his path was blocked by the zealous young Nazi. Teichmüller told Landau that his Jewish way of presenting calculus was fundamentally incompatible with the Aryan way of thinking. Landau crumbled under the pressure, resigned his position and retired to Berlin. It hurt him greatly to be denied the chance to teach. Hardy invited him to Cambridge to give some lectures. 'It was quite pathetic to see his delight when he found himself again in front of a blackboard and his sorrow when his opportunity came to an end,' Hardy recalled. Unable to contemplate abandoning his homeland, Landau returned to Germany, where he died in 1938.

That year, Siegel – who had no Jewish connections – was moved from Frankfurt to Göttingen to try to rescue the reputation of the mathematics department. In 1940 he went into self-imposed exile in America to protest at the horrors of the war. After his terrible childhood experiences during the First World War, he had vowed never to remain in Germany if his country ever went to war again. He spent the war years at the Institute of Advanced Study in Princeton. Of the mathematicians who had forged Göttingen's great reputation, only Hilbert remained in Germany. He had always been rather obsessed with Göttingen's mathematical dominance. An old man now, he couldn't comprehend the devastation that was happening around him. Siegel tried to explain to Hilbert why many members of the faculty had left. Siegel recalled, 'I felt that he had the impression we were trying to play a bad joke on him.'

Within the space of a few weeks, Hitler had destroyed the great Göttingen tradition forged by Gauss, Riemann, Dirichlet and Hilbert. One commentator wrote that it was 'one of the greatest tragedies experienced by human culture since the time of the Renaissance'. Göttingen (and, some might add, German mathematics itself) has never fully recovered from its decimation by Nazi Germany during the thirties. Hilbert died on St Valentine's Day in 1943 after suffering a fall in Göttingen's medieval streets. His death marked the end of the city's position as the Mecca of mathematics.

Across the whole of Europe, mathematics was plunged into crisis. As nations prepared themselves for the inevitable confrontation, it became more and more impossible to justify the pursuit of abstract ideas for their own sake. Once again, European science became geared to giving nations a military edge. Many mathematicians would follow Siegel and emigrate from Europe to America. Most found the prosperity and support they encountered on the other side of the Atlantic the perfect environment for pure research. While America benefited from this academic migration, Europe has never regained its position as the world's mathematical powerhouse.

Some mathematicians did return from exile. Once the war had finished, Siegel made his way back to Germany. Sheltered in Princeton, he had been totally cut off from mathematical developments in Europe and thought that little could have happened during his absence. He was to get a surprise. Whilst most mathematicians had fled or stopped doing mathematics, it turned out that there was indeed one piece of news. Siegel met up with his friend Harald Bohr, Hardy's collaborator in attempts at the Riemann Hypothesis in Copenhagen. 'So, has anything happened since my exile in Princeton?' Siegel asked his old colleague. Bohr simply replied, 'Selberg!'

Selberg, the solitary Scandinavian

In 1940 Siegel had made his way to Princeton via Norway. He had been invited to deliver a lecture at the University of Oslo. The German authorities approved the visit, ignorant of the fact that Siegel was using the lecture as a front. The principal purpose of the trip was to make his escape from Europe on a ship that was leaving Oslo bound for America. As the ship pulled out of the harbour, Siegel watched a fleet of German merchant ships approaching. Later he learnt that those ships were the advance party for the German invasion force. He had escaped, but left behind in the mathematics department at the University of Oslo was a young mathematician by the name of Atle Selberg. This young gun was burying his head in the mathematical sands trying to ignore the turmoil around him.

Even before war engulfed the region, Selberg was happy to spend his

Atle Selberg, professor at the Institute for Advanced Study, Princeton.

working days in self-imposed isolation. A cloistered existence often forces the mathematician to head off in a completely new direction. Selberg had already decided to work in a field of mathematics that no one else in the region was particularly familiar with. The fact that he had no help from colleagues in his mathematical endeavours did not deter him. Far from it – he seemed to revel in the isolation. As war approached and Norway became increasingly cut off, with foreign scientific journals failing to get through, Selberg found the silence an inspiration. 'It was like being in a kind of prison. You were cut off. You certainly got the opportunity to concentrate on your own ideas. You were not distracted by what other people do. In that sense I considered that the situation in many ways was rather a good one for doing my work.'

This self-sufficiency was to characterise Selberg's mathematical life. It had been cultivated during his youth, when he would sit undisturbed for hours in his father's personal library delving into the many mathematics books on its shelves. It was during those long hours that Selberg came across an article about Ramanujan in a journal of the Norwegian Mathematical Society. Selberg recalls how the 'strange and beautiful formulas . . . made a very deep and lasting impression on me'. The work of Ramanujan became one of Selberg's main inspirations. 'It seemed like a revelation – a completely new world to me, with much more appeal to the imagination.' As a present, his father gave him Ramanujan's *Collected Papers*, which Selberg still carries with him today. Self-taught with the help of his father's large collection of mathematics books, he was already producing original work by the time he entered the University of Oslo in 1935.

He was particularly fascinated by Ramanujan's formula for the sequence of partition numbers, which the Indian mathematician had discovered with Hardy. Although Ramanujan's formula was recognised as a stunning achievement, there was something slightly unsatisfactory about it. The formula generated an answer which was not a whole number; the nearest whole number was the partition number itself. Surely there was a formula that produced exactly the number of partitions. Selberg was delighted when in the autumn of 1937 he succeeded in going one better than Ramanujan, coming up with an exact formula. Shortly after his discovery he was reading a review of his very first paper when his eyes were drawn to the review alongside it. To his great disappointment, he'd been beaten to the finishing line by Hans Rademacher, in a paper published the year before. Rademacher had fled his native Germany to America in 1934 after the Nazis terminated his employment at Breslau because of his pacifist sympathies. 'I felt it was a bit of a blow at the time, later I got more used to such things!' That Selberg had not heard of Rademacher's contribution illustrates the degree to which Norway was now isolated from mathematical developments further afield.

It was always something of a surprise to Selberg that Hardy and Ramanujan had missed the exact formula. 'I believe firmly that the responsibility for this rests with Hardy . . . Hardy did not fully trust Ramanujan's insight and intuition . . . I think that if Hardy had trusted Ramanujan more, they should have inevitably ended with the Rademacher series. There is little doubt about that.' Maybe, but the route they took did result in Hardy and Littlewood's contribution to the Goldbach Conjecture, which otherwise might not have happened.

Selberg began to read as much as he could of the work of the Cambridge trio – Ramanujan, Hardy and Littlewood. He was particularly taken by their work on primes and connections to the zeta function. There was a statement in one of Hardy and Littlewood's papers that especially intrigued him. They had written that their current methods seemed to offer no hope of proving that most zeros, the points at sea level in Riemann's landscape, would be on Riemann's ley line. Hardy had taken the great step of proving that there were at least an infinite number of zeros on the line, but he had failed to show that this infinite number amounted to even a fraction of the total number of zeros. Despite some improvements that he made with Littlewood, the number of zeros they could prove were on the line was swamped by the zeros they couldn't catch. They stated boldly that their result could not be improved with the methods they had developed.

But Selberg was not as pessimistic as Hardy and Littlewood. He thought there was still some mileage to be had from their ideas. 'I was looking at that passage in Hardy and Littlewood's original paper where they explain at the end why their method would not give more than they could prove. I read that and I thought about it. And then I realised that what they had there was complete nonsense.' Selberg's hunch that he could go further than Hardy and Littlewood proved right. Although he still could not prove that all the zeros were on the line, he was able to show that the percentage captured by his method would not tail off to zero as he counted farther north. He wasn't too sure what fraction of the total number of zeros he had caught, but this was the first substantial bite out of the pie which left some tooth marks. In retrospect, it looks as if he managed to prove that about 5 to 10 per cent of the zeros were on the line. As you counted north, then, at least this proportion of zeros obeyed the Riemann Hypothesis.

Even if it wasn't a proof of the Riemann Hypothesis, Selberg's bite was still a psychological breakthrough. But no one yet knew about it. Selberg himself wasn't sure whether he might have been beaten to the result. The war ended, and he was invited to speak at the Scandinavian Congress of Mathematicians in Copenhagen in the summer of 1946. He had already been disappointed at being beaten to his discovery of the exact formula for the number of partitions, so he decided he had better check whether his result about the zeros was old news or not. But the University of Oslo had still not received copies of the journals which had failed to get through during the war. 'I had heard that the library at the Institute of Technology in Trondheim had received copies. So I went up to Trondheim specifically for that. I spent about a week in the library.'

He needn't have worried. He found himself way ahead of anyone else's appreciation of the zeros in Riemann's zeta landscape. His lecture in Copenhagen confirmed Bohr's declaration to the visitors from America that the mathematical news in Europe amounted to 'Selberg!' Selberg spoke about his views on the Riemann Hypothesis. Although he had made a major contribution on the way to a proof, he stressed that there was still very little to support its truth. 'I think the reason that we were tempted to believe the Riemann Hypothesis then was essentially that it is the most beautiful and simple distribution that we can have. You have this symmetry down the line. It would lead also to the most natural distribution of primes. You think that at least something should be right in this universe.'

Some misinterpreted his comments, thinking that Selberg was casting doubt on the validity of the Riemann Hypothesis. Yet he was not as pessimistic as Littlewood who believed the lack of evidence meant the Hypothesis was false. 'I have always been a strong believer in the Riemann Hypothesis. I would never bet against it. But at that stage I maintained that we didn't really have any results either numerical or theoretical that pointed very strongly to its truth. What the results pointed to was that it was mostly true.' In other words, most zeros were probably on the line, just as Riemann had claimed he could prove nearly a century before.

Selberg's wartime breakthrough was the death-rattle of European dominance in mathematics. Following his success he was headhunted by Hermann Weyl, a professor at the Institute for Advanced Study in Princeton, who in 1933 had escaped the worsening situation in Göttingen. This lone mathematician who had remained in Europe and endured the privations of the Second World War succumbed to the attractions that beckoned from the other side of the Atlantic. Selberg took up his invitation to visit the Institute, excited by the prospect of new inspiration. He arrived at the bustling port of New York and made his way to the sleepy town of Princeton, a short drive south of Manhattan.

The United States was to benefit immensely from the influx from overseas of talented mathematicians such as Selberg. Once a backwater of mathematical activity, America was now becoming the major power it still is today. It is the home of mathematics, drawing mathematicians from across the globe. Although Göttingen's reputation as the Mecca of mathematics had been smashed by the devastation wrought by Hitler and the Second World War, it would rise phoenix-like at the Institute in Princeton.

The Institute had been founded in 1932 with the help of a five million dollar endowment from Louis Bamberger and his sister Caroline Bamberger Fuld. Its aim was to attract the world's best scholars by offering them a peaceful haven and a handsome salary – indeed, the place gained the nickname of the Institute for Advanced Salaries. It sought to reproduce the collegiate atmosphere of Oxford and Cambridge where scholars from all disciplines could benefit by interacting with one another.

In contrast, however, to the musty atmosphere of those ancient seats of learning, Princeton had the air of somewhere young and fresh, bursting with life and ideas. Whilst it was considered a faux pas at Oxford or

Cambridge to talk shop at high table, Princeton knew no such niceties. Members of the Institute talked openly about their work, whenever they wanted to. Einstein described it as a pipe yet unsmoked. 'Princeton is a wondrous spot, a quaint and ceremonious village of puny demigods on stilts. Yet by ignoring certain social conventions I have been able to create for myself an atmosphere conducive to study and free from distraction. Into this small university town the chaotic voices of human strife barely penetrate.'

Although the Institute was founded to cater for all disciplines, it began life in the old mathematics building of Princeton University. The mathematics department would later move to the only skyscraper in Princeton and take its name – Fine Hall – with it. The Institute's first home probably

The Institute for Advanced Study in Princeton.

influenced its particular strengths, in mathematics and physics. Inscribed above the fireplace of the faculty lounge were some words which Einstein would often quote: 'Raffiniert ist der Herr Gott, aber boshaft ist Er nicht' ('The Lord God is subtle, but malicious he is not'). The mathematicians, though, were rather more sceptical about the truth of such a statement. As Hardy had explained to Ramanujan, there is 'diabolical malice inherent in the primes'.

The Institute moved to its new premises in 1940. Situated on the outskirts of Princeton and surrounded by woodland, it was insulated from the horrors happening in the outside world. Einstein described his exile to Princeton as 'a banishment to paradise. I wished for this isolation all my life and now I have finally achieved it here in Princeton.' In many ways the Institute would mirror its ancestor in Göttingen. It thrived on its isolation. People came from far and wide and were sucked into its self-sufficient community. Some would say that Princeton's self-sufficiency grew into self-satisfaction. Not only had they accepted Göttingen's mathematicians, but they appeared to have appropriated the German town's motto: for members of the Institute, there was no life outside Princeton. Isolated in the woods, the Institute provided the perfect working environment for banished and fleeing Europeans.

Erdős, the wizard from Budapest

There was another mathematical émigré from Europe at the Institute whose life was to become intertwined with Selberg's. While Ramanujan's story had been inspiring the young Selberg in Norway, its magic was also working on another young mind. Paul Erdős, a Hungarian, was to become one of the most intriguing mathematicians of the second half of the twentieth century. But Ramanujan would not be the only thing to link these two young mathematicians. There was also controversy.

Whereas Selberg liked to work alone, Erdős thrived on collaboration. His stooped figure, clad in sandals and a suit, was familiar in mathematical common rooms across the world. He could be seen hunched over a notepad with a new collaborator at his side as they indulged his passion for creating and solving problems about numbers. He wrote over fifteen hundred papers in his lifetime, a phenomenal achievement. The only mathematician to have written more papers is Euler. Erdős was a mathematical monk who shed all his personal possessions lest they distract him from his mission. He gave away any money he earned to students or as rewards for answers to the many questions he posed. Like Hardy before him, God played a leading albeit unconventional role in his view of the world. The 'Supreme Fascist' was the name he gave to the custodian of the 'Great Book', which contained details of all the most elegant proofs of mathematical problems, both solved and unsolved. Erdős's highest compliment for a proof was 'that's straight from the Book!' He believed that all babies – or 'epsilons' as he called them, after the Greek letter that mathematicians use for a very small number – are born with knowledge of the Great Book's proof of the Riemann Hypothesis. The trouble was that, after six months, they had forgotten it.

Erdős enjoyed doing his mathematics to music, and was often to be seen at concerts scribbling away in a notebook, unable to contain the excitement of a new idea. Although he was a great collaborator and hated to be alone, he found physical contact quite abhorrent. It was mental pleasure that sustained him, fuelled by a diet of coffee and caffeine tablets. As he once famously explained, 'A mathematician is a machine for turning coffee into theorems.'

Erdős, like so many great mathematicians, was lucky to have a father who could expose him to ideas that would stimulate his passion for numbers. On one occasion his father had shown Erdős Euclid's proof that there were infinitely many prime numbers. But Erdős was fascinated when his father twisted Euclid's argument to prove that you could find stretches of numbers of arbitrary length where there were no primes.

If you want a sequence of 100 consecutive numbers where there are no primes, just take all the numbers up to 101 and multiply them together. The result is a number called the *factorial* of 101 and written as 101!. Then 101! is certainly divisible by all the numbers from 1 to 101. But if N is any of these numbers, then $101! + N$ will also be divisible by N, since 101! and N are both divisible by N. So all the numbers

$$101! + 2, 101! + 3, \ldots, 101! + 101$$

are not prime. Here, then, is a list of 100 consecutive numbers, none of which are prime.

Erdős's interest was piqued. How long would he have to count from 101! or some other number before he was guaranteed to find a prime number? Euclid had made sure there must be a prime somewhere, but would you have to wait an arbitrarily long time before finding the next prime? After all, if primes were being selected by the tossing of Nature's coin, there's no knowing how long it will be from one 'heads' to another. Of course, getting 1,000 tails in a row is very unlikely – but not impossible. As Erdős explored further, he learnt that in this respect the primes were *not* like the tossing of a coin. They may look a chaotic bunch of numbers, but their behaviour isn't completely random.

In fact it was a French mathematician, Joseph Bertrand, who in 1845 first guessed how far you need to go before you're guaranteed to find a prime. He believed that if you take any number, for example 1,009, and you count up to twice that number, then you should be guaranteed to find a prime on the way. There are actually quite a lot of primes between 1,009 and 2,018, the first being 1,013. Would this be true if Bertrand chose any number N? He couldn't prove that you'll always find a prime between any number N and its double, $2N$. But the striking prediction he made, when he was only twenty-three, that this would always be the case became known as Bertrand's Postulate.

It didn't hold out as long Riemann's Hypothesis as an unsolved problem. Within seven years the Russian mathematician Pafnuty Chebyshev had come up with a proof. Chebyshev used ideas similar to those he had employed in making the first inroads into the Prime Number Theorem, when he proved that Gauss's guess was never more than 11 per cent away from the true number of primes. His methods were not as sophisticated as the powerful ones that Riemann had developed, but they were effective. So, unlike the tossing of a coin where there are no guarantees when the next head will appear, Chebyshev proved that there was a small measure of predictability to the primes.

One of the first results that Erdős published in 1931, when he was only eighteen, was a new proof of Bertrand's Postulate. But to his dismay, someone pointed him to Ramanujan's work and he discovered that his proof wasn't as new as he had hoped. One of Ramanujan's last achievements was an argument that greatly simplified Chebyshev's proof of Bertrand's Postulate. Although the young Erdős was rather upset, this was more than outweighed by the joy of discovering Ramanujan.

Erdős decided to see whether he could do better than Ramanujan and Chebyshev. He began to look at how big the gap between primes might be. The problem of the difference between primes was one that would continue to fascinate Erdős throughout his life. He was famous for offering prizes as rewards for proving his own conjectures. The second-largest he ever offered, $10,000, was for a proof of his conjecture about how big the gap between consecutive primes really is. The problem remains unsolved to this day, and the money is still there to be claimed, even though Erdős is no longer alive to appreciate the proof. But, as he liked to joke, the work that had to be done to earn one of his prizes probably violated the minimum wage law. He once rashly offered 10 billion factorial dollars for a proof of a conjecture which generalised Gauss's Prime Number Theorem (10 billion factorial is the product of all the numbers from 1 to 10 billion). 100 factorial is already a number which exceeds the number of atoms in the universe, and Erdős expressed his relief when, in the 1960s, the mathematician who produced a proof didn't claim his reward.

As soon as Erdős arrived at the Institute for Advanced Study in the late 1930s, he made his mark. Mark Kac was an émigré from Poland who was sheltering from the storm in Europe. Although Kac was interested in the theory of probability, he had announced a lecture which aroused Erdős's interest. Kac was going to discuss a function that kept track of how many different primes divide each number as one counted higher. For example, $15 = 3 \times 5$ is divisible by two different prime numbers, whilst $16 = 2 \times 2 \times 2 \times 2$ is divisible by only one. So each number gets a score according to how many different primes divide into it.

Erdős recalled that Hardy and Ramanujan had been interested in how these scores varied. But it required a statistician like Kac to see that these scores were behaving in a completely random fashion. Kac could see that if you plotted a graph recording the scores made by each number as you counted higher, the shape of the graph was the familiar bell-shaped curve known to statisticians as the signature of randomness. Although Kac had recognised the behaviour of the function counting the number of prime building blocks, he didn't possess the tricks from number theory that were needed to prove his hunch about this randomness. 'I first stated the conjecture during a lecture in Princeton in March 1939. Fortunately for me and possibly for mathematics, Erdős was in the audience and he immediately perked up. Before the lecture was over he had completed the proof.'

This success began Erdős's lifelong passion for mixing number theory and probability theory. At first sight the subjects look like chalk and cheese. Hardy once dismissively declared, 'Probability is not a notion of pure mathematics but of philosophy or physics.' The objects studied by number theorists have been set in stone since the beginning of time, immovable and unchanging. As Hardy said, 317 is a prime whether we like it or not. Probability theory, on the other hand, is the ultimate slippery subject. You're never quite sure what's going to happen next.

Orderly zeros mean random primes

Although Gauss had used the idea of tossing a prime number coin to guess at the number of primes, it wasn't until the twentieth century that mathematicians were happy to contemplate a union of the diverse disciplines of probability and number theory. In the first few decades of the century, physicists were proposing that chance was an integral part of the subatomic world. An electron might behave as though it were a tiny billiard ball, but you can never be too sure where this ball is located. Many physicists were reluctant to admit it, but it seems that the role of a quantum dice dictates where you will find the electron. Perhaps the unsettling effect of the emerging theory of quantum physics and its probabilistic model of the world helped to challenge the view that chance had no role to play in something as deterministic as the primes. While Einstein was trying to deny that God played dice with Nature, down the corridor at the Institute Erdős was proving that the throw of the dice lay at the heart of number theory.

Indeed, it was during this period that mathematicians began to understand how the Riemann Hypothesis, which was about the regimented behaviour of the zeros in the zeta landscape, explained why the primes look so wild and random. The best way to understand this tension between the order of the zeros and the chaos of the primes is to take a deeper look at the quintessential model of randomness – the tossing of a coin.

If you toss a coin a million times, you should get half heads and half tails. But you wouldn't expect to get an exact score. With a fair coin – one that behaves randomly, without any bias – you shouldn't be surprised to see an error of about 1,000 either side of 500,000 heads. The theory of probability has provided a measure of how big this error can be if the experiment has its source in some random process. If the coin is tossed N times, there will be some deviation from $\frac{1}{2}N$ heads – the 'error' – one way or the other. This error has been analysed for a fair coin and is expected to be in the order of the square root of N. Thus, for example, out of 1,000,000 tosses of a fair coin, the number of heads is most likely to lie somewhere between 499,000 and 501,000 (1,000 being the square root of 1,000,000). If the coin were biased, you would expect the error to be consistently more than the square root of N.

Gauss had modelled his guess at the number of primes by tossing a coin. The probability that it would land heads on the Nth toss was only $1/\log(N)$ rather than $\frac{1}{2}$. However, in the same way that a conventional coin doesn't come up exactly half heads, half tails, Nature's prime number coin was not indicating exactly the number of primes that Gauss had predicted. But what was the error like? Was it within the limits for a coin behaving randomly, or was there a strong bias for producing primes in certain regions of numbers and leaving other regions barren?

The answer lies in the Riemann Hypothesis and its prediction about the location of the zeros. These points at sea level control the errors made by Gauss's guess for the number of primes. Each zero with east–west coordinate equal to $\frac{1}{2}$ produces an error of $N^{1/2}$ (which is another way of writing the square root of *N*). So if Riemann was correct about the location of the zeros, then the error between Gauss's guess for the number of primes less than *N* and the true number of primes is at most of the order of the square root of *N*. This is the error margin expected by the theory of probability if the coin is fair, behaving randomly with no bias.

If the Riemann Hypothesis is false and there are zeros farther to the east of Riemann's critical line, these zeros will produce an error which is much bigger than the square root of *N*. It would be like the coin producing many more heads than the fifty–fifty split expected from the fair coin. If the Riemann Hypothesis is false, that would imply that the prime number coin is far from fair. The farther east one finds zeros off Riemann's ley line, the more biased is the prime number coin.

A fair coin produces truly random behaviour, whereas a biased coin produces a pattern. The Riemann Hypothesis therefore captures why the primes look so random. Riemann's brilliant insight had turned this randomness on its head by finding the connection between the zeros of his landscape and the primes. To prove that the primes are truly random, one has to prove that on the other side of Riemann's looking-glass the zeros are ordered along his critical line.

Erdős liked this probabilistic interpretation of the Riemann Hypothesis. For one thing, it reminded mathematicians why they had entered Riemann's looking-glass world in the first place. Erdős wanted to encourage a return to what number theory was fundamentally about: numbers. It was striking that, ever since Riemann's wormhole had opened up and sucked mathematicians through to a new world, fewer number theorists were talking about numbers. They were much more concerned with navigating the geometry of Riemann's zeta landscape, on the lookout for points at sea level, than with talking about the primes themselves. Erdős initiated an about-turn, to studying primes for their own sake. He soon found out that he was not alone on this return journey.

Mathematical controversy

Although Selberg had been fascinated primarily with Riemann's zeta landscape, at Princeton his interest was shifting away from the zeta function and becoming more directly focused on the primes themselves. His mathematical exodus to America was combined with a return to the solid side of Riemann's looking-glass.

Since de la Vallée-Poussin and Hadamard's proof of the Prime Number Theorem, mathematicians were frustrated at not being able to find an easier way to prove Gauss's connection between logarithms and prime numbers. Was it only with highly sophisticated tools such as Riemann's zeta function and this imaginary landscape that mathematicians would be able to prove Gauss's estimate of the number of primes? Mathematicians were prepared to admit that such tools might be necessary to prove that the estimate was as good as Riemann's Hypothesis would imply, namely that the error would always be as little as the square root of *N*, but they believed there had to be a simpler way to get the first rough estimate that Gauss had predicted. They had hoped they could extend Chebyshev's elementary approach which proved that Gauss was at least within 11 per cent of the correct answer. But as time went by, and fifty years of attempts to find a simpler proof had failed, people were beginning to believe that the sophisticated tools that Riemann had introduced, and de la Vallée-Poussin and Hadamard had exploited, were simply unavoidable.

Hardy didn't believe there was an elementary proof. Not that he didn't wish for one; mathematicians constantly strive not only for proof but also simplicity. Hardy was simply becoming pessimistic and doubting that such a thing existed. He would have appreciated the contribution made by Erdős and Selberg who, just a few months after he died in 1947, found an elementary argument that linked primes and logarithms. However, the controversy that surrounded the credit for this elementary proof would have appalled him. The story has been told in various places, not least in two biographies of Erdős. Given the huge network of collaborators and correspondents that Erdős cultivated, combined with Selberg's reticence, it is not surprising that the majority of these stories have been told from Erdős's viewpoint. It is worth, however, recording something of Selberg's side of the affair.

The first to wield the sophisticated tool of the zeta function was Dirichlet, who used it to confirm one of Fermat's hunches. Dirichlet proved that if you take a clock calculator with *N* hours on the clock face, and you feed in the primes, the calculator will hit one o'clock infinitely often. In other words, there are infinitely many primes that have remainder 1 after dividing by *N*. Dirichlet's proof had relied on the sophisticated use of the zeta function. His proof was the catalyst for Riemann's great discoveries.

However, in 1946, nearly 110 years after Dirichlet's discovery, Selberg came up with an elementary proof of Dirichlet's Theorem that was closer in spirit to Euclid's proof that there are infinitely many primes. His proof, avoiding the zeta function, was a psychological breakthrough at a time when many believed it was impossible to make any headway in the theory of prime numbers without using Riemann's ideas. The proof, though subtle, required no sophisticated nineteenth-century mathematics and could possibly even have been understood by the ancient Greeks themselves.

Paul Turán, a Hungarian mathematician visiting the Institute at Princeton, had become friendly with Selberg during the time they spent together. He was also a good friend of Erdős. In fact, a joint paper that he had written with Erdős was the only ID he could produce when a Soviet military patrol stopped him in the streets of liberated Budapest in 1945. The patrol was suitably impressed, and Turán was saved from a trip to the gulag. As Turán later joked, this was 'a surprising application of number theory'.

Turán was keen to understand something of the ideas behind Selberg's proof of Dirichlet's result, but he was due to leave the Institute after spending the spring there. Selberg was happy to show him some of the details, and even suggested that Turán lecture on the proof while Selberg went to renew his visa during a trip to Canada. But during his discussion with Turán, Selberg showed slightly more of his hand than he had meant to.

During the lecture Turán mentioned a rather extraordinary formula Selberg had proved, one not directly related to the proof of Dirichlet's Theorem. Erdős was in the audience and saw that this formula was just what he needed to improve on Bertrand's Postulate, that there will always be a prime between N and $2N$. What Erdős was trying to do was to see whether you needed to go as far as 2 times N. For example, could you always find a prime between N and 1.01 times N? He realised that this wouldn't work for every N. After all, if N is 100, there are no whole numbers, let alone prime numbers, between 100 and 101 (which is 100 times 1.01). But Erdős believed that once N was big enough, then, in the spirit of Bertrand's Postulate, there would be a prime between N and $1.01N$. There was nothing special about 1.01. Erdős believed this would work for any choice of number between 1 and 2. Having listened to Turán's lecture, Erdős could see that Selberg's formula provided the missing link in his proof.

'Erdős asked me when I got back if I had anything against him using this to give an elementary proof of this generalisation of Bertrand's

Postulate.' It was a result that Selberg had himself thought about, but he hadn't got anywhere. 'I was not working on that so I said that I had no objection.' Selberg was distracted by a multitude of practical problems at the time. He had to renew his visa, find somewhere to live in Syracuse, where he had accepted a position for the coming academic year, and prepare lectures to teach a summer school for engineers. 'At any rate, Erdős was always rather quick at things and he managed to find a proof.'

Now there were certain things that Selberg had not let on to Turán. In particular, the reason Selberg had been thinking about this generalisation of Bertrand's postulate was that he could see how to fit it into a jigsaw to complete the picture of an elementary proof of the Prime Number Theorem. With Erdős's result, Selberg now had that final piece that gave him the proof.

He told Erdős how he had used his result to complete an elementary proof of the Prime Number Theorem. Erdős suggested they present the work to the small group that had been present at Turán's lecture. But Erdős couldn't contain his excitement, and he busily began issuing invitations to what he promised would be a very interesting lecture. Selberg had not expected such a large audience.

> When I arrived there in the late afternoon, around 4 or 5, the room was packed. So I went up and I went through the argument and then I asked Erdős to go through his part. Then I went through the rest that was needed to complete the proof. So the first proof was obtained by using this intermediary result that he had got.

Erdős proposed that they write a paper together explaining the proof. But as Selberg explains,

> I had never published joint papers. I really wanted to publish separate papers but Erdős insisted that one should do things the same way that Hardy and Littlewood had done things. But I had never agreed to cooperate. When I came to the States I had done all my mathematics in Norway. It was done alone, even without talking to anybody . . . no, I have never been a collaborator in that sense. I talk with people but I work alone, this is what suits my temperament.

The truth is that here were two mathematicians with completely different temperaments. One was an entirely self-sufficient loner who wrote only one joint paper in his life, with the Indian mathematician Saravadam Chowla, and that somewhat against his will. The other took collaboration to such an extreme that mathematicians talk of their Erdős number, the number of co-authors that link them to a paper with Erdős. Mine is 3, which means I've written a paper with someone who's written a paper with someone who's written a paper with Erdős. Since Chowla was one of Erdős's 507 co-authors, Selberg's one joint paper that he ever wrote gave him an Erdős number of 2. Over five thousand mathematicians have an Erdős number of 2.

After this refusal, as Selberg admits, 'things got out of hand'. By 1947 Erdős had built up an extensive network of collaborators and correspondents. He would keep them up to date with his mathematical progress by firing off postcards. The story goes that the nail in the coffin for Selberg was being greeted on his arrival in Syracuse by a faculty member who asked, 'Have you heard the news? Erdős and some Scandinavian mathematician have produced an elementary proof of the Prime Number Theorem.' By then Selberg had found an alternative argument that avoided the need for the intermediary step that Erdős had provided him. Selberg went ahead and published alone. His paper appeared in the *Annals of Mathematics*, the Princeton-based publication generally regarded as one of the three leading mathematical journals in the world, and where Andrew Wiles eventually published his proof of Fermat's Last Theorem.

Erdős was furious. He asked Hermann Weyl to adjudicate the issue. Selberg recounts, 'I take pleasure in the fact that Hermann Weyl essentially came down on my side in the end after he had heard both sides.' Erdős published his proof acknowledging Selberg's role. But it was a very unfortunate episode. Despite the unworldly nature of mathematics, mathematicians still have egos that need massaging. Nothing acts as a better drive to the creative process than the thought of the immortality bestowed by having your name attached to a theorem. The story of Selberg and Erdős highlights the importance in mathematics – indeed, in all of science – of credit and priority. That is why Wiles spent seven years alone in his attic working on Fermat's Last Theorem in secret, lest he have to share the glory.

Although mathematicians are like runners in a relay team, passing the baton from one generation to the next, they still yearn all the while for the individual glory that crossing the finishing line will bestow. Mathematical research is a complex balance between the need for collaboration in projects which can span centuries, and the longing for immortality.

After a while it became clear that Selberg's elementary proof of the Prime Number Theorem wasn't the striking breakthrough that had been hoped for. Some thought that the insight might provide an elementary path to proving the Riemann Hypothesis. After all, it might have shown that the difference between Gauss's guess and the actual number of primes would never be more than the square root of *N* off its mark. And people knew that this was equivalent to the zeros falling onto Riemann's regimented straight line.

By the end of the 1940s, Selberg still held the record for proving how many zeros lay on Riemann's line. That was one of the achievements for which he was awarded a Fields Medal in 1950. Hadamard, who was then eighty, was due to attend the International Congress of Mathematicians in Cambridge, Massachusetts, to celebrate Selberg's award. He was particularly looking forward to meeting the explorer who'd found an elementary route to the base camp that he and de la Vallée-Poussin had established fifty years before. However, both Hadamard and Laurent Schwartz, the other mathematician due to receive a Fields Medal, were denied visas because of their Soviet connections – McCarthyism had just begun to raise its ugly head. It required President Truman's intervention before they were allowed to enter America, just days before the congress.

People have subsequently extended Selberg's arguments about the percentage of zeros that we can prove are really on Riemann's ley line, adding their own ingenious twists. Some proofs of mathematical theorems evolve very naturally once you have an idea of the general direction in which to head. Finding the first bit of the path is the hard part. Improving Selberg's estimate is very different, however. The proofs require very delicate analysis. They are not susceptible to one great idea but require tremendous perseverance to see them through to their end. The path is littered with traps. One false move, and a number you thought was bigger than zero can suddenly turn negative on you. Each step needs to be taken with great care, and mistakes can easily creep in.

In the 1970s, Norman Levinson improved on Selberg's estimates and thought at one stage that he'd managed to capture as many as 98.6 per cent of the zeros. Levinson gave his colleague Gian-Carlo Rota at MIT a copy of the manuscript of the proof, joking that he'd proved that 100 per cent were on the line – the manuscript did 98.6 per cent of the zeros, and the other 1.4 per cent were left to the reader. Rota thought he was serious, and started spreading the word that Levinson had proved the Riemann Hypothesis. Of course, even if he had got to 100 per cent, that did not necessarily mean that all the zeros were on the line because we are dealing with the infinite. But that did not stop the rumour spreading.

Eventually a mistake was found in the manuscript which brought the located zeros down to 34 per cent. Still, this was a record that stood for some time, and was all the more impressive because Levinson was in his sixties when he did his best work. As Selberg says, 'He had to have a great deal of courage to carry out such numerical calculation because in advance you wouldn't know whether it would lead anywhere.' It was said that Levinson had great ideas for how to extend his methods, but he died from a brain tumour before he could bring them to fruition. The record currently belongs to Brian Conrey of Oklahoma University, who proved in 1987 that 40 per cent of the zeros must lie on the line. Conrey has some ideas about how to improve on his estimate, but the huge amount of work it would take doesn't seem worth it for the extra few per cent. 'It would be worth it if I could get the estimate above 50 per cent because then at least you could say most of the zeros were on the line.'

Erdős was very hurt by the controversy surrounding credit for the elementary proof, but he remained prolific throughout his life, defying the myths of ageing and mathematical burn-out. When he failed to secure a permanent position at Princeton, he chose instead the life of the itinerant mathematician. With no home and no job, he preferred to descend suddenly on one of his many friends around the world to indulge his love for collaboration, often staying with them for several weeks before moving on just as suddenly. He died in 1996, the centenary year of the first proof of the Prime Number Theorem. Erdős was still collaborating on joint papers at the age of eighty-three. He said, shortly before he died, 'It will be another million years, at least, before we understand the primes.'

Now silver-haired and in his nineties, Selberg is still reading the latest about the Riemann Hypothesis and attending conferences, at which he offers pearls of wisdom to young delegates. In his gentle voice you can still hear the singing tones of his home of Norway, but underneath are often penetrating and cutting commentaries on the work he is appraising. He does not suffer fools gladly. In 1996 his talk at a meeting in Seattle celebrating the centenary of the proof of the Prime Number Theorem was greeted with a standing ovation by six hundred mathematicians.

Selberg believes that despite much progress, we still have no real idea how to prove the Hypothesis:

> I think it is anybody's guess whether we are close to a solution or not. There are some people who think we are getting closer. Of course as time progresses if we ever get a solution then we are getting closer. But some believe that we have very essential elements of a solution. I don't really see that. It's very different from Fermat. There has been no corresponding breakthrough. It may very well survive a bicentennial by 2059 but of course I will not see that. How long the problem will last it is impossible to say. I do think that a solution will eventually be found. I don't think it is a

> result which is unprovable. Maybe though the proof will be so involved that the human brain will not catch up with it.

In the lecture he delivered in Copenhagen after the war, Selberg had cast doubt on whether there was any evidence that the Riemann Hypothesis was true. Then it seemed like wishful thinking, but today his view has changed. The evidence that has emerged in the fifty years since the war has in Selberg's view become quite overwhelming. But it was the war, and in particular the code-breakers at Bletchley Park, that were responsible for the development of the machine that would generate this new evidence: the computer.

CHAPTER EIGHT

Machines of the Mind

I propose to consider the question, 'Can machines think?' Alan Turing, *Computing Machinery and Intelligence*

Alan Turing's name will always be associated with the cracking of Germany's wartime code, Enigma. From the comfort of the country house of Bletchley Park, halfway between Oxford and Cambridge, Churchill's code-breakers created a machine which could decode the messages sent each day by German intelligence. The story of how Turing's unique combination of mathematical logic and determination helped save many lives from the threat of the German U-boats is the stuff of novels, plays and movies. Yet the inspiration for the creation of his 'bombes', the code-cracking machines, can be traced back to Turing's mathematical days in Cambridge, when Hardy and Hilbert were still in the ascendancy.

Before the Second World War engulfed Europe, Turing was already planning machines that would blow two of Hilbert's twenty-three problems out of the water. The first was a theoretical machine, existing only in the mind, which would demolish any hope that the secure basis of the foundations of the mathematical edifice could be checked. The second was very real, made of cogs and dripping with oil, and with this machine Turing intended to challenge another mathematical orthodoxy. He dreamt that this spinning contraption might have the power to disprove the eighth and Hilbert's favourite of the twenty-three problems: the Riemann Hypothesis.

After years of his colleagues failing to prove the Riemann Hypothesis, Turing believed that perhaps it was time to investigate whether Riemann might have been wrong. Perhaps there actually was a zero off Riemann's critical line, which would thus force some pattern onto the sequence of primes. Turing could see that machines would become the most powerful tools in a search for the zeros that might disprove Riemann's conjecture. Thanks to Turing, mathematicians would now have the help of a new, mechanical partner in their investigation of Riemann's Hypothesis. But it wasn't just Turing's physical machines that would have an impact on mathematicians' exploration of the primes. His machines of the mind, originally created to attack Hilbert's second problem, would lead in the late twentieth century to the most unexpected offshoot: a formula for generating all the primes.

Turing's fascination with machines was stimulated by a book that he was given in 1922 when he was ten years old. *Natural Wonders Every Child Should Know* by Edwin Tenney Brewster was packed with nuggets which fired the young Turing's imagination. Published in 1912, it explained that there were explanations for natural phenomenon, and it didn't rely on feeding its young readers with passive observations. Given Turing's later passion for artificial intelligence, Brewster's description of living things is particularly enlightening:

> For of course the body is a machine. It is a vastly complex machine, many, many times more complicated than any machine ever made with hands; but still after all a machine. It has been likened to a steam machine. But that was before we knew as much about the way it works as we know now. It really is a gas engine; like the engine of an automobile, a motor boat or a flying machine.

Even at school, Turing was obsessed with inventing and building things: a camera, a refillable ink-pen, even a typewriter. It was a passion that he would take with him when he went to Cambridge in 1931 to read mathematics as an undergraduate at King's College. Although Turing was shy and something of an outsider, like many before him he found reassurance in the absolute certainty that mathematics provided. But his passion for building things stayed with him. He would always be on the lookout for the physical machine that would lay bare the mechanism of some abstract problem.

Turing's first piece of research as an undergraduate was an attempt to understand one of the borders where abstract mathematics rubs up against the vagaries of nature. His starting point was the practical problem of tossing a coin. The outcome was a sophisticated theoretical analysis of the scores produced by any random experiment. Turing was a little upset when he presented his proof only to find that, like Erdős and Selberg before him, his first piece of research had duplicated what had been achieved some ten years before by a Finnish mathematician, J.W. Lindeberg, and was called the Central Limit Theorem.

Number theorists would later find that the Central Limit Theorem provides new insights into estimating the number of primes. The Riemann Hypothesis would confirm that the deviation between the true number of primes and Gauss's estimate is the same as the deviation we expect from tossing a fair coin. But the Central Limit Theorem revealed that the distribution of primes cannot be perfectly modelled by the tossing of a coin. The more refined measure of randomness made possible by the Central Limit Theorem is not obeyed by the primes. Statistics is all about

Alan Turing (1912–54).

the different angles from which to judge collections of data. From the viewpoint of Turing and Lindeberg's Central Limit Theorem, mathematicians could see that although the primes and the tossing of a coin had much in common, they were not one and the same thing.

Turing's proof of the Central Limit Theorem, although not original, was proof enough of his potential, and he was elected to a Fellowship at King's at the precocious age of twenty-two. Turing remained something of a loner in the Cambridge mathematical community. Whilst Hardy and Littlewood battled with classical problems in number theory, Turing preferred to work outside the mathematical canon. Rather than read the mathematical papers of his peers, he preferred to come up with his own conclusions. Like Selberg, he cut himself off from the distractions of a conventional academic life.

Despite this self-imposed isolation, Turing could not fail to be aware of a crisis that was sweeping through mathematics. People in Cambridge were talking about the work of a young Austrian mathematician who had put uncertainty at the heart of the subject that had promised security for Turing.

Gödel and the limitations of the mathematical method

In his second problem, Hilbert had challenged the mathematical community to produce a proof that mathematics did not contain contradictions. It was the ancient Greeks who began the development of mathematics as a subject of theorems and proofs. They started with basic statements about numbers which seemed to be self-evident truths. These statements, the axioms of mathematics, are the seeds from which the rest of the mathematical garden has grown. Since Euclid's first proofs about the primes, mathematicians have been using deduction to extend our knowledge of numbers beyond these axioms.

But a worrying question had been thrown up by Hilbert's study of different sorts of geometry. Are we sure that we could never prove that a statement was both true and false? How certain are we that there doesn't exist one sequence of deductions from the axioms which would prove that the Riemann Hypothesis was true, whilst an alternative sequence would prove it false? Hilbert felt certain that mathematical logic could be used to prove that mathematics contained no such contradictions. In his view, the second of his twenty-three problems was just a matter of putting the mathematical house in order. The question became slightly more urgent after a number of people, including Bertrand Russell, the philosopher friend of Hardy and Littlewood, had produced what appeared to be mathematical paradoxes. Although Russell's monumental work *Principia Mathematica* found a way to resolve these paradoxes, it alerted many to the serious nature of Hilbert's question.

On September 7, 1930, Hilbert was accorded the privilege of being made an honorary citizen of Königsberg, his beloved home town. It was the year of his retirement from Göttingen. He ended his acceptance speech with a clarion call to all mathematicians: 'Wir müssen wissen. Wir werden wissen.' ('We must know, we shall know.') After making his speech he was whisked off to a recording studio to record the last part of it for a radio broadcast. One can detect in the crackle of the recording Hilbert laughing after he declares, 'We must know.' Unknown to Hilbert, though, the last laugh had already been had the day before at a conference held down the road at the University in Königsberg. A twenty-five-year-old Austrian logician, Kurt Gödel, had made an announcement which struck at the heart of Hilbert's world-view.

As a child, Gödel had been known as 'Herr Warum' – Mr Why – because of his incessant stream of questions. A bout of rheumatic fever in his childhood left him with a weak heart and incurable hypochondria.

Kurt Gödel (1906–78) with Albert Einstein in 1950.

Towards the end of his life, his hypochondria turned to outright paranoia. He became so convinced that people were trying to poison him that he literally starved himself to death. But at twenty-five he was the one poisoning Hilbert's dream and inducing a rush of paranoia throughout the mathematical community.

For his dissertation, Gödel had turned his inquisitiveness to Hilbert's question which went to the heart of mathematical endeavour. Gödel proved that mathematicians could never prove that they had the secure foundations Hilbert had craved. It was impossible to use the axioms of mathematics to prove that these axioms would never lead to contradictions. Perhaps the problem could be rectified by changing the axioms or adding more axioms? That wouldn't work. Gödel showed that whatever axioms one chose for mathematics, they could never be used to prove that contradictions would never arise.

Mathematicians call a collection of axioms *consistent* if they don't lead to contradictions. It might be true that the axioms one has chosen never produce contradictions, but one can never prove that by using those same axioms. It might be possible to prove the consistency from some other collection of axioms, but that would be only a partial victory because then the consistency of that other choice of axioms is equally questionable. It is like Hilbert's attempt to prove that geometry is consistent by turning geometry into a theory of numbers. It only led to the question about the consistency of arithmetic.

Gödel's realisation is reminiscent of the description of the universe provided by a little old lady at the beginning of Stephen Hawking's *A Brief History of Time*. The lady stands up at the end of a popular astronomy lecture and declares, 'What you have told us is rubbish. The world is really a flat plate supported on the back of a giant tortoise.' Her reply to the lecturer's question as to what the tortoise is sitting on would have brought a smile to Gödel's face: 'You're very clever, young man, very clever. But it's turtles all the way down.'

Gödel had provided mathematics with a proof that the mathematical universe was built on a tower of turtles. One can have a theory without contradictions but one can't *prove* that within that theory there won't be contradictions. All one can do is to prove the consistency within another system whose own consistency could not be proved. It was ironic that mathematics could be used to prove that proof itself is limited in what it can deliver. The French mathematician André Weil summed up the situation post-Gödel with the following nugget: 'God exists since mathematics is consistent, and the Devil exists since we cannot prove it.'

Hilbert had declared in 1900 that in mathematics there was no 'unknowable'. Thirty years later, Gödel had proved that ignorance is an integral part of mathematics. Hilbert learnt about Gödel's bombshell some months after his day in Königsberg. He was apparently 'somewhat angry' upon hearing the news. Hilbert's declaration 'Wir müssen wissen. Wir werden wissen', made the day after Gödel's announcement, found its appropriate resting place. It was inscribed on Hilbert's gravestone, an idealistic dream from which mathematics had finally awoken.

At a time when physicists were learning from Heisenberg's Uncertainty Principle that there were limits to what physicists could know, Gödel's proof meant that mathematicians would always have to live with their own uncertainty: that they might suddenly discover that the whole of mathematics was a mirage. Of course, for most mathematicians the fact that this hasn't happened yet is the best justification for why it won't happen. We have a working model that seems to justify the consistency.

But since the model is ultimately infinite, we can't be sure that somewhere along the line the model will not contradict our axioms. And as we have discovered, things as innocent as prime numbers can hide surprises in distant stretches of the universe of numbers, surprises which would never have been stumbled upon by experiment and observation alone.

Gödel did not stop there. His dissertation contained a second bombshell. If the axioms of mathematics *are* consistent, then there will always be true statements about numbers which cannot formally be proved from the axioms. This went against the whole ethos of what mathematics had meant since the time of the ancient Greeks. Proof had always been considered the pathway to mathematical truth. Now Gödel had blown this faith in the power of proof out of the water. Some hoped that by adding new axioms it might be possible to patch up the mathematical edifice. But Gödel showed that such efforts were in vain. However many new axioms one adds to the mathematical foundations, there will always remain some true statements without proofs.

This was named Gödel's Incompleteness Theorem – any consistent axiom system is necessarily incomplete in that there will be true statements that can't be deduced from the axioms. And to assist him in this act of mathematical terrorism he enlisted the help of none other than the prime numbers. Gödel used the primes to give each mathematical statement its own individual code number, called the Gödel number. By analysing these numbers Gödel could show that for any given choice of axioms there would always exist true statements that could not be proved.

Gödel's result was a major body blow to mathematicians everywhere. There were so many statements about numbers, and especially prime numbers, which appeared to be true but we had no idea how to prove. Goldbach: that every even number is the sum of two prime numbers; Twin Primes: that there are infinitely many primes differing by 2, such as 17 and 19. Were these going to be statements that we couldn't prove from the existing axiomatic foundations?

There is no denying that it was an unnerving state of affairs. Maybe the Riemann Hypothesis was simply unprovable within our current axiomatic description of what we think we mean by arithmetic. Many mathematicians consoled themselves with the belief that anything that is really important should be provable, that it is only tortuous statements with no valuable mathematical content that will end up being one of Gödel's unprovable statements.

But Gödel was not so sure. In 1951, he questioned whether our current axioms were sufficient for many of the problems of number theory:

> One is faced with an infinite series of axioms which can be extended further and further, without any end being visible . . . It is true that in the mathematics of today the higher levels of this hierarchy are practically never used . . . it is not altogether unlikely that this character of present-day mathematics may have something to do with its inability to prove certain fundamental theorems, such as, for example, Riemann's Hypothesis.

Gödel believed that mathematics had failed to prove the Riemann Hypothesis because its axioms are not sufficient to explain the Hypothesis. We might have to broaden the base of the mathematical edifice to discover a mathematics in which we can solve this problem. Gödel's Incompleteness Theorem drastically altered people's mindsets. If problems were so impossible to answer, like Goldbach's and Riemann's, maybe they were simply unprovable with the logical tools and axioms that we were bringing to bear on them.

At the same time, we should be careful not to overemphasise the significance of Gödel's result. This was not the death knell of mathematics. Gödel had not undermined the truth of anything that had been proved. What his theorem showed was that there's more to mathematical reality than the deduction of theorems from axioms. Mathematics was more than a game of chess. There would be an ongoing evolution of the foundations of mathematics in parallel with the continued construction of the edifice above. In contrast to the formal nature of the construction above the base, the evolution of the foundations would rely on the intuition of mathematicians as to what new axioms they believed best described the world of mathematics. Many were happy to celebrate Gödel's Theorem as confirmation of the superior nature of the mind over the mechanistic spirit that had emerged from the Industrial Revolution.

Turing's miraculous machine of the mind

Gödel's revelation had opened up a whole new question which began to fascinate both Hilbert and the young Turing. Was there some way to tell the difference between true statements which have proofs and those Gödel statements which, although true, had no proofs? Turing, in his pragmatic manner, began to contemplate the possibility of a machine which could rescue mathematicians from the uncertainty of trying to prove an unprovable statement. Was it possible to conceive of a machine that would decide whether any statement fed into it could be deduced from the axioms of mathematics even if it didn't actually produce the proof? We could use such a machine like some Delphic oracle to reassure ourselves that it is at least worth trying to find a proof of Goldbach's Conjecture or the Riemann Hypothesis.

The question of the existence of such an oracle was not so dissimilar to the tenth question that Hilbert had posed at the dawn of the century. In that problem, Hilbert had speculated that there might exist a universal method or algorithm which could decide whether any equation had solutions or not. He was getting at the idea of a computer program before the idea of a computer had really been put forward. He envisaged a mechanical procedure which could be applied to the equation and answer 'yes' or 'no' to the question 'Does this equation have solutions?' without the need for any personal intervention by an operator.

All this talk of machines was purely theoretical. No one was yet contemplating an actual physical object. These were machines of the mind – methods or algorithms for producing answers. It was like coming up with the idea of software before there was any hardware on which to implement it. Even if Hilbert's machine existed it would still be useless in practice, because the amount of time the machine would need to establish whether any equation had solutions would most likely far exceed the lifetime of the universe. For Hilbert, the existence of this machine was of philosophical importance.

The idea of such theoretical machines horrified many mathematicians. They would effectively put the mathematician out of business. No longer would we need to rely on the imagination, on the cunning intuition of the human mind to produce clever arguments. The mathematician could be replaced by a mindless automaton which would bludgeon its way through new problems without the slightest need for subtle new modes of thought. Hardy was adamant that no such machine could exist. The very thought of one threatened his whole existence:

> There is of course no such theorem and this is very fortunate, since if there were we should have a mechanical set of rules for the solution of all mathematical problems, and our activities as mathematicians would come to an end. It is only the very unsophisticated outsider who imagines that mathematicians make discoveries by turning the handle of some miraculous machine.

Turing's fascination with the complexities of Gödel's ideas stemmed from a series of lectures given in the spring of 1935 by Max Newman, one of the mathematics dons in Cambridge. Newman had been captivated by Hilbert's questions when he heard the great Göttingen mathematician speaking during the International Congress of Mathematicians in Bologna in 1928. It was the first time since the First World War that a delegation from Germany had been invited to an International Congress. Many German mathematicians refused to attend, still outraged that they had been excluded from the previous Congress in 1924. But Hilbert rose above such political divisions, and headed a delegation of sixty-seven German mathematicians. As he entered the hall to hear the opening session, the audience rose to its feet and applauded him. He responded with a view shared by many mathematicians: 'It is a complete misunderstanding of our science to construct differences according to peoples and races and the reasons for which this has been done are very shabby ones. Mathematics knows no races . . . for mathematics, the whole cultural world is a single country.'

As soon as Newman learnt in 1930 that Hilbert's programme had been comprehensively demolished by Gödel, he was keen to explore some of the complexities of Gödel's ideas. Five years later he felt confident enough to announce a series of lectures on Gödel's Incompleteness Theorem. Turing sat in the lectures, transfixed by the twists and turns of Gödel's proof. Newman ended with the question that would act as a catalyst for both Hilbert and Turing's imagination. Could one in some way distinguish between statements that had proofs and those that didn't? Hilbert christened the question 'the Decision Problem'.

As he listened to Newman lecturing on Gödel's work, Turing became convinced that it was impossible to construct a miraculous machine that could make these distinctions. But it was going to be difficult to prove there could never be any such machine. After all, how can you know what the limitations of human ingenuity are going to be in the future? You might be able to prove that one particular machine won't produce answers, but to extend that to all possible machines was to deny the unpredictability of the future. Yet Turing did.

This was Turing's first great breakthrough. He came up with the idea of special machines that could effectively be made to behave like any person or machine that was doing arithmetic computations. They would later be known as *Turing machines*. Hilbert had been rather vague about what he meant by a machine that could tell whether statements could be proved. Now, thanks to Turing, Hilbert's question had been put into focus. If one of Turing's machines could not distinguish the provable from the unprovable, then no other machine could. So were his machines powerful enough to meet the challenge of Hilbert's Decision Problem?

While he was out one day, running along the banks of the River Cam, Turing experienced the second flash of enlightenment that told him why none of these Turing machines could be made to distinguish between statements that had proofs and those that didn't. As he paused for a breather, lying on his back in a meadow near Granchester, he saw that an idea which had been used successfully to answer a question about irrational numbers might be applicable to this question about the existence of a machine to test for provability.

Turing's idea was based on a startling discovery made in 1873 by Georg Cantor, a mathematician from Halle in Germany. He had found that there were different sorts of infinities. It may seem a strange proposition, but it is actually possible to compare two infinite sets and say that one is bigger than the other. When Cantor announced his discovery, in the 1870s, it was considered almost heretical or at best the ramblings of a madman. To compare two infinities, imagine a tribe that has a counting system that goes 'one, two, three, lots'. They can still judge who is the richest member of the tribe, even though the exact numerical value of that wealth cannot be discerned. If chickens are the sign of a person's wealth, then two people just need to pair up each other's chickens. Whoever runs out of chickens first is clearly the poorer of the two. They don't need to be able to count their chickens to see that one collection outstrips another.

Using this idea of pairing objects, Cantor showed that if you compare all the whole numbers against all the fractions (such as $\frac{1}{3}$, $\frac{3}{4}$, $\frac{5}{101}$) the members of the two sets can be paired off exactly. This seems counter-intuitive, since there would appear to be many more fractions than whole numbers. Yet Cantor found a way to establish a perfect match, with no fractions left without partners. He also produced a cunning argument to show that, in contrast, there was no way to match all fractions with all *real* numbers, which include irrational numbers such as π and $\sqrt{2}$, and numbers with a non-repeating decimal expansion. Cantor showed that any attempt to pair the fractions with the real numbers would necessarily miss some number with an infinite decimal expansion. Here, then, were two infinite sets which Cantor could show were of different sizes.

Hilbert recognised that Cantor was creating a genuinely new mathematics. He declared Cantor's ideas on infinities to be 'the most astonishing product of mathematical thought, one of the most beautiful realisations of human activity in the domain of the purely intelligible . . . no one shall expel us from the paradise which Cantor has created for us.' In recognition of Cantor's pioneering ideas he dedicated the first problem on his list of twenty-three to a question posed by Cantor: Is there an infinite set of numbers bigger than the set of fractions but smaller than the set of all real numbers?

It was Cantor's demonstration that infinite decimals outnumber fractions that had raced through Turing's mind as he lay soaking up the Cambridge sun. He suddenly realised why this fact could be used to show that Hilbert's dream of a machine that could check whether a statement had a proof was pure fantasy.

Turing started by supposing that one of his machines could decide whether any true statement had a proof. By an elegant method, Cantor had shown that some decimal numbers would always be left over, however the fractions were matched with the real numbers. Turing took this technique and used it to produce a 'left-over' true statement for which the Turing machine could not possibly decide whether a proof existed. The beauty of Cantor's argument was that if you tried to adapt the machine to include this missing statement, there would always be another statement that had been missed, just as Gödel's proof of the Incompleteness Theorem showed that adding another axiom would only lead to some new unprovable statement.

Turing could see that it was rather a slippery argument that he was proposing. As he ran back to his rooms in King's College, he turned it over in his mind, probing for any weaknesses. There was one point that worried him. He had shown that none of his Turing machines could answer Hilbert's Decision Problem. But how could he convince people that there wasn't some other machine that could answer Hilbert's problem? This was his third breakthrough: the idea of a *universal machine*. He drew up a blueprint for a single machine that could be taught to behave like all of his Turing machines or like any other machine that might answer Hilbert's problem. He was already beginning to understand the power of a program that could teach this universal machine to behave like any other machine capable of answering Hilbert's question. The brain was also a machine that might be able to decide between the provable and unprovable, and this stimulated Turing's later investigations into whether a machine was capable of thinking. For now, he focused on checking all the details of his proposed solution to Hilbert's question.

For a year Turing laboured away, making sure his argument held water. He knew that, when he presented it, it would be subjected to the utmost scrutiny. He decided that the best person to try it out on was the man who had explained the problem to him in the first place: Newman. At first, Newman felt rather uneasy about the argument. It looked as though there was the potential for tricking oneself into thinking something was true when it wasn't. But as Newman turned and turned the argument he became more convinced that Turing had got it. But they were to discover that Turing wasn't the only one to have got it.

Turing learned that he had been pipped at the post by one of the mathematicians in Princeton. Alonzo Church had reached the same conclusion at almost the same time as Turing, but had been fastest to publicise his discovery. Turing was naturally concerned that his bid for recognition in the tough academic jungle was going to be thwarted by Church's announcement. But with the support of Newman, his mentor in Cambridge, Turing's own proof was accepted for publication. To Turing's dismay, it received little recognition at the time it was published. However, his idea of a universal machine was more tangible than Church's method, and much more far-reaching in its consequences. Turing's addiction for real-life inventions had infused his theoretical considerations. Although the universal machine was only a machine of the mind, his description of it sounded like the plan for an actual contraption. A friend of his joked that if it were ever built it would probably fill the Albert Hall.

The universal machine marked the dawn of the computer age, which would equip mathematicians with a new tool in their exploration of the universe of numbers. Even during his lifetime, Turing appreciated the impact that real computing machines might have on investigating the primes. What he could not have foreseen was the role that his theoretical machine would later play in unearthing one of the Holy Grails of mathematics. Turing's very abstract analysis of Hilbert's Decision Problem would become the key, decades later, to the serendipitous discovery of an equation that generates all the primes.

Cogs and pulleys and oil

The next step for Turing was a trip across the Atlantic to visit Church. He hoped he might also get the chance to meet Gödel, who was visiting the Institute for Advanced Study in Princeton. Although theoretical machines were on his mind as he crossed the Atlantic, he hadn't lost his passion for real equipment. He whiled away the week on the ship by using a sextant to chart its progress.

Turing was disappointed when he arrived in Princeton to find that Gödel had gone back to Austria. Gödel would return two years later to take up a permanent position at the Institute, having fled persecution in Europe. One person Turing did meet in Princeton was Hardy, who happened to be visiting at the same time. Turing wrote to his mother about his encounter with Hardy, 'At first he was very stand-offish or possibly shy. I met him in Maurice Pryce's rooms the day I arrived, and he didn't say a word to me. But he is getting much more friendly now.'

Once he had written up his proof of Hilbert's Decision Problem for publication, Turing looked around for another big problem to attack. Cracking the Decision Problem would be a hard act to follow. But if you were going to go for another big problem, why not go for the ultimate prize, the Riemann Hypothesis? Turing got his colleague in Cambridge, Albert Ingham, to send him the most recent papers on the Hypothesis. He also began talking to Hardy to see what he thought about it.

By 1937 Hardy was getting more pessimistic about the validity of the Riemann Hypothesis. He'd spent so long trying to prove it, only to keep failing, that he was beginning to think it might actually be false. Turing, influenced by Hardy's mood in Princeton, believed he could build a machine to prove that Riemann was wrong. He had also heard about Siegel's discovery of Riemann's fantastic method for calculating zeros. The formula that Siegel had discovered had made clever use of adding up sines and cosines to efficiently estimate the height of the Riemann landscape. In Cambridge, Turing's approach to Hilbert's Decision Problem was regarded as positively industrial in its proposal to create a machine to solve the problem. But Turing realised that machines might also shed light on Riemann's secret formula. He recognised that there were strong similarities between Riemann's formula and the ones that were used to predict periodic physical phenomena such as the orbits of planets. In 1936, Ted Titchmarsh, an Oxford mathematician, had already adapted a machine, built originally to calculate celestial motions, to prove that the first 1,041 zeros in the zeta landscape were indeed on Riemann's ley line. But Turing had seen an even more sophisticated piece of machinery for predicting another periodic natural phenomenon: the tides.

The tides posed a complicated mathematical problem because they depended on calculating the daily cycle of the Earth's rotation, the monthly cycle of the Moon's orbit around the Earth, and the yearly cycle of the Earth's orbit around the Sun. Turing had seen a machine in Liverpool that carried out these calculations automatically. Adding up all the periodic sine waves was replaced by manipulating a system of strings and pulleys, and the answer was indicated by the length of certain sections of the string in the contraption. Turing wrote to Titchmarsh, admitting that when he had first seen the Liverpool machine he had no inkling that it could be used for investigating prime numbers. His mind was now racing. He would build a machine that would calculate the height of the Riemann landscape. This way he might be able to find a point at sea level off Riemann's critical line and prove the Riemann Hypothesis false.

Turing was not the first to contemplate the use of machinery to speed up tedious computations. The grandfather of the idea of computing machines was another Cambridge graduate, Charles Babbage. Babbage was an undergraduate at Trinity in 1810 and had been as fascinated as Turing with mechanical devices. In his autobiography he recalls the genesis of his idea for a machine to calculate the mathematical tables which were central to England's abilities to navigate the sea so skilfully:

> One evening I was sitting in the rooms of the Analytical Society, at Cambridge, my head leaning forward on the Table in a kind of dreamy mood, with a Table of logarithms lying open before me. Another member, coming into the room, and seeing me half asleep, called out, 'Well Babbage, what are you dreaming about?' to which I replied 'I am thinking about how all these Tables (pointing to the logarithms) might be calculated by machinery.'

It was not until 1823 that Babbage was able to begin to realise his dream of building his 'Difference Engine'. But the project foundered in 1833 when he fell out with his chief engineer over money. Part of the machine was eventually completed, but it took until 1991, the bicentenary of Babbage's birth, for his vision to be fully realised. That was when, at a cost of £300,000, a Difference Engine was constructed at the Science Museum in London, where it is still on display.

Turing's idea for a zeta machine was similar to Babbage's plan for calculating logarithms with the Difference Engine. The mechanism was specifically tuned to the individual problem it would calculate. This was not one of Turing's theoretical universal machines that could be made to imitate any computation. The physical properties of the device mirrored the problem, making it useless for attacking other questions. Turing admitted as much in an application he made to the Royal Society for money to start building his zeta machine: 'Apparatus would be of little permanent value . . . I cannot think of any application that would not be connected with the zeta-function.'

Babbage himself had appreciated the disadvantages of building a machine that could only calculate logarithms. In the 1830s he was dreaming of a grander machine which could be made to perform a range of tasks. He was inspired by the French Jacquard weaving looms that were being used in mills across Europe. Skilled operators had been replaced by cards punched with holes which, when fed into the loom, controlled the loom's operation. (Some have grandly referred to these cards as the first computer software.) Babbage was so impressed by Jacquard's invention that he bought a portrait of the inventor on a silk tapestry woven with the aid of one of these punched cards. 'The loom is capable of weaving any design which the imagination of man may conceive,' he marvelled. If this machine could produce any pattern, then why couldn't he build a machine that could be fed a card to tell it to perform any mathematical computation? His blueprint for the Analytical Engine, as he named it, was a forerunner of Turing's plan for a universal machine.

It was the poet Lord Byron's daughter, Ada Lovelace, who recognised the enormous programming potential of Babbage's machine. While translating into French a copy of Babbage's paper describing the machine, she couldn't resist adding some extra notes to extol the machine's capability. 'We may say most aptly that the Analytical Engine weaves Algebraic patterns, just as the Jacquard loom weaves flowers and leaves.' Her notes included many different programs that could be implemented on Babbage's new machine, even though the machine was purely theoretical and had never been built. By the time she had finished the translation, her additions had become so out of hand that the French version was three times the length of the English edition. Lovelace is generally regarded today as the world's first computer programmer. She died in great pain of cancer in 1852, aged just thirty-six.

While Babbage was working away on his ideas for machines in England, Riemann was developing his theoretical mathematical concepts in Germany. Eighty years later, Turing hoped to unite these two themes. He had already cut his teeth on the abstract computability of Gödel's Incompleteness Theorem, which had formed the basis of his thesis. Now he was to turn to the job of cutting physical teeth for the cogs for his zeta machine. He was successful in his application for a grant of £40 from the Royal Society to help build his creation, thanks to Hardy and Titchmarsh's support.

By the summer of 1939, Turing's room was 'a sort of jigsaw puzzle of gear wheels across the floor', according to Turing's biographer, Andrew Hodges. But Turing's dream of a zeta machine that would unite the nineteenth-century English passion for machines with German theory was to be rudely interrupted. The onset of the Second World War saw this burgeoning intellectual unity between the two countries replaced by military conflict. British intellectual forces were rallied at Bletchley Park, and minds turned from finding zeros to cracking codes. Turing's success in designing machines to crack Enigma owes something to his apprenticeship calculating the zeros of the Riemann zeta function. His complex mesh of interlocking gear wheels hadn't uncovered the secrets of the primes, but Turing's new contraptions would prove to be spectacularly successful in revealing the secret movements of the German war machine.

Bletchley Park was a strange mix of the ivory tower and the real world. It was like a Cambridge College, with games of cricket played on the front lawn; for Turing and company, cosseted in their country retreat, the encoded messages that arrived each day were a substitute for doing the *Times* crossword in a Cambridge common room. A theoretical puzzle – yet they all knew that lives depended on its solution. Given such an atmosphere, it is not surprising that Turing should have continued to think about mathematics while he was helping to win the war.

It was while at Bletchley that Turing came to understand, as Babbage had some hundred years before, that it was better to construct a single machine that could be told to do different tasks rather than build a completely new one for each new problem. Although he already knew this in principle, he was to learn the hard way that this should be implemented in practice too. When the Germans changed the designs of the Enigma machines being used in the field, Bletchley Park was plunged into weeks of silence. Turing realised that the code-breakers needed a machine that could be adapted to cope with any change the Germans might make to their machines.

After the war had finished, Turing began to explore the possibility of building a universal computing machine that could be programmed to perform a multitude of tasks. After several years at Britain's National Physics Laboratory, he went to work with Max Newman in Manchester at the newly created Royal Society Computing Laboratory. Newman had been alongside Turing in Cambridge during the development of the theoretical machine that had smashed Hilbert's hope of devising an algorithm to tell whether a true statement had a proof. Now they were to work together to design and build a real machine.

In Manchester, Turing had the time to exploit the technical expertise he had developed cracking codes in Bletchley, even though his wartime activities would remain classified for decades. He returned to the idea that had obsessed him in the days before the war: using machines to explore Riemann's landscape in the search for counter-examples to the Riemann Hypothesis – zeros off the critical line. But this time, rather than build a machine whose physical properties reflected the problem he was trying to solve, Turing sought to create a program that could be implemented on the universal computer he and Newman were constructing from cathode ray tubes and magnetic drums.

Naturally, theoretical machines run smoothly and effortlessly. Real machines, as Turing had discovered at Bletchley Park, are far more temperamental. But by 1950 he had his new machine up and running and ready to start navigating the zeta landscape. The pre-war record for the number of zeros located on Riemann's line was held by Hardy's former student, Ted Titchmarsh. Titchmarsh had confirmed that the first 1,041 points at sea level fulfilled the Riemann Hypothesis. Turing went further and managed to make his machine check as far as the first 1,104 zeros and then, as he wrote, 'unfortunately at this point the machine broke down'. But it wasn't only his machines that were breaking down.

Turing's personal life was beginning to collapse around him. In 1952 he was arrested when the police investigated his homosexuality. He had been burgled and had called the police; the burglar turned out to be an acquaintance of one of Turing's lovers. The police, as well as chasing the burglar, homed in on Turing's admission of (as the law then described it) an 'act of gross indecency'. He was distraught. This could mean jail. Newman testified on his behalf that Turing was 'completely involved in his work and is one of the most profound and original mathematical minds of his generation'. Turing was spared a prison sentence, on condition that he voluntarily subject himself to treatment by drugs to control his sexual behaviour. He wrote to one of his old tutors in Cambridge, 'It is supposed to reduce sexual urge whilst it goes on, but one is supposed to return to normal when it is over. I hope they are right.'

On June 8, 1954, Turing was found dead in his room from cyanide poisoning. His mother could not come to terms with the possibility that he had committed suicide. Her son had experimented with chemicals ever since he was a boy, and never washed his hands. She insisted it had been an accident. But by Turing's bedside lay an apple, with several bites out of it. Although the apple was never analysed, there is little doubt that it was steeped in cyanide. One of Turing's favourite film scenes was when the witch in Disney's *Snow White and the Seven Dwarfs* creates the apple that will put Snow White to sleep: 'Dip the apple in the brew, let the Sleeping Death seep through.'

Forty-six years after his death, at the dawn of the twenty-first century, rumours began to spread through the mathematical community that Turing's machines had indeed turned up a counter-example to the Riemann Hypothesis. Yet because the discovery had been made at Bletchley Park during the Second World War on the same machines that had cracked Enigma, British Intelligence had insisted it remain classified. Mathematicians clamoured for the records to be declassified so they could locate Turing's discovery of a zero off the line. It turned out that the rumour was no more than that, and had been started by one of Bombieri's friends who shared the Italian's taste for wicked April Fool's jokes.

Turing's machine may have broken down only a short way beyond the pre-war record for zeros, but it had taken the first step into an era in which the computer would take over from the human mind in the exploration of Riemann's landscape. It would take some time to develop efficient 'Riemann rovers', but soon these unmanned explorers would travel farther and farther north along Riemann's ley line, sending us back more and more evidence – if not final proof – that, contrary to Turing's belief, Riemann was right.

Although Turing's real machines had yet to make an impact on the Riemann Hypothesis, his theoretical ideas were to contribute to a strange twist in the story of the primes: the discovery of an equation for generating all the prime numbers. He could never have guessed that this equation would emerge from the devastation that he and Gödel had wrought on Hilbert's programme to provide mathematics with solid foundations.

From the chaos of uncertainty to an equation for the primes

Turing had proved that his universal machine could not answer all the questions of mathematics. But what about something less ambitious: could it say anything about the existence of solutions to equations? This was the heart of Hilbert's tenth problem, which in 1948 was beginning to obsess Julia Robinson, a talented mathematician based in Berkeley.

With very few notable exceptions, women have made very few appearances in mathematical history until the last few decades. The French mathematician Sophie Germain corresponded with Gauss, but she pretended to be a man, fearing that if she did not, her ideas would be dismissed out of hand. She had discovered special kinds of prime numbers related to Fermat's Last Theorem now called Germain primes. Gauss was very impressed with the letters he received from 'Monsieur le Blanc' and was amazed when after a lengthy correspondence he discovered that monsieur was a mademoiselle. He wrote back to her:

> The taste for the mysteries of numbers is rare . . . the charms of this sublime science in all their beauty reveal themselves only to those who have the courage to fathom them. But when a woman, because of her sex, our customs, and prejudices . . . overcomes these fetters and penetrates that which is most hidden, she doubtless had the most noble courage, extraordinary talent and superior genius.

Gauss tried to persuade Göttingen to award her an honorary degree, but Germain died before he could succeed.

In Hilbert's Göttingen, Emmy Noether was an exceptionally talented algebraist. Hilbert fought on her behalf to overturn archaic rules that denied women positions in German academic institutions. 'I do not see that the sex of the candidate is an argument against her admission,' he objected. The University, he declared, was not 'a bath-house'. Noether, who was Jewish, eventually had to flee Göttingen for the USA. Certain algebraic structures which permeate mathematics are named in her honour.

Julia Robinson was always seen as more than just a gifted mathematician. She was also a woman in the 1960s, and her success encouraged more women to pursue careers into mathematics. Later she recalled how, because she was one of very few women in academia, she would always get asked to complete surveys. 'I am in everyone's scientifically selected sample.'

Robinson's childhood was spent in the Arizona desert. It was a solitary life, with little but a sister and the land for company. Even at an early age she found patterns hidden in the desert. She recalled: 'One of my earliest memories is of arranging pebbles in the shadow of a giant saguaro, squinting because the sun was so bright. I think I have always had a basic liking for the natural numbers. To me they are the only real thing.' At the age of nine she fell ill with rheumatic fever and was bedridden for a couple of years.

Such isolation can be a source of inspiration for budding young scientists. Cauchy and Riemann had both escaped into the mathematical world from their real-world physical and emotional problems. Although Robinson didn't spend her hours confined to bed concocting theorems, she did learn skills that prepared her well for the mathematical battles that lay ahead. 'I am inclined to think that what I learned during that year in bed was patience. My mother said that I was the stubbornest child she had ever known. I would say that my stubbornness has been to a great extent responsible for whatever success I have had in mathematics. But then, it is a common trait among mathematicians.'

By the time she had recovered from her illness, Robinson had missed two years of school. After a year of private tuition, however, she found herself way ahead of her classmates. On one occasion her tutor explained that the ancient Greeks of two thousand years ago knew that the square root of 2 could not be written exactly as a fraction. Unlike the decimal expansion of a fraction, the decimal expansion of the square root of 2 did

Julia Robinson (1919–85).

not have a pattern which repeated itself. It seemed remarkable to Robinson that someone could prove such a thing. How could you be sure that after millions of decimal places a pattern didn't emerge? 'I went home and utilized my newly acquired skills at extracting square roots to check it but finally, late in the afternoon, gave up.' Despite her failure, she began to appreciate the power of mathematical argument to show convincingly that, however far you calculated the decimal expansion of the square root of 2, no pattern would ever emerge.

It is this power of simple argument that captivates many who turn to mathematics. Here is a problem that brute-force calculation can never solve, even with the aid of the most powerful computer, yet stringing together a few cleverly chosen mathematical ideas will open up the mystery of this infinite decimal expansion. Instead of the impossible task of checking an infinite number of decimals, the task is reduced to a cunning little argument.

When she was fourteen, Robinson started to look for anything about mathematics that might relieve the dry tedium of school arithmetic. She listened eagerly to a radio programme called the *University Explorer*. She was particularly intrigued by one broadcast which told the story of the mathematician D.N. Lehmer and his son, D.H. Lehmer. The broadcast explained how this mathematical team were attacking mathematical problems with computing machines they were building from bicycle sprockets and chains. The younger Lehmer would be the first to snatch the baton from Turing's hand, using modern computing machines to show in 1956 that the first 25,000 zeros satisfy Riemann's Hypothesis. The elder Lehmer described how their pre-war machine 'would run happily and sweetly for a few minutes and then suddenly become incoherent. Then it would suddenly pull itself together before having another tantrum.' They eventually traced the cause of the tantrums – a neighbour listening to the radio. Their favourite mathematical problem was finding the prime building blocks of large numbers. Robinson was so excited by the description of these machines that she sent for the transcript of the broadcast.

She found a small item in a newspaper about the supposed discovery of a record prime number which she eagerly cut out. Under the headline FINDS LARGEST NUMBER BUT NOBODY CARES, it reported:

> Dr Samuel I. Krieger wore out six pencils, used 72 sheets of legal size note paper and frazzled his nerves quite badly but he was able to announce today that 231,584,178,474,632,390,847,141,970,017,375,815,706,539,969, 331,281,128,078,915,826,259,279,871 is the largest known prime number. He was unable to say offhand who cared.

Perhaps the lack of interest reflects the fact that this number is actually divisible by 47 (as the newspaper would have discovered, had it checked). Robinson kept the clipping for the rest of her life, along with the script about the Lehmers' calculating machine and a pamphlet she acquired about the mysteries of the fourth dimension.

The groundwork had been laid for Robinson's mathematical career. She took her degree in mathematics at San Diego State College, then made her way to the University of California at Berkeley where her passion for number theory was awakened by a lecturer whom she would later marry, Raphael Robinson. Whilst they were still dating, Raphael found that mathematics was the way to Julia's heart. He started bombarding her with explanations of all the latest mathematical breakthroughs.

One discovery that particularly intrigued Julia was Raphael's description of Gödel and Turing's results. 'I was very impressed and excited by the fact that things about numbers could be proved by symbolic logic,' she said. Despite the unsettling nature of Gödel's results, she still held firm to the sense of reality of numbers that she had acquired as a child playing with pebbles in the desert. 'We can conceive of a chemistry that is different from ours, or a biology, but we cannot conceive of a different mathematics of numbers. What is proved about numbers will be a fact in any universe.'

Although she was blessed with a great mathematical ability, she admitted that without the support of her husband she would have found it hard to continue doing mathematics at a time when most women found it difficult to sustain an academic career. University rules at Berkeley meant that husband and wife could not be members of the same department. In recognition of her research ability, a position was created for her in statistics. The job description she submitted to personnel to accompany her application is a classic description of most mathematicians' working week: 'Monday – tried to prove theorem, Tuesday – tried to prove theorem, Wednesday – tried to prove theorem, Thursday – tried to prove theorem, Friday – theorem false.'

Her interest in Gödel and Turing's work was fuelled by the chance to study with one of the great logicians of the twentieth century, Alfred Tarski, a Pole who had found himself stranded by the onset of war during a visit in 1939 to Harvard. Robinson, however, didn't want to lose sight of her passion for number theory. The tenth of Hilbert's problems presented the perfect blend of both subjects: Is there an algorithm – in computing terms, a program – that could be used to show whether any equation has a solution?

In the light of Gödel and Turing's work, it was becoming clear that, contrary to Hilbert's initial belief, there probably wasn't such a program. Robinson felt sure that there should be a way to exploit the groundwork laid by Turing. She understood that each Turing machine gives rise to a sequence of numbers. For example, one of the Turing machines could be made to produce a list of the square numbers 1, 4, 9, 16, . . ., whilst another generated the primes. One of the steps in Turing's solution of Hilbert's Decision Problem had been to prove that no program existed that could decide whether, given a Turing machine and a number, that number is the output of the machine. Robinson was looking for a connection between equations and Turing machines. Each Turing machine, she believed, would correspond to a particular equation.

If there were such a connection, Robinson hoped that asking whether a number was the output of a particular Turing machine would translate into asking whether the equation corresponding to that machine had a solution. So if she could establish such a connection, she would be home. If there existed a program to test equations for solutions, as Hilbert was hoping for when he posed his tenth problem, then the same program could be used, via Robinson's as yet hypothetical connection between equations and Turing machines, to check which numbers were outputs of Turing machines. But Turing had shown that no such program – one that could decide about the outputs of Turing machines – existed. Therefore there could be no program that could decide whether equations had solutions. The answer to Hilbert's tenth problem would be 'no'.

Robinson set about understanding why each Turing machine might have its own equation. She wanted an equation whose solutions were connected to the sequence of numbers output by the Turing machine. She was rather amused by the question she had set herself. 'Usually in mathematics you have an equation and you want to find a solution. Here you were given a solution and you had to find the equation. I liked that.' As Robinson grew older, the interest that had been sparked in 1948 turned to obsession. After her illness at the age of nine, doctors predicted that her heart had been weakened to such an extent that she might not live beyond forty. Every birthday, 'when it came time for me to blow out the candles on my cake, I always wished, year after year, that the Tenth Problem would be solved – not that I would solve it, but just that it would be solved. I felt that I couldn't bear to die without knowing the answer.'

As each year passed she made more progress. She was joined in her quest by two other mathematicians, Martin Davis and Hilary Putnam. By the end of the 1960s they had reduced the problem to something simpler. Rather than having to find all equations for all outputs of Turing machines, they discovered that if they found an equation for one particular sequence of numbers, they would have proved Robinson's hunch. It was a remarkable achievement. Everything came down to finding the equation for this one sequence of numbers. Their whole theory now relied on them being able to confirm the existence of a single brick in their mathematical wall. If it turned out that this sequence didn't have its own Robinson equation, then the whole wall they had spent so long building would collapse.

There was growing scepticism that Robinson's approach was the right way to attack Hilbert's tenth problem. A number of mathematicians were complaining that it was misguided. Then suddenly, on February 15, 1970, Robinson got a phone call from a colleague who had just returned from a conference in Siberia. There had been a very exciting talk which he thought Robinson would be interested in. A twenty-two-year-old Russian mathematician called Yuri Matijasevich had found the last piece of the jigsaw and solved Hilbert's tenth problem. He had shown that there was an equation that would produce the sequence as Robinson had predicted. That was the brick upon which the whole of Robinson's approach relied.

The solution to Hilbert's tenth problem was complete: a program that could decide whether equations had solutions did not exist.

'That year when I went to blow out the candles on my cake, I stopped in mid-breath, suddenly realizing that the wish I had made for so many years had actually come true.' Robinson saw that the solution had been under her nose all the time, but it had taken Matijasevich to spot it. 'There are lots of things, just lying on the beach as it were, that we don't see until someone else picks one of them up. Then we all see that one,' she explained. She wrote to congratulate Matijasevich: 'I am especially pleased to think that when I first made the conjecture you were a baby and I just had to wait for you to grow up.'

It is striking how mathematics has the ability to unite people across political and historical boundaries. Despite the difficulties posed by the Cold War, these American and Russian mathematicians would forge a strong friendship upon their obsession with Hilbert's inspirational problem. Robinson described this strange bond between mathematicians as being like 'a nation of our own without distinction of geographical origins, race, creed, sex, age or even time (the mathematicians of the past and of the future are our colleagues too) – all dedicated to the most beautiful of the arts and sciences'.

Matijasevich and Robinson would fight over credit for the proof, but not for self-aggrandisement – rather, each insisted that the other had done the hardest bit. It is true that, because Matijasevich ended up putting in the last piece of the jigsaw, the solution of Hilbert's tenth problem is often attributed to him. The reality of course is that many mathematicians contributed to the long journey from Hilbert's announcement in 1900 to the final solution seventy years later.

Although the problem had been solved in the negative – namely, there was no program that could be used to tell whether any equations had solutions – there was a silver lining. Robinson had been proved right in her belief that lists of numbers produced by Turing machines could be described by equations. Mathematicians knew that there was one Turing machine that could reproduce the list of prime numbers. So, thanks to the work of Robinson and Matijasevich, in theory there had to be a formula that could output all the primes.

But could mathematicians find such a formula? In 1971, Matijasevich devised an explicit method for arriving at such a formula, but he did not follow it through to produce an answer. The first explicit formula to be written out in detail used 26 variables, from A to Z, and was discovered in 1976:

$$(K + 2)\{1 - [WZ + H + J - Q]^2 - [(GK + 2G + K + 1)(H + J) + H - Z]^2 - [2N + P + Q + Z - E]^2 - [16(K + 1)^3(K + 2)(N + 1)^2 + 1 - F^2]^2 - [E^3(E + 2)(A + 1)^2 + 1 - O^2]^2 - [(A^2 - 1)Y^2 + 1 - X^2]^2 - [16R^2Y^4(A^2 - 1) + 1 - U^2]^2 - [((A + U^2(U^2 - A))^2 - 1) \times (N + 4DY)^2 + 1 - (X + CU)^2]^2 - [N + L + V - Y]^2 - [(A^2 - 1)L^2 + 1 - M^2]^2 - [AI + K + 1 - L - I]^2 - [P + L(A - N - 1) + B(2AN + 2A - N^2 - 2N - 2) - M]^2 - [Q + Y(A - P - 1) + S(2AP + 2A - P^2 - 2P - 2) - X]^2 - [Z + PL(A - P) + T(2AP - P^2 - 1) - PM)]^2\}$$

This formula works like a computer program. You randomly change the letters $A, \ldots, Z$ into numbers and then use the formula to perform a calculation on those numbers; for example, you might choose $A = 1$, $B = 2$, $\ldots$, $Z = 26$. If the answer is bigger than zero, then the result of the calculation is prime. You can go on repeating the process, assigning different choices of numbers to the letters and redoing the calculation. By systematically going through every possible choice of numbers for the letters A, $\ldots$, Z, you will be guaranteed to produce all prime numbers. No prime number is missed by the formula: there is always some way to choose the numbers $A, \ldots, Z$ so that the equation will produce the prime in question. There is one piece of small print: some choices of $A, \ldots, Z$ give a negative answer, and we just have to ignore them. Our choice of $A = 1$, $B = 2$, $\ldots$, $Z = 26$, for example, is one such choice that we must sweep under the carpet.

Wasn't this the Holy Grail at the end of the quest – the discovery that this extraordinary polynomial will generate the primes? Had this equation been found in Euler's day, it would certainly have been great news. Euler had discovered an equation which had successfully produced lots of primes, but he had been quite pessimistic about the chance of finding an equation which could produce all the primes. But since Euler's day, mathematics had moved on from the mere study of equations and formulas to embrace Riemann's belief in the importance of the underlying structures and themes running through the mathematical world. Mathematical explorers were now busy charting passageways to new worlds. The discovery of this prime number equation was a result born into the wrong age. To the new generation of mathematicians, the equation was like a very technical survey of a land explored years ago and now abandoned. Mathematicians were rather surprised that such an equation existed, but Riemann had moved the study of the primes onto a different plane. A classical symphony in the style of Mozart written and performed in the time of Shostakovich would not have impressed audiences, even if it had attained some perfection of style.

But it wasn't just the new mathematical aesthetic that muted the reception given to this miraculous equation. The truth is, it was practically useless. Most of the values of this equation are negative. Even theoretically it is somewhat lacking in significance. Robinson and Matijasevich had shown that any list of numbers that can be produced by a Turing machine will have such an equation, so in that sense there is nothing special about the primes as opposed to any other list of numbers. This was a view shared by many at the time. When someone told the Russian mathematician Yu. V. Linnik about Matijasevich's result on primes, he replied, 'That's wonderful. Most likely we soon shall learn a lot of new things about primes.' When it was explained to him how the result had been proved and that it applied to many sequences of numbers, he retracted his earlier enthusiasm. 'It's a pity. Most likely we shall not learn anything new about primes.'

If the existence of such an equation is universally true for any sequence of numbers, it tells us nothing specific about the primes. This is what makes Riemann's interpretation so much more intriguing. The existence of Riemann's landscape and the notes he created from each point at sea level is music utterly unique to the primes. You won't find this harmonic structure underlying just any sequence of numbers.

While Robinson was putting paid to the tenth of Hilbert's problems, a friend of hers in Stanford was demolishing Hilbert's belief that in mathematics there is no unknowable. As a student in 1962, Paul Cohen had rather arrogantly asked his professors in Stanford which of Hilbert's problems would make him famous if he could solve it. They thought for while, then told him that the first was one of the most important. In crude terms, it was asking how many numbers there are. To head his list, Hilbert had chosen Cantor's question about different infinities. Is there an infinite collection of numbers which is bigger in size than the set of all fractional numbers, yet small enough that they can't be paired with all the real numbers – including irrational numbers such as π, $\sqrt{2}$ or any number with an infinite decimal expansion?

Hilbert probably turned in his grave when Cohen returned a year later with the solution: both answers are possible! Cohen proved that this most basic of questions was one of Gödel's unprovable sentences. Gone, then, was any hope that only obscure problems were undecidable. What Cohen had proved was this: it was impossible to prove, on the basis of the axioms we currently use for mathematics, that there is a set of numbers whose size is strictly between the number of fractions and the number of all real numbers; equally, it can't be proved that there isn't such a set. Indeed, he had managed to build two different mathematical worlds which satisfied the axioms we are using for mathematics. In one of those worlds the answer to Cantor's question was 'yes'; in the other world it was 'no'.

Some people compare Cohen's result to Gauss's realisation that there are different geometries, not just the geometry of the physical world around us. In some sense this is true. But the point is that mathematicians have a strong sense of what they mean by numbers. Sure, the axioms used for proving things about these numbers might also be satisfied by other 'supernatural' numbers. Nevertheless, most mathematicians still believe that Cantor's question has just one answer that is true for the numbers with which we construct our mathematical edifice. Robinson summed up the response of most mathematicians to Cohen's proof when she exclaimed in a letter she was writing to him, 'For heaven's sake there is only one *true number theory*! It's my religion.' But she crossed out the last sentence before sending the letter to Cohen.

Cohen's groundbreaking work, unsettling as it was to the mathematical orthodoxy, earned him a Fields Medal. After his remarkable discovery that we can't decide the answer to Cantor's question from the classical axioms of mathematics he decided that he would move on to what he considered to be the next most challenging problem on Hilbert's list: the Riemann Hypothesis. Cohen has been one of the few mathematicians to admit that he was actively working on this notoriously difficult problem. So far, though, it has held firm against his attack.

Intriguingly, the Riemann Hypothesis is in a different category to Cantor's question. If Cohen has the same success and proves that the Riemann Hypothesis is undecidable from the axioms of mathematics, he will have shown that the Hypothesis is in fact true! If it is undecidable, then either it is false and we can't prove it, or it is true and we can't prove it. But if it is false then there is a zero off the critical line which we can use to *prove* it is false. It can't be false without us being able to prove it is false. So, the only way the Riemann Hypothesis can be undecidable is if it is true but we still can't find a proof that all the zeros are on the critical line. Turing was one of the first to see the possibility of such a strange confirmation of Riemann's Hypothesis. But few believe that logistic trickery of this nature will be successful in answering Hilbert's eighth problem.

Computers of the mind had played a critical role in our understanding of the mathematical world, thanks to Turing's universal machine. But it was the physical machines he had tried to build that would be in the ascendancy in the second half of the twentieth century. The tide changed in favour of computers made of valves, wires and ultimately silicon, not of neurones and infinite memories. All around the world, machines were built that would allow mathematicians to stare deep into the universe of numbers.

CHAPTER NINE

The Computer Age: From the Mind to the Desktop

I will offer you a bet. When the Riemann Hypothesis is proved it will be proved without the use of a computer. Gerhard Frey, inventor of the crucial link between Fermat's Last Theorem and elliptic curves

Once they've left school, most people's sole encounter, if at all, with prime numbers is via perennial news items about large computers making the latest discovery of the biggest known prime number. Julia Robinson's treasured FINDS LARGEST NUMBER newspaper cutting illustrates that, as early as the 1930s, even incorrect discoveries were making the news. Thanks to Euclid's proof that there are an infinite number of primes, this is a news story that will run and run. By the end of the Second World War, the largest known prime number was one with thirty-nine digits that had held the record since its discovery in 1876. Today, the record prime has more than a million digits. The number would take more pages than this book to print and several months to read aloud. It is the computer that has allowed us to reach such heady heights. In Bletchley, Turing had already been thinking about how to use his machines to find record-breaking prime numbers.

Although Turing's theoretical universal machine was blessed with an infinite amount of memory in which to store information, the machines that he and Newman built in Manchester after the war were very limited in terms of what they could remember. They could only carry out calculations which didn't require too much memory. For example, all it requires to generate the Fibonacci sequence (1, 1, 2, 3, 5, 8, 13, . . .) is to remember the previous two numbers in the sequence, and their computers had no trouble with such a simple sequence. Turing knew of a clever trick that had been developed by the younger member of the Lehmer family duo to find the special primes made famous by the seventeenth-century monk Marin Mersenne. Turing realised that, as with the Fibonacci numbers, Lehmer's test wouldn't need a lot of memory. Searching for Mersenne's primes would be the perfect task for the machines Turing was conceiving.

Mersenne had hit on the idea of generating prime numbers by multiplying 2 together many times and then subtracting 1. For example, $2 \times 2 \times 2 - 1 = 7$ is a prime number. He saw that if $2^n - 1$ was going to stand a chance of being prime, then he would have to choose n to be prime. Yet, as he discovered, this doesn't guarantee that $2^n - 1$ is prime. $2^{11} - 1$ is not a prime, although 11 is. Mersenne had made the prediction that

2, 3, 5, 7, 13, 19, 31, 67, 127, 257

were the only choices of n up to 257 that would make $2^n - 1$ a prime number.

A number the size of $2^{257} - 1$ is so huge that the human mind could not possibly check Mersenne's claim. Perhaps that's why he felt safe in making his bold statement. He had believed that 'all time would not suffice to determine whether they are prime'. His choice of numbers had been driven by Euclid's proof of the infinity of primes. Take a number like 2^n that is divisible by lots of numbers, and then shift it by 1 in the hope of making it indivisible.

Although it didn't guarantee primes, Mersenne's intuition about his numbers was right in one respect. Because Mersenne's numbers were close to this highly divisible number 2^n, there is a very efficient way to check whether such numbers are prime. The method was devised in 1876, when the French mathematician Édouard Lucas discovered how to confirm that Mersenne was correct about $2^{127} - 1$ being prime. This thirty-nine digit prime remained the largest known until the dawn of the computer age. Armed with his new method, Lucas succeeded in revealing Mersenne's list of 'primes' for what they really were. The monk's own list of those numbers n for which $2^n - 1$ is prime turned out to be well off the mark: he'd missed out 61, 89 and 107, and mistakenly included 67. But $2^{257} - 1$ was still beyond Lucas's reach.

Mersenne's mystical insight turned out to be blind guesswork. Mersenne's reputation might have taken a knock, but his name lives on as king of the big primes. The record primes that make the news are invariably one of Mersenne's numbers. Although Lucas could confirm that $2^{67} - 1$ wasn't prime, his method couldn't crack this number into its prime building blocks. As we shall see, cracking such numbers is considered such a tough problem that it is now at the heart of the present-day cryptographic security systems, successors to the Enigma code Turing cracked with his Bletchley bombes.

Turing wasn't the only one thinking about primes and computers. As Robinson had discovered as a child listening to the radio, the Lehmer family had also been fascinated by the idea of using machines to explore prime numbers. The elder Lehmer had already, at the turn of the century, produced a table of primes up to 10,017,000. (No one has since published a table of primes beyond this.) His son had made a more theoretical contribution. In 1930, aged only twenty-five, he discovered a refinement of Lucas's idea for testing whether Mersenne's numbers were prime.

To prove that a Mersenne number is prime and not divisible by any smaller number, Lehmer showed that you could turn the problem on its head. The Mersenne number $2^n - 1$ will be prime only when $2^n - 1$ *divides* a second number, called a Lucas–Lehmer number and denoted by L_n. These numbers are built up, as are the Fibonacci numbers, from the previous numbers in the sequence. To get L_n you square the previous number, L_{n-1}, and subtract 2:

$$L_n = (L_{n-1})^2 - 2$$

The test gets going when $n = 3$ and the corresponding Lucas–Lehmer number is $L_3 = 14$. From there the sequence continues with $L_4 = 194$ and $L_5 = 37{,}634$. What gives this test its power is that all you need to do is generate the number L_n and test whether the Mersenne number $2^n - 1$ divides this number, a computationally easy task. For example, since $2^5 - 1 = 31$ divides the Lucas–Lehmer number $L_5 = 37{,}634$, Mersenne's number $2^5 - 1$ is prime. This simple test allowed Lehmer to finish off Mersenne's list and prove that Mersenne had been wrong about $2^{257} - 1$ being prime.

How did Lucas and Lehmer discover their test for Mersenne primes? It's not the most obvious idea to come to mind. This sort of discovery is very different to the thunderbolt discovery of the Riemann Hypothesis or Gauss's discovery of a connection between primes and logarithms. The Lucas–Lehmer test is not a pattern that will emerge through experiment or numerical observation. They discovered this by playing around with what it means for $2^n - 1$ to be prime, continually turning the statement like a Rubik's cube until suddenly the colours come together in a new way. Each turn will be like a step in the proof. Unlike other theorems where the destination is clear from the outset, the Lucas–Lehmer test ultimately emerged by following the proof without quite knowing where it was going. Lucas had begun turning the cube but Lehmer successfully brought it into the simple form used today.

While he was cracking the German Enigma codes in Bletchley, Turing discussed with his colleagues the potential for machines, similar to the bombes they had built, to find large prime numbers. Thanks to the method developed by Lucas and Lehmer, Mersenne's numbers are particularly amenable to having their primality checked. The method was perfectly suited to being automated on a computer, but the pressures of the war effort soon pushed Turing's ideas on to the back burner. But after the war,

Turing and Newman could return to the idea of finding more Mersenne primes. It would be a perfect test of the machine they were proposing to build at the research laboratory in Manchester. Despite its tiny storage capacity, the Lucas–Lehmer method for determining primes would not require too much memory at each stage. To calculate the *n*th Lucas–Lehmer number, the computer just had to remember what the $(n - 1)$th number had been.

Turing had been unlucky with Riemann's zeros, and his fortune didn't change when he turned his attention to finding Mersenne primes. His computer in Manchester failed to surpass the seventy-year-old record prime, $2^{127} - 1$. The next Mersenne prime wouldn't appear until $2^{521} - 1$, which was just beyond the reach of Turing's machine. By a strange twist of fate it was Julia Robinson's husband, Raphael, who claimed the discovery of this new record prime. He had somehow obtained the manual of a machine that Derrick Lehmer had built in Los Angeles. By this time, Lehmer had moved on from the bicycle gears and chains of his pre-war machine. He was now director of the National Bureau of Standards' Institute for Numerical Analysis and had created a machine called the Standards Western Automatic Computer (SWAC). From the comfort of his office in Berkeley, and never having set eyes on the machine, Raphael wrote a program that would run on the SWAC to hunt for Mersenne primes. On January 30, 1952, the computer discovered the first prime numbers beyond the reach of the computational abilities of the human mind. Just a few hours after the record $2^{521} - 1$ had been set, SWAC spat out a new big prime, $2^{607} - 1$. Within the year, Raphael Robinson had broken his record another three times. The biggest prime number was now $2^{2,281} - 1$.

The hunt for record primes came to be dominated by those with access to the biggest computers. Through to the mid-1990s, new records were set using Cray computers, the giants of the computing world. Cray Research, founded in 1971, had exploited the fact that a computer doesn't have to finish one operation before it starts the next. This simple idea was what lay behind the creation of machines acknowledged for decades to be the fastest calculators in the world. Since the 1980s, the Cray machine based at the Lawrence Livermore Laboratory in California, under the watchful eye of Paul Gage and David Slowinski, had been grabbing the records and headlines. In 1996 they announced their seventh record-breaking prime, $2^{1,257,787} - 1$, a number with 378,632 digits.

Recently, though, the tide has turned in favour of the little guys. Like David conquering Goliath, the records are now falling to humble desktop computers. And the catapult that gives them the power to challenge the Cray computers? The Internet. With the combined strength of countless small computers networked together, the potential is there to set this ant colony of machines hunting for big primes. This is not the first time the Internet has been used to empower the amateur in doing real science. Astronomy has benefited greatly from assigning to thousands of amateur astronomers their own small piece of the night sky to scour; the Internet had provided the network to coordinate this astronomical effort. Inspired by the success of the astronomers, an American programmer, George Woltman, published a piece of software on the Internet which, once downloaded, assigns your desktop a tiny part of the infinite expanse of numbers. Instead of training their telescopes on the night sky looking for a new supernova, amateur scientists are using their computers' idle time to scan corners of the galaxy of numbers for new record primes.

The search was not without its dangers. One of Woltman's recruits worked for a US telephone company and had enlisted the help of 2,585 of the company's computers in his search for Mersenne primes. The company grew suspicious that something was amiss when the computers in Phoenix were taking five minutes as opposed to five seconds to retrieve telephone numbers. When the FBI eventually tracked the source of the slowdown, the employee admitted that 'All that computational power was just too tempting for me.' The telephone company showed little sympathy with their employee's pursuit of science, and fired him.

The first discovery of a new Mersenne prime by this band of Internet hunters came a few months after the Cray announcement in 1996. Joel Armengaud, a Paris-based computer programmer, struck gold in the small seam of numbers he had been mining for Woltman's project. For the media, the discovery came a little too soon after the previous find. When I contacted *The Times* about the latest biggest prime, they told me they would run this story only every other year. Slowinski and Gage, the Cray twins, had matched supply to demand with discoveries coming on average every two years since 1979.

But there was more to all of this than the discovery of new primes. It marked a turning point in the role of the computer in the search for primes, and it wasn't missed by the Internet-savvy *Wired* magazine. *Wired* ran a story about what is now known as the Great Internet Mersenne Prime Search, or GIMPS. Woltman has successfully enrolled over two hundred thousand computers from all around the world to create what is in effect one huge parallel-processing machine. Not that the big guns with

Crays are out of business. They are now equal partners, checking the discoveries of the little guys.

By 2002 there have been five lucky winners in the search for Mersenne primes. The Paris find was followed by one in England and a third in California. But it was Nayan Hajratwala from Plymouth, Michigan who truly struck gold in June 1999. His prime, $2^{6,972,593} - 1$, contains 2,098,960 digits and was the first to pass the one-million-digit milestone. A symbolic prize in itself, this achievement also won him a cash prize of $50,000 from the Electronic Frontier Foundation. This Californian organisation is the self-styled protector of the civil liberties of Netizens – those who use the Internet. If your appetite has been whetted by Hajratwala's success, the Foundation still has half a million dollars in prize money to reward the discoverers of more big primes, with the next milestone set at 10 million digits. Hajratwala's prime was beaten in November 2001 by Canadian student Michael Cameron, who used his PC to prove the primality of $2^{13,466,917} - 1$, a number with over 4 million digits. Mathematicians believe that there are infinitely many of these special Mersenne primes waiting to be discovered.

The computer – the death of mathematics?

If the computer can out-count us, doesn't that make the mathematician redundant? Fortunately, it doesn't. Rather than heralding the end of mathematics, it highlights the true difference between the mathematician as creative artist and the computer as tedious calculator. The computer is certainly a powerful new ally in mathematicians' attempts to navigate their world, and a sturdy sherpa in our ascent of Mount Riemann, but it can never replace the mathematician. Although the computer can outstrip the mathematician in any finite computation, it lacks the imagination (as yet) to embrace an infinite picture and unmask the structure and patterns underlying mathematics.

For example, does the computer search for big primes give us any better understanding of the primes? We might be able to sing higher and higher notes, but the music remains hidden. Euclid has already assured us that there will always be a bigger prime to find. It isn't known, though, whether Mersenne's special numbers will produce primes infinitely often. It may be that Michael Cameron has discovered the thirty-ninth and final Mersenne prime. When I talked to Paul Erdős, he ranked the task of proving that there are infinitely many Mersenne primes as one of the greatest unsolved problems of number theory. It is generally believed that there are infinitely many choices of n that make $2^n - 1$ prime. But it is very unlikely that a computer will prove it.

That's not to say that computers can't prove things. Given a set of axioms and rules of deduction, you can program a computer to start churning out mathematical theorems. The point is that, like the monkey on the typewriter, the computer won't be able to distinguish Gaussian theorems from primary-school sums. The human mathematician has developed the critical faculties to differentiate between theorems which are important and those which aren't. The aesthetic sensibilities of the mathematical mind are tuned to appreciate proofs that are beautiful compositions and shun proofs which are ugly. Although the ugly proof is just as valid, elegance has always been recognised as an important criterion in charting the best course through the mathematical world.

The first successful use of a computer to prove a theorem was in connection with a challenge called the Four-Colour Problem, which began as an amateur curiosity. The problem is about something we probably all discovered as kids: if you want to colour a map so that no two neighbouring countries have the same colour, you can always get away with just four colours. Despite the most creative redrawing of national boundaries it seems you just can't force the map of Europe to need any more colours. The current boundaries of France, Germany, Belgium and Luxembourg prove that you do need at least four colours:

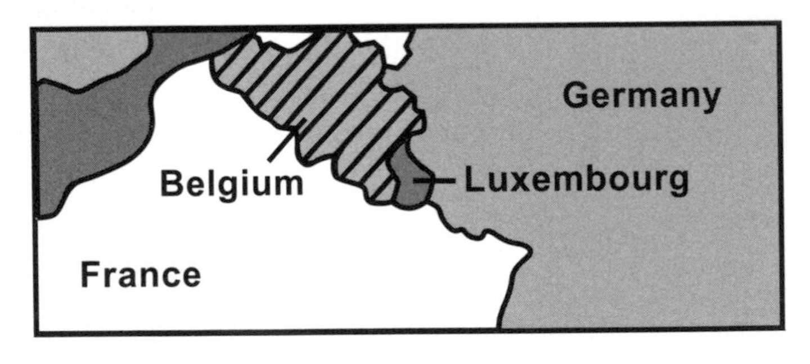

At least four colours are needed to colour this map so that no two neighbouring countries share the same colour.

But could you prove that four colours would be enough for any map?

The question was first aired in 1852 when a law student, Francis Guthrie, wrote to his brother, a mathematician at University College, London, asking if anyone had proved that four colours would always suffice. Admittedly it was not then a question that many people felt was important. A number of minor mathematicians tried their hand at providing Guthrie with a proof. But, as a proof remained elusive, the problem gradually worked its way up the ladder of mathematical ability. Even Hilbert's best friend in Göttingen, Hermann Minkowski, had his fingers burnt by it. The question of the Four-Colour Problem arose in a lecture course he was giving. 'This theorem has not yet been proved but that is because only mathematicians of the third rank have occupied themselves with it,' he announced. 'I believe I can prove it.' He spent several lectures battling with his ideas on the blackboard. One morning, as he entered the lecture hall, there was a clap of thunder. 'Heaven is angered by my arrogance,' he conceded. 'My proof is defective.'

The more people tried and failed, the greater the problem grew in stature, especially as it was so simple to pose. It resisted all attempts to prove it until 1976, over a hundred years after Guthrie's letter to his brother. Two mathematicians, Kenneth Appel and Wolfgang Haken at the University of Illinois, showed that instead of the impossible task of colouring the infinite number of all possible maps, the problem could be reduced to considering 1,500 different basic maps. This was a crucial breakthrough. It was like the discovery of a cartographic Periodic Table of elementary maps from which all others could be built. But checking each of these 'atomic' maps by hand would mean that even if they had started back in 1976, Appel and Haken would still be colouring away today. Instead, a computer was used for the first time to complete the proof. It took 1,200 hours of computing time, but eventually it came back with the answer that, yes, every map can be coloured with four colours. The combination of human ingenuity to prove that you only needed to consider these 1,500 basic maps to understand all maps, coupled with the brute force of the computer, confirmed what Guthrie had conjectured in 1852: that any map needs only four colours.

Knowing that the Four-Colour Theorem is true is of no practical use. Cartographers did not breathe a collective sigh of relief on hearing the news that they wouldn't need to go out and buy a fifth pot of paint. Mathematicians were not eagerly awaiting confirmation of the result before they could explore beyond this problem. They simply couldn't see much on the other side that invited much investigation. This wasn't the Riemann Hypothesis, with thousands of results depending on its proof. The significance of the Four-Colour Problem was that it illustrated that we still didn't understand two-dimensional space sufficiently to be able to answer it. For as long as it remained unsolved, it spurred mathematicians to seek a deeper understanding of the space around us. That is why many were unsatisfied by Appel and Haken's proof. The computer had given us an answer but it had not deepened our understanding.

Whether Appel and Haken's computer-aided proof of the Four-Colour Problem really captures the true spirit of 'proof' has been a matter of hot debate. Many felt unsettled by the computer's role, even if most knew that the proof was more likely to be correct than many human-generated proofs. Shouldn't a proof offer understanding? 'A mathematical proof should resemble a simple and clear-cut constellation, not a scattered Milky Way' was how Hardy liked to express it. The computer proof of the Four-Colour Problem resorted to laboriously mapping out the chaos of the heavens rather than offering a deep understanding of why the heavens look the way they do.

The computer-aided proof highlighted that the pleasure of mathematics comes not just from the end result. We don't read mathematical mystery stories just to find whodunnit. The pleasure comes from seeing how the convolutions of the plot unfold themselves in the build-up to the moment of revelation. Appel and Haken's proof of the Four-Colour Problem had deprived us of the 'Aha – now I get it!' feeling that we crave when we read mathematics. We like to share the eureka moment that the creator of the proof first experienced. Debate will rage for decades over whether computers will ever feel emotions, but the proof of the Four-Colour Problem certainly didn't offer us the chance to share any exhilaration the computer might have felt.

Aesthetic sensibilities notwithstanding, the computer has gone on to serve the mathematical community in proving theorems. Whenever a problem has been reduced to checking a finite number of things, a computer can and does help. Can the computer therefore help us in our ascent on the Riemann Hypothesis? By the time of Hardy's death at the end of the Second World War there was some suspicion that the Riemann Hypothesis might be false. As Turing recognised, if it is false, then a computer can help. A machine can be programmed to search for zeros until it finds a zero off the line. But if the Hypothesis is true, then the computer is useless in actually proving that this infinite number of zeros are all on the line. The best it can do is to provide more and more evidence to support our faith in Riemann's hunch.

The computer also satisfied another need. By the time of Hardy's death, mathematicians were rather stuck. The theoretical advances on the Riemann Hypothesis had dried up. It seemed that, given the available techniques, Hardy, Littlewood and Selberg had obtained the best results possible about the location of the points at sea level in Riemann's landscape. They had squeezed these techniques for all they were worth. Most mathematicians agreed that new ideas were needed if they were to get any closer to proving the Riemann Hypothesis. In the absence of new ideas the computer gave the impression of progress. But it was just an impression – the participation of the computer was covering up the distinct lack of progress then being made with the Riemann Hypothesis. Computing became a substitute for thought, chewing gum for the mind that lulled us into thinking that we were really doing something when in fact we were facing a brick wall.

Zagier, the mathematical musketeer

The secret formula that Siegel had discovered in 1932 amongst Riemann's unpublished notes was a formula for accurately and efficiently calculating the location of zeros in Riemann's landscape. Turing had tried to speed up the calculation using his complex system of cogs, but it has taken more modern machines to realise the formula's full potential. Once this secret formula was programmed into a computer, one could start exploring this landscape in regions that were previously unimaginable. In the 1960s, as humankind began to probe the distant universe with unmanned spacecraft, mathematicians began to send computers calculating their way into the outer reaches of Riemann's landscape.

The farther the mathematicians went north in search of zeros, the more evidence they gathered. But what use is that evidence? How many zeros did you have to show were on the line before you could feel confident that the Riemann Hypothesis was true? The trouble is that, as Littlewood's work proved, in mathematics evidence is rarely grounds for confidence. That is why many dismissed the computer as a useful instrument to probe the Riemann Hypothesis. However, there was a surprise in store that would begin to convince even the most diehard sceptics that there was a good chance the Riemann Hypothesis was true.

At the beginning of the 1970s, one mathematician stood at the head of this small band of sceptics. Don Zagier is one of the most energetic mathematicians on today's mathematical circuit, cutting a dashing figure as he sweeps through the corridors of the Max Planck Institute for Mathematics in Bonn, Germany's answer to the Institute for Advanced Study in Princeton. Like some mathematical musketeer, Zagier flourishes his razor-sharp intellect, ready to slay any passing problem. His enthusiasm and energy for the subject whisk you away in a whirlwind of ideas delivered in

Don Zagier, professor at the Max Planck Institute for Mathematics, Bonn.

a rat-a-tat-tat voice and at a speed that leaves you breathless. He has a playful approach to his subject and is always ready with a mathematical puzzle to spice up lunch at the Institute in Bonn.

Zagier had become exasperated by some people's desire to believe in the Riemann Hypothesis on purely aesthetic grounds, ignoring the dearth of real evidence to support it. Faith in the Hypothesis was probably based on little more than a reverence for simplicity and beauty in mathematics. A zero off the line would be a blot on this beautiful landscape. Each zero contributed a note to the prime number music. Enrico Bombieri expresses what it would mean if the Riemann Hypothesis turns out to be false: 'Say you go to a concert and you are listening to the musicians playing together in a very harmonious way. Then suddenly there is a big tuba which plays with an extremely strong sound which drowns all the rest.' So much beauty abounds in the mathematical world that we cannot – dare not – believe that Nature would choose a discordant universe in which the Riemann Hypothesis was false.

Whereas Zagier was the leading sceptic at this point, Bombieri represented the archetypal believer in the Riemann Hypothesis. In the early

1970s he had yet to move to the Institute in Princeton, and was still a professor in his native Italy. As Zagier explained, 'Bombieri believes as an absolute article of faith that the Riemann Hypothesis is true. It is a religious belief that it has to be true or else the whole world would be wrong if it weren't.' Indeed, as Bombieri elaborated, 'When I was in the eleventh grade I studied several of the medieval philosophers. One of them, William of Occam, elevated the idea that when one must choose between two explanations, one should always choose the simpler. Occam's razor as the principle is called cuts out the difficult and chooses the simple.' For Bombieri, a zero off the line would be like an instrument in the orchestra 'that drowns out the others – an aesthetically distasteful situation. As a follower of William of Occam, I reject that conclusion, and so I accept the truth of the Riemann Hypothesis.'

Things came to a head when Bombieri visited the Institute in Bonn, and discussion over tea turned to the Riemann Hypothesis. Zagier, ever the mathematical swashbuckler, had his chance to challenge Bombieri to a duel. 'I told him over tea that there wasn't enough evidence yet to convince me either way. So I would be willing to take an evens money bet against it. Not that I thought it was necessarily false, but I was willing to play devil's advocate.'

Bombieri responded, 'Well, I'll certainly be prepared to take you up on that,' and Zagier realised that he was stupid to offer evens – Bombieri was such a firm believer he would have taken odds of a billion to one. The stakes were agreed upon: two bottles of very good Bordeaux to be chosen by the winner.

'We wanted the bet to be settled in our lifetime,' Zagier explains. 'However, there was a very good chance we would go to our graves with the bet still open. We didn't want to put a time limit on it so that if ten years passed then we would drop it. That seemed silly. What does ten years have to do with the Riemann Hypothesis? We wanted something mathematical.'

So Zagier offered the following. Although Turing's machine had broken down after calculating the first 1,104 zeros, by 1956 Derrick Lehmer had been more successful. He had got his machines in California to check that the first 25,000 zeros were on the line. By the beginning of the 1970s, a famous calculation confirmed that the first three and a half million zeros were indeed on the line. The proof was a fantastic tour de force which exploited some brilliant theoretical techniques to push the computation to the absolute limits of the computer technology available at the time. As Zagier explains:

> So I said OK, right now three million zeros have been computed, I'm not yet convinced, even though most people would say *what do you want . . . blood . . . my god, three million zeros*. Most people would say *what's going to change, what's the difference three million . . . three trillion*. That's the whole point of what I'm telling you. That's not the case. At three million I still wasn't convinced. I wish I'd made the bet a bit early because I was already beginning to be convinced. I wish I'd made the bet at a hundred thousand because at that point there was absolutely no reason to believe the Riemann Hypothesis. It turns out that when you analyse the data a hundred thousand zeros is totally useless. It's essentially zero evidence. Three million is beginning to be interesting.

But Zagier recognised that 300 million zeros represented an important watershed. There were theoretical reasons why the first several thousand zeros had to be on Riemann's ley line. However, as one advanced farther north, the reasons why early zeros had to be on Riemann's line began to be outweighed by even stronger reasons why zeros should start falling off the line. By 300 million zeros, Zagier realised, it would be a miracle if zeros weren't pushed off the line.

Zagier based his analysis on a graph he knew kept track of the behaviour of the gradient in the hills and valleys along Riemann's ley line. Zagier's graph represented a new perspective from which to view the cross-section of Riemann's landscape along the critical line. What was interesting was that it facilitated a new interpretation of the Riemann Hypothesis. If this new graph ever crosses Riemann's critical line, there has to be a zero off Riemann's line in this region, making the Riemann Hypothesis false. To begin with, the graph is nowhere near the critical line, and in fact climbs away. But as one marches farther and farther north the graph starts coming down, edging towards the critical line. Every now and again Zagier's graph attempts to crash through the critical line, but as the figure opposite illustrates something seems to be preventing it from crossing.

So, the farther north one advances, the more likely it seems that this graph might cross the critical line. Zagier knew that the first real weakness was around the 300-millionth zero. This region of the critical line would be the real test. By the time you had gone this far north, if the graph still did not cross the critical line, then there really had to be a reason why it didn't. And that reason, Zagier reasoned, would be that the Riemann Hypothesis was true. And that is why Zagier set the threshold for the bet at 300 million zeros. Bombieri would win the bet either if a proof was found or if 300 million zeros were calculated and no counter-example was found.

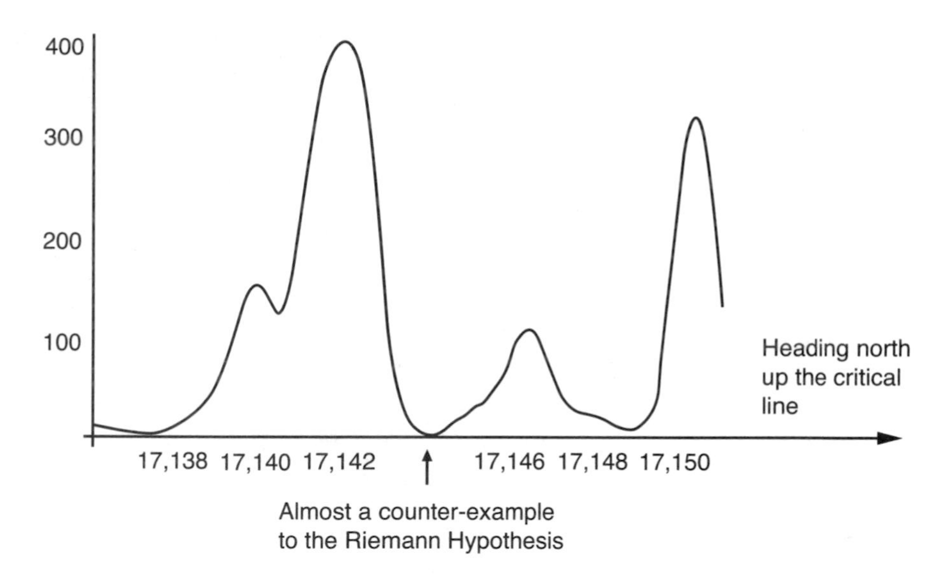

Zagier's auxiliary graph – the Riemann Hypothesis is false if the graph crosses the horizontal axis.

Zagier could see that the computers of the 1970s were still far too weak to explore this region of Riemann's ley line. Computers had managed to calculate three and a half million zeros. Zagier estimated, considering the growth of computer technology at the time, that it might be as long as thirty years before 300 million zeros could be calculated. But he hadn't counted on the computer revolution that was just around the corner.

For about five years nothing happened. Computers grew slowly more powerful, but to compute even twice as many zeros let alone a hundred times as many would have required such a huge amount of work that nobody bothered. After all, in this business there was no point expending huge amounts of energy on merely doubling the amount of evidence. But then, about five years later, computers suddenly got a lot faster, and two teams took up the challenge of exploiting this new power to calculate more zeros. One team was in Amsterdam, run by Herman te Riele, and the second was in Australia, led by Richard Brent.

Brent was the first to make an announcement in 1978, that the first 75 million zeros were still on the line. The team in Amsterdam then joined forces with Brent, and after a year's work they produced a big paper, carefully written up and beautifully presented, everything finished and in place. How far had they calculated . . . 200 million! Zagier laughs:

> I breathed a sigh of relief because this was such a huge project. Thank God they had luckily stopped at 200 million. Obviously they could have gone to 300 million, but thank God they didn't. Now I'll have a reprieve of many years. They won't go on just another 50 per cent. People will wait until they can go to a billion. So it's going to be many years. Unfortunately I didn't count on my good friend Hendrik Lenstra, who knew about the bet and was in Amsterdam.

Lenstra went to te Riele and asked him, 'Why did you people stop at 200 million? You know if you go to 300 million Don Zagier will lose a bet.' So the team went on to 300 million. Naturally they didn't find a zero off the line, so Zagier had to pay up. He took the two bottles to Bombieri, who shared the first with Zagier. As Zagier likes to point out, that was probably the most expensive bottle ever drunk, because

> two hundred million had nothing to do with my bet. They'd done that independently. But the last hundred million, they did that computation at that time only because they'd heard about my bet. It took about a thousand hours of CPU time approximately to calculate the extra 100 million. The going price for an hour of CPU time at the time was $700. Since they did this computation solely that I should lose the bet and have to pay up my two bottles of wine, I claim that these bottles were 350 thousand dollars each – and that's a lot more than the most expensive bottle of wine ever sold at auction.

More importantly though, the evidence, in Zagier's view, was now overwhelmingly in support of the Hypothesis. The computer as a calculating tool was finally powerful enough to navigate far enough north in Riemann's zeta landscape for there to be every chance of throwing up a counter-example. Despite numerous attempts by Zagier's auxiliary graph to storm across Riemann's critical line, it was clear that something was acting like a huge repulsive force stopping the graph from crossing the line. And the reason? The Riemann Hypothesis.

'That is what makes me a firm believer in the Riemann Hypothesis,' Zagier now admits. He compares the role of the computer to the way a particle accelerator is used in support of theoretical physics. Physicists have a model of what makes up matter, but to test that model requires the generation of sufficient energy to blow apart the atom. For Zagier, 300 million zeros was finally enough energy to test whether the Riemann Hypothesis had a good chance of being true:

> This, I believe, is 100 per cent convincing evidence that there is something preventing the graph from crossing, and the only thing I can imagine that

is possibly happening is yes, indeed, the Riemann Hypothesis is true. And now I am absolutely as firm a believer in the Riemann Hypothesis as Bombieri, not a priori because it is too beautiful and is such an elegant thing or because of the existence of God, but because of this evidence.

One of te Riele's team in Amsterdam, Jan van de Lune, has now retired. But mathematicians never entirely lose the mathematical bug, even though they might have to give up their offices. Using the program the team had used decades before, he has now confirmed, using three PCs he has running at home, that the first 6.3 billion zeros are all obeying Riemann's Hypothesis. However long his computers keep calculating, there is no chance that they will provide a proof in this manner. But if there is a zero off the line, the computer does have the potential to play a role in exposing the Riemann Hypothesis as pure fantasy.

This is where the computer is in its element – as a destroyer of conjectures. In the 1980s, calculations of zeros were used to bring down a close cousin of the Riemann Hypothesis – something called the Mertens Conjecture. But the calculations weren't made from the comfort of a mathematics department. Instead, interest switched to calculations of zeros made by a rather unexpected source: the AT&T telephone company.

Odlyzko, the calculating maestro of New Jersey

In the heart of New Jersey, near the sleepy town of Florham Park, an unlikely powerhouse of mathematical talent thrives under the commercial aegis of AT&T's research laboratories. Once inside their building, you could be mistaken for believing you were in the mathematics department of a university. But this is the home of a major telecommunications business. The origins of the laboratory go back to the 1920s, when AT&T Bell Labs was first created. Turing had spent some time in the Bell Laboratory in New York during the war. He had been involved in a project to design a voice encryption system to enable Washington and London to talk securely on the telephone. Turing claimed that his time at Bell Labs was more exciting than Princeton, though that might have had something to do with the Village nightlife in Manhattan. Erdős would often make visits to the New Jersey base on his mathematical wanderings.

With the explosion of technology that hit the telecommunications industry in the 1960s, it was clear that to stay ahead of the game AT&T would need more and more sophisticated mathematical expertise. After the rapid expansion of the universities in that decade, the seventies by contrast were lean years for mathematicians trying to find academic jobs. By expanding their research facilities, AT&T were able to attract some of this excess. Although they were ultimately hoping that the research would translate into technological innovation, they were happy for their mathematicians to continue pursuing their mathematical passions. It sounds altruistic, but it was in fact good business: because of the monopoly it enjoyed in the 1970s, certain restrictions were put on how the company spent its profits. Investing in its research laboratory was seen as a canny way of soaking up some of the profits.

Whatever the reasons behind AT&T's move, mathematics has a lot to be thankful for. Some of the most interesting theoretical advances have their source in ideas coming out of their laboratory. It is a fascinating mix of academia and the hard-nosed world of business. On a visit to talk to mathematicians at the laboratory, I witnessed this mix at first hand. Faced with the job of maximising AT&T's bids in auction for mobile phone bandwidth, several mathematicians were presenting over a working lunch a theoretical model that would provide the company with the best strategy to negotiate the complex bidding process. For the mathematicians, it could have been a strategy for a game of chess rather than spending millions of the company's dollars. But the two weren't inconsistent.

The head of the research laboratory until 2001 was Andrew Odlyzko. Originally from Poland, Odlyzko still retains a strong but gentle eastern European accent. His time in the commercial sector has made him a very good communicator of difficult mathematical ideas. He has an engaging, inclusive attitude which encourages you to join him on his mathematical journey. Nevertheless he is very precise and always the consummate mathematician: each step must leave no room for ambiguity. Odlyzko's interest in the zeta function had been sparked during his doctorate under the supervision of Harold Stark at MIT. One of the problems he had been looking at required him to know as accurately as possible the location of the first few zeros in the zeta landscape.

High-precision calculations are just the sort of thing that a computer does much better than a human. Shortly after Odlyzko joined AT&T Bell laboratories, he got his break. In 1978 the lab purchased their first supercomputer – a Cray 1. It was the first Cray to be owned by a private company as opposed to the government or a university. Since AT&T was a commercial organisation, with accounting and budgets controlling most things, each department had to pay for time on the mainframe computers. However, it took a while for people to acquire the skills to program the Cray, and to start with it saw little use. So the computer section decided to

Andrew Odlyzko, head of AT&T's research laboratory until 2001.

offer free chunks of five hours on the Cray for worthy scientific projects that had no funding.

The opportunity to exploit the power of the Cray was too much for Odlyzko to resist. He contacted the teams in Amsterdam and Australia who had proved that the first 300 million zeros were on the line. Had any of them accurately located the positions of these zeros along Riemann's ley line? None of them had. They had focused on proving that the east–west coordinate of each zero was $\frac{1}{2}$, as Riemann predicted it should be. They hadn't been too concerned with the exact north–south location.

Odlyzko applied for time on the Cray to determine the exact location of the first million zeros. AT&T agreed, and for decades Odlyzko has been using whatever computer time the company can spare to compute more and more zeros. These calculations aren't just an unmotivated exercise in computing. His supervisor, Stark, had applied the knowledge gained about the location of the first few zeros to prove one of Gauss's conjectures about how certain sets of imaginary numbers factorise. Odlyzko, on the other hand, employed his accurate location of the first 2,000 zeros to disprove a conjecture which had been on the mathematical circuit since the beginning of the twentieth century: the Mertens Conjecture. He was joined in the demolition of this conjecture by te Riele in Amsterdam, the mathematician who had helped Zagier to lose his bet by proving that the first 300 million zeros were on Riemann's line. The Mertens Conjecture is very closely related to the Riemann Hypothesis, and its disproof showed mathematicians that if the Riemann Hypothesis were true, it was only just true.

The Mertens Conjecture is best understood as a variation on the game of tossing the prime number coin. The Mertens coin lands heads on the Nth toss if N is built from an even number of prime building blocks. For example, the result is 'heads' for $N = 15$, since 15 is the product of two primes, 3 and 5. If, on the other hand, N is built from an odd number of primes, for example $N = 105 = 3 \times 5 \times 7$, the result is 'tails'. There is, however, a third possibility. If N uses one of the building blocks twice, then the result is counted as zero: 12, for example, is built from two 2's and a 3 ($12 = 2 \times 2 \times 3$), so it scores 0. Think of zero results as corresponding to the coin landing out of sight or on its side. Mertens made a conjecture about the behaviour of this coin as N gets larger. It is very similar to Riemann's Hypothesis, which says that the prime number coin isn't biased. But the Mertens Conjecture was slightly stronger than what Riemann was predicting for the primes. It was predicting that the error was slightly smaller than what one might expect from a fair coin. If the conjecture were true, then so was the Riemann Hypothesis – but not vice versa.

In 1897 Mertens produced tables of calculations up to $N = 10{,}000$ to support his conjecture. By the 1970s experimental evidence had reached numbers up to a billion. But, as Littlewood had shown, experimental evidence in the billions is peanuts. There was now growing scepticism that the Mertens Conjecture would indeed be true. It took Odlyzko and te Riele's accurate calculations of the locations of the first 2,000 zeros calculated to 100 decimal places to finally disprove Mertens Conjective. As yet another warning, though, to those impressed by numerical experimental evidence, Odlyzko and te Riele estimated that even if Mertens had analysed the tosses of his coin up to 10^{30}, his conjecture would still appear to be true.

Odlyzko's computers at AT&T have continued to help mathematicians in their quest to unearth the mysteries of the primes. But it isn't one-way traffic. Prime numbers are now making their own contribution to the ever-expanding computer age. In the 1970s the primes suddenly became the key, literally, to securing the privacy of electronic communication. Hardy had always been very proud of how useless mathematics and especially number theory was in the real world:

> The 'real' mathematics of the 'real' mathematicians, the mathematics of Fermat and Euler and Gauss and Abel and Riemann, is almost wholly 'useless' (and this is true of 'applied' as of 'pure' mathematics). It is not possible to justify the life of any genuine professional mathematician on the ground of the 'utility' of his work.

Hardy could not have got it more wrong. The mathematics of Fermat, Gauss and Riemann was to be put to use at the heart of the commercial world. This is why AT&T were to recruit even more mathematicians during the 1980s and 1990s. The security of the electronic village stands or falls on our understanding of prime numbers.

CHAPTER TEN

Cracking Numbers and Codes

If Gauss were alive today, he would be a hacker. Peter Sarnak, professor at Princeton University

In 1903, Frank Nelson Cole, a professor of mathematics at Columbia University in New York, gave a rather curious talk to a meeting of the American Mathematical Society. Without saying a word, he wrote one of Mersenne's numbers on one blackboard, and on the next blackboard wrote and multiplied together two smaller numbers. In the middle he placed an equals sign, and then sat down.

$$2^{67} - 1 = 193{,}707{,}721 \times 761{,}838{,}257{,}287$$

The audience rose to its feet and applauded – a rare outburst for a roomful of mathematicians. But surely, multiplying together two numbers was not so difficult, even for mathematicians at the turn of the century? In fact, Cole had done the opposite. It had been known since 1876 that $2^{67} - 1$, a twenty-digit Mersenne number, was not itself prime but the product of two smaller numbers. However, no one knew which ones. It had taken Cole three years of Sunday afternoons to 'crack' this number into its two prime components.

It was not only Cole's 1903 audience who appreciated his feat. In 2000 an esoteric off-Broadway show called *The Five Hysterical Girls Theorem* paid homage to his calculation by having one of the girls crack Cole's number. Prime numbers are a recurrent theme in this play about a mathematical family's trip to the seaside. The father laments his daughter's coming of age, not because she will be old enough to run off with her lover but because 17 is a prime number, whereas 18 can be divided by *four* other numbers!

Over two thousand years ago the Greeks proved that every number can be written as a product of prime numbers. A fast and efficient way to find which prime numbers have been used to build up other numbers has eluded mathematicians ever since. What we are missing is a mathematical counterpart of chemical spectroscopy, which tells chemists which elements of the Periodic Table make up a chemical compound. A discovery of a mathematical analogue that would crack numbers into their constituent primes would earn its creator more than just academic acclaim.

In 1903, Cole's calculation was regarded as an interesting mathematical curiosity – the standing ovation he received was in recognition of his extraordinary hard labour rather than any intrinsic importance the problem had. Such number-cracking is no longer a Sunday afternoon pastime but lies at the heart of modern code-breaking. Mathematicians have devised a way to wire this difficult problem of cracking numbers into the codes that protect the world's finances on the Internet. This innocent-sounding task is sufficiently tough for numbers with 100 digits that banks and e-commerce are prepared to stake the security of their financial transactions on the impossibly long time it takes – at present – to find the prime factors. At the same time, these new mathematical codes have been used to solve a problem that dogged the world of cryptography.

The birth of Internet cryptography

For as long as we have been able to communicate, we have needed to deliver secret messages. To prevent important information from falling into the wrong hands, our ancestors devised ever more intriguing ways of disguising the content of a message. One of the first methods used to hide messages was devised by the Spartan army over two and a half thousand years ago. The sender and recipient each had a cylinder of exactly the same dimensions, called a scytale. To encode a message, the sender would first wrap a narrow strip of parchment around the scytale so that it spiralled down the cylinder. He would then write his message on the parchment, along the length of the scytale. Once the parchment was unwound, the text looked meaningless. Only when it was wrapped around an identical cylinder would the message reappear. Since then, successive generations have concocted ever more sophisticated cryptographic methods. The ultimate mechanical encoding device was the German Enigma machine used by German forces in the Second World War.

Before 1977, anyone who wanted to send a secret message faced an inherent problem. Before the message was transmitted, sender and receiver would have to meet in advance to decide which cipher – the method of encryption – to use. The Spartan generals, for example, would have needed to agree on the dimensions of the scytale cylinder. Even with the mass-produced Enigma machine, Berlin would still have to dispatch agents to deliver to U-boat captains and tank commanders the books detailing the machine settings for encoding each day's messages. Of course, if an enemy got their hands on the code book, the game was up.

Imagine the logistics of using such a system of cryptography to do business on the Internet. Before we could send our banking details safely, we would have to receive secure letters from the company running each website at which we wanted to shop telling us how to encode the details. Given the huge traffic on the Internet, there would be a very high chance that many of these letters would be intercepted. A crypto-system suited to the emerging era of rapid global communication needed to be developed. And just as it was mathematicians at Bletchley Park who cracked the wartime Enigma, it would be mathematicians who created a new generation of codes that took cryptography out of the spy novel and into the global village. These mathematical codes underpinned the birth of what is known as *public-key cryptography*.

Think of encoding and decoding as locking and unlocking a door. With a conventional door the same key is used both to lock it and to unlock it. With the Enigma machine, the setting used to encode a message is the same as the setting used to decode it. The setting – call it the *key* – must be kept secret. The farther the recipient is from the sender, the more logistically difficult it becomes to deliver the key being used to lock and unlock the message. Suppose a spymaster wanted to receive reports securely from a number of different agents in the field, but didn't want them to be able to read one another's messages. Different keys would need to be delivered to each agent. Now replace a few agents with millions of eager Internet shoppers. An operation on this scale, although not impossible, is a logistical nightmare. For a start, a customer visiting a website couldn't place an order immediately but would have to wait for the delivery of a secure encryption key. The World Wide Web then really would be the World Wide Wait.

The system known as public-key cryptography is like a door with two different keys: key A locks the door, but a different key, B, opens it. Suddenly there is no need for any secrecy surrounding key A. Being in possession of this key does not compromise security. Now, imagine this door at the entrance to the secure part of a company's website. The company can freely distribute key A to any visitor to its website who wants to send a secure message, such as the number of their credit card. Even though everyone is using the same key to encode their data – to lock the door and secure their secret – no one can read anyone else's encoded message. In fact, once data is encoded, customers are unable to read it, even if it is their own. Only the company running the website has key B, to unlock the door and read those credit card numbers.

Public-key cryptography was first openly proposed in 1976 in a seminal paper by two mathematicians based at Stanford University in California, Whit Diffie and Martin Hellman. The duo sparked a counter-culture in the cryptographic world that would challenge the governmental agencies' monopoly on cryptography. Diffie in particular was the archetypal anti-establishment, long-haired child of the 1960s. Both were keen that cryptography should not remain a subject discussed solely behind government doors, and that their ideas should be made public, for the benefit of individuals. It transpired much later that such a system had been put forward by a number of government security agencies, but rather than finding its way into a published journal the proposal was stamped TOP SECRET and hidden away.

The Stanford group's paper, entitled 'New directions in cryptography', heralded a new era in encryption and electronic security. Public-key encryption, with its two keys, sounded great in theory. But was it possible to put theory into practice and create a code that worked this way? After several years of trying, some cryptographers began to doubt that such a lock could be made. They feared this academic lock just wouldn't be able to hack it in the real world of espionage.

RSA, the MIT trio

One of the many who were inspired by Diffie and Hellman's paper was Ron Rivest at the Massachusetts Institute of Technology. Rivest, in contrast to the rebellious style of Diffie and Hellman, is a man of convention. He is reserved and softly spoken, and has a measured response to the world around him. At the time he read 'New directions in cryptography', his ambition was to become part of the academic establishment. He was dreaming of professorships and theorems rather than spies and codes. He had no idea that reading this paper would send him on a journey that would lead to one of the most powerful and commercially successful cryptographic systems ever created.

Rivest had joined the department of computer science at MIT in 1974, after doing research in Stanford and Paris. Like Turing he was interested in the interaction between abstract theory and real machines. In Stanford he had spent some time making smart robots, but his mind was turning towards the more theoretical side of computer science.

In Turing's day the main question in computing, inspired by Hilbert's second and tenth problems, was whether, in theory, programs existed to solve certain types of problems. As Turing had shown, no program could ever decide which mathematical truths had proofs or not. By the 1970s a different theoretical question was all the rage in the academic world of computer science. Suppose a program did exist to solve a particular mathematical problem. Was it possible to analyse how fast that program would take to solve the problem? This was obviously important if the program was going to be implemented. The question demanded highly theoretical analysis but was nevertheless rooted in the real world. It was this combination that made the challenge perfect for Rivest. He left his robots at Stanford and moved to MIT to pursue the burgeoning subject of computational complexity.

'One day a graduate student handed me this article and said, "You might be interested in this",' Rivest recalls. It was Diffie and Hellman's paper, and immediately he was captivated. 'It laid out this broad vision of what cryptography was and what it might be. If only you could come up with an idea.' The challenge of the paper brought together all of Rivest's interests: computing, logic and mathematics. Here was a problem which clearly had practical implications for the real world, but which also had direct links with the theoretical concerns that were very much on Rivest's mind. 'What you care about in cryptography is distinguishing between the problems that are easy and the problems that are hard,' he explained. 'That was what computer science was all about.' If a code was going to be hard to break, it had to be based on a mathematical problem whose solution was difficult to calculate.

Rivest started his attempt to build a public-key cryptography by plundering the wealth of problems he knew computers would take a long time to solve. He also needed someone he could bounce ideas off. MIT was already beginning to break the mould of a traditional university, loosening the boundaries between departments in the hope of encouraging interdisciplinary interactions. Rivest, a computer scientist, worked on the same floor as members of the mathematics department. In offices nearby were two mathematicians, Leonard Adleman and Adi Shamir.

Adleman was a more gregarious character than Rivest but still a classic academic with wild and wonderful ideas about things that seemed to have nothing much to do with reality. Adleman recalls coming into Rivest's office one morning: 'Ron is sitting there with this manuscript. "Did you see this thing from Stanford about this crypto, secret code, scrambling, blah, blah, blah . . ." My reaction was, "Well, that's nice, Ron, but I have something important to talk about. I couldn't care less." But Ron got very interested in this.' What Adleman cared about was the abstract world of Gauss and Euler. Cracking Fermat's Last Theorem was what mattered to him, not some trendy subject like cryptography.

Rivest found more receptive ears down the corridor in the office of Adi Shamir, an Israeli mathematician visiting MIT. Together, Shamir and

Adi Shamir, Ron Rivest and Leonard Adleman.

Rivest began to search for some idea they could use to implement Diffie and Hellman's dream. Though Adleman wasn't too interested, it was hard to ignore Rivest and Shamir's obsession with this problem. 'Every time I'd go into their office they'd be talking about it. Most of the systems they came up with sucked, and since I was there I would just join in with their discussions to see if what they were proposing today made sense.'

As they explored the range of 'hard' mathematical problems, their embryonic crypto-systems began to use more ideas from number theory. This was right up Adleman's street: 'Since that was my area of expertise, I could be more useful in analysing their systems – and mostly dispensing with them.' He thought he'd met his match when Rivest and Shamir proposed a very secure-looking system. But after a sleepless night spent working through all the number theory he knew, he could see how to crack their latest code. 'This went on and on. We would go for a ski trip, and on the ride up that's what we would talk about . . . Even as we headed to the top of the ski slopes in a gondola, that's what we would talk about . . .'

The breakthrough came one evening when all three had been invited to dine at a graduate's house to celebrate the first night of Passover. Adleman doesn't drink, but he remembers Rivest knocking back the Seder wine.

Adleman got home at midnight. Soon after, the phone rang. It was Rivest. 'I've got another idea . . .' Adleman listened carefully. 'Ron, I think this time you've got it. This one sounds right to me.' They had been thinking for a while about the difficult problem of factorising numbers. There had been no clever proposals for programs which could crack numbers into their prime building blocks. This problem had the right flavour to it. Under the influence of the Seder wine, Rivest had seen how to program this problem into his new code. Rivest recalls, 'It had a good feel to it at first. But we knew from experience that things that have a good feel initially can still fall apart. So I put it aside till the morning.'

When Adleman arrived at the department in MIT late next morning, Rivest greeted him with a handwritten manuscript with the names Adleman, Rivest and Shamir blazoned across the top. As Adleman read through it he recognised what Rivest had told him the previous night on the telephone. 'So I tell Ron, "Take my name off this, this is your stuff," and we proceed to have a fight about whether I should stay on the paper.' Adleman agreed to think about it. At the time he thought it probably didn't matter either way, as the paper would probably be the least read of all his publications. But he did remember the early crypto-system that had kept him up all night. He had saved them from rushing into print with an insecure code which would have left them with egg on their faces. 'So I went back to Ron. "Make me third on the list." So that's how it became RSA.'

Rivest decided they had better find out how difficult factorising really was. 'The problem of factorising was an obscure art form at that time. There was little literature on it. It was difficult to get good estimates on the time that the algorithms that had already been proposed would take.' Someone who knew more than most happened to be Martin Gardner, one of the world's great popularisers of mathematics. Gardner was intrigued by Rivest's proposal and asked if it would be all right to run an article about the idea in his regular column for *Scientific American*.

The reaction to Gardner's article finally convinced Adleman that they were on to something big:

> That summer I was out at Berkeley in a bookstore. A customer and the man behind the desk are discussing something and the customer's saying, 'Did you see that crypto thing in *Scientific American*?' So I go, 'Hey, I was involved in that thing.' And the guy turns to me and says, 'Can I get your autograph?' How many times do we get asked for autographs? Zero. Wow, what is this . . . maybe something's going on here!

Gardner had said in his article that the three mathematicians would send a preprint of their paper to anyone who sent them a stamped addressed envelope. 'When I get back to MIT there are thousands, literally thousands, of these things from all over the world, including Bulgarian National Security blah, blah, blah.'

People started telling the trio that they were going to be rich. Even in the 1970s, when e-commerce was hardly dreamt of, people understood the potency of these ideas. Adleman thought the money would start to pour in within a few months, and went straight out and got himself a little red sports car to celebrate. Bombieri was not the only mathematician for whom the reward for mathematical success was a fast car.

Adleman's car was eventually paid off with instalments from his regular income from MIT. It took a little time for security agencies and business to fully appreciate the security and the power of RSA. While Adleman was speeding round in his sports car still thinking of Fermat, Rivest was the one whose head was tuned to the real-world implications of their proposal:

> We thought there might be some business potential to the scheme. We went through the MIT patent office, and then we tried to see if there might be some company that would be interested in marketing the product. But there was really no market in the early eighties. There was really very little interest at that stage. The world was not well networked. People didn't have computers on their desks.

The people who were interested were, of course, the security agencies. 'The security agencies became very concerned about the development of all this technology,' says Rivest. 'They did their best to see that our proposal wouldn't go very fast.' It seems that the same idea had been suggested behind the closed doors of the world of intelligence. But the security agencies weren't too sure whether to place their agents' lives in the hands of a few mathematicians who thought that cracking numbers was difficult. Ansgar Heuser of the BSI, the German National Security Agency, recalls how in the 1980s they considered using RSA in the field. They asked the mathematicians whether the West was stronger than the Russians at number theory. When the answer came back 'No', the idea was shelved. But in the following decade RSA proved its worth not just for protecting the lives of spies but also in the public world of business.

A cryptographic card trick

RSA cryptography is now what safeguards most of the transactions on the Internet. Remarkably, the mathematics that goes into making possible such a system of public-key cryptography harks back to the clock calculators of Gauss and to a theorem proved by one of Adleman's heroes, Pierre de Fermat, known as Fermat's Little Theorem.

Addition on Gauss's clock calculators is something we are all familiar with when working with the time on a 12-hour clock. We know that 4 hours after 9 o'clock it will be 1 o'clock. This is the principle of addition on the clock calculator: add the numbers together and work out the remainder after division by 12. We write this, as Gauss did some two hundred years ago, as

$$4 + 9 = 1 \text{ (modulo 12)}$$

Multiplication or raising a number to a power on Gauss's clock calculator works in a similar way: calculate the answer on a conventional calculator, and take the remainder on division by 12.

Gauss had realised that one didn't need to stick to clocks with the conventional 12 hours on them. Even before Gauss had explicitly formulated the concept of his clock arithmetic, Fermat had made a fundamental discovery, his so-called Little Theorem, about a clock calculator with a *prime* number of hours, say p, on it. If one takes a number on this calculator and raises it to the power p, one always gets the number one started with. For example, on a clock with 5 hours, multiply 2 together 5 times gives you 32, which indeed is 2 o'clock on the calculator with 5 hours. What seemed to be happening was that each time Fermat multiplied by 2, the arm on the clock would map out a pattern. After 5 moves the arm had returned to its original position ready to repeat the pattern.

Power of 2	2^1	2^2	2^3	2^4	2^5	2^6	2^7	2^8	2^9	2^{10}
On a conventional calculator	2	4	8	16	32	64	128	256	512	1,024
On a 5-hour clock calculator	2	4	3	1	2	4	3	1	2	4

If we take a clock with 13 hours, and go through the process with powers of 3 from 3^1, 3^2, . . . up to 3^{13}, we get

3, 9, 1, 3, 9, 1, 3, 9, 1, 3, 9, 1, 3

This time the hand doesn't visit all the hours on the clock face, but it still makes a repeating pattern which brings it back to 3 o'clock after multiplying 3 together 13 times. Whatever value Fermat chose for the prime p, the same magic seemed to happen. Written in Gauss's notation for clock or modular arithmetic, Fermat had discovered that for any prime number p and any time x on the clock with p hours,

$$x^p = x \text{ (modulo } p)$$

Fermat's discovery is the sort of thing that gets a mathematician's blood racing. What is it about prime numbers that makes this kind of magic? Not content with experimental observation, Fermat wanted a proof that, whatever prime number clock he chose, the primes wouldn't let him down.

It was in a letter he wrote in 1640 to a friend, Bernard Frenicle de Bessy, rather than in the margins of a book that Fermat declared he had found a proof – but as with his Last Theorem, the proof was too long to set out in the space available. Despite promising to send it to Frenicle, he kept the proof a secret from the world. We had to wait another hundred years for it to be rediscovered. In 1736 Leonhard Euler found why the hand on Fermat's prime number clocks always returned to its starting point after the hour is multiplied by itself a prime number of times. Euler also managed to generalise Fermat's discovery to clocks with N hours, where $N = p \times q$ is the product of two primes p and q. Euler discovered that on such a clock the pattern would start repeating itself after $(p - 1) \times (q - 1) + 1$ steps.

It was Fermat's discovery of the magic of prime number clocks and Euler's generalisation that raced through Rivest's mind as he sat thinking late into the night after that fateful Seder dinner. Rivest began to see that he could use Fermat's Little Theorem as the key to a mathematical code that could make a credit card number disappear and then magically reappear.

Encrypting a credit card number is something like the beginning of a card trick. But this is no ordinary pack: the number of cards in this pack will be so huge that you need over a hundred digits to write it. The customer's credit card number is one of these playing cards. The customer places the credit card on top of the pack. The website shuffles the pack so that the location of the customer's card seems to have been completely lost. Any hacker is faced with the impossible task of extracting that single card from the scrambled pack. The website, however, knows a cunning trick. Thanks to Fermat's Little Theorem, the credit card can be made to reappear at the top of the pack after another sequence of shuffles. This second sequence of shuffles is the secret key known only to the company that owns the website.

The mathematics Rivest used to design this cryptographic trick is quite simple. The shuffling of the cards is done by a mathematical calculation. When a customer places an order at a website, the computer takes the customer's credit card number and performs a calculation on it. The calculation will be easy to perform but will be almost impossible to undo without knowledge of the secret key. That is because the calculation will be done not on a conventional calculator, but on one of Gauss's clock calculators.

The Internet company tells its customers when they place an order on its website how many hours to use on the clock calculator. It decides how many hours to choose by first taking two large prime numbers, p and q, of around 60 digits each. The company then multiplies the primes together to get a third number, $N = p \times q$. The number of hours on the clock will be huge, up to 120 digits long. Every customer will use the same clock to encode their credit card number. The security of the code means that the company can use the same clock for months before they need to consider changing the number of hours on the clock face.

Selecting the number of hours on the website's clock calculator is the first step in choosing a public key. Although the number N is made public, the two primes p and q are kept secret. They are two ingredients of the key that is used to unscramble the encrypted credit card number.

Next, every customer receives a second number, E, called the encoding number. This number E is the same for everyone and is as public as the number N of hours on the clock face. To encrypt their credit card number, C, the customer raises it to the power E on the website's public clock calculator. (Think of the number E as the number of times the magician shuffles the pack of cards to hide the one you've chosen.) The result, in Gauss's notation, is C^E (modulo N).

What makes this so secure? After all, any hacker can see the encrypted credit card number as it travels through cyberspace and can look up the company's public key, which consists of the N-hour clock calculator and the instruction to raise the credit card number to the power E. To crack this code all the hacker has to do is find a number which, when multiplied together E times on the N-hour clock calculator, gives the encrypted credit card number. But that is very difficult. An extra twist comes from the way powers are computed on a clock calculator. On a conventional calculator the answer grows in proportion to the number of times we multiply the credit card number together. That doesn't happen on the clock calculator.

There, you very quickly lose sight of the starting place because the size of the answer bears no relationship to where you start from. The hacker is completely lost after E shuffles of the pack of cards.

What if the hacker tries working through every possible hour on the clock calculator? No chance. Cryptographers are now using clocks on which N, the total number of hours, has over a hundred digits – in other words, there are more hours on the clock face than there are atoms in the universe. (In contrast, the encoding number E is usually quite small.) If it is impossible to solve this problem, how on earth does the Internet company recover the customer's credit card number?

Rivest knew that Fermat's Little Theorem guaranteed the existence of a magic decoding number, D. When the Internet company multiplies the encrypted credit card number together D times, the original credit card number reappears. The same idea is used by magicians in card tricks. After a certain number of shuffles it looks as though the card order is completely random, but the magician knows that a few more shuffles will bring the pack back to its original order. For example, the perfect shuffle – where the pack is equally divided, then interleaved one card at a time from each half – takes eight shuffles to bring the pack into its original position. Of course, the art of the magician is being able to perform a perfect shuffle eight times in a row. Fermat had discovered the analogue for clocks of how many perfect shuffles it takes to return a pack of 52 cards to its original order. It was Fermat's trick that Rivest had exploited to decode messages in RSA.

Although the pack of cards with your credit card number has been shuffled by the website a number of times to lose it, the Internet company knows that shuffling it another D times will make your card appear as if by mathematical magic at the top of the pile. But you can work out what D is only if you know the secret primes p and q. Rivest used the generalisation of Fermat's Little Theorem discovered by Euler which works on clock calculators built from two primes rather than just one. Euler showed that on these clocks the pattern repeats itself after $(p - 1) \times (q - 1) + 1$ shuffles. So the only way you can know how long it takes to see the repeating pattern on the clock with $N = p \times q$ hours is to know both the primes p and q. Knowledge of these two primes therefore becomes the key to unlocking the secrets of RSA. The number of shuffles required to recover the lost credit card is known only by the Internet company, which has kept the prime numbers p and q very secret.

Although the two numbers p and q have been kept secret, their product, $N = p \times q$, has been made very public. The security of Rivest's RSA code therefore relies on the difficult task of factorising *N*. A hacker is faced with the same problem that occupied Professor Cole at the beginning of the last century: find the two prime building blocks for the number *N*.

Throwing down the gauntlet of RSA 129

To persuade business that the problem of factorising had a respectable heritage, the MIT three would quote what one of the big guns, Gauss, had to say about factorising: 'The dignity of the science itself seems to require that every possible means be exploited for the solution of a problem so elegant and so celebrated.' Although Gauss acknowledged the importance of the factorisation problem, he made no headway with its solution. If Gauss had tried and failed to crack it, surely corporate security was safe in the hands of RSA.

Despite Gauss's 'endorsement' of the RSA system, before they wired it into their new code the problem of factorising large numbers lay on the margins of mathematics. Most mathematicians showed little interest in the nitty-gritty of cracking numbers. What if it did take the lifetime of the universe to find the prime building blocks of large numbers – surely that was of no theoretical importance. But with Rivest, Shamir and Adleman's discovery, the problem of factorising assumed a significance way beyond what it had in Cole's day.

So just how difficult is it to crack a number into its prime constituents? Cole had no access to electronic computers, so it would have taken him a good number of Sundays before he stumbled on 193,707,721 or 761,838,257,287 as one of the two building blocks of the Mersenne number $2^{67} - 1$. Armed with our computers, can't we just check one number after another until we find one that divides the number we are trying to crack? The trouble is that to crack a number with over a hundred digits entails checking more numbers than there are particles in the observable universe.

With so many numbers to check, Rivest, Shamir and Adleman felt confident enough to issue a challenge: to crack a number with 129 digits that they had built from two primes. The number, along with an encoded message, was published in Martin Gardner's *Scientific American* article that brought the code to the world's attention. They were not yet the millionaires they were to become, so they offered only $100 as a prize for unmasking the two primes used to build the number dubbed 'RSA 129'. In the article they estimated that it would take as many as 40 quadrillion years to crack RSA 129. They soon realised they'd made an arithmetical slip in estimating the time it would take. Nevertheless, given the techniques for cracking numbers into primes then available, it should still have taken thousands of years.

RSA seemed to be the code-maker's dream come true: an unbreakable code. With so many primes to check, confidence in the impregnability of the system seemed justified. But the Germans had thought that Enigma was invincible since it had more possible combinations than there are stars in the universe – but the Bletchley mathematicians had showed that one cannot always place one's faith in large numbers.

The gauntlet of RSA 129 was thrown down. Never ones to shirk a challenge, mathematicians around the world began beavering away. In the years that followed, they began to devise ever more cunning plans to find Rivest, Shamir and Adleman's two secret primes. Instead of 40 quadrillion years, as the MIT three had estimated, their number was eventually cracked in a paltry seventeen years. That is still long enough for a credit card encrypted with RSA 129 to be well out of date. Nevertheless, it begs the question how much longer it will be for a mathematician to emerge with ideas that bring seventeen years down to seventeen minutes.

New tricks on the block

The interaction between cryptography and mathematics introduced modern mathematicians to a new culture more akin to the experimental and practical sciences. It was a culture not experienced since the nineteenth-century German school had snatched the baton from the mathematicians of Revolutionary France. The French had regarded their subject as a practical tool, as a means to an end, whereas Wilhelm von Humboldt believed in the pursuit of knowledge for its own sake. Those theoreticians still steeped in the German tradition were quick to condemn the study of methods for factorising numbers as a 'pig in the rose garden', in Hendrik Lenstra's words. In contrast to the pursuit of watertight proofs, this truffling for primes was seen as minor work of little mathematical importance. But as RSA grew commercially more important, it became impossible to ignore the practical implications of finding an efficient technique for unveiling the prime building blocks behind large numbers. Gradually, more mathematicians were drawn into the challenge of cracking RSA 129. The final breakthrough came about not so much from the development of faster computers as from unexpected theoretical advances. The new problems that sprang out of these forays into code-breaking have led to the development of some deep and difficult mathematics.

One of the mathematicians attracted to this emerging subject was Carl Pomerance. Pomerance is happy to split his time between the academic corridors of the University of Georgia and the commercial environment of Bell Laboratories in Murray Hill, New Jersey. As a mathematician he has never lost that childish love of playing with numbers and looking for new connections between them. He came to Paul Erdős's attention when the Hungarian read an intriguing article by Pomerance on the numerology of baseball scores. Stimulated by a curious question raised in the article, Erdős descended on Pomerance in Georgia to begin a collaboration that would produce over twenty joint papers.

Factorising numbers had fascinated Pomerance ever since he had been asked to factorise the number 8,051 in a high-school mathematics competition. There was a time limit of five minutes, and in the 1960s pocket calculators didn't exist. Although Pomerance was excellent at mental arithmetic, he decided first to look for a quick route to the solution rather than just testing one number after another. 'I spent a couple of minutes looking for the clever way, but grew worried that I was wasting too much time. I then belatedly started trial division, but I *had* wasted too much time, and I missed the problem.'

His failure to crack 8,051 fuelled Pomerance's lifelong quest for a fast way to factorise numbers. Eventually he learnt about the trick his schoolteacher had had in mind. Before 1977, the smartest way to crack numbers still, amazingly, belonged to the man whose Little Theorem was the catalyst for the invention of RSA's prime number code. Fermat's Factorisation Method is a fast way to factorise special choices of numbers by exploiting some simple algebra. Using Fermat's method, Pomerance took just seconds to crack 8,051 into 83×97. Fermat, who loved the idea of secret codes, would probably have been delighted to find his work at the heart of making and breaking codes some three centuries later.

When Pomerance heard of Rivest, Shamir and Adleman's challenge, he knew immediately that cracking the 129-digit number was the way to exorcise the memories of his childhood failure. In the early 1980s it dawned on him that there was a way to exploit Fermat's Factorisation Method. By implementing the method on a variety of different clock calculators, it could provide a powerful factorisation machine. Now it was no longer just the outcome of a high-school mathematics competition that was at stake. This new discovery, called the *quadratic sieve*, had very serious implications for the emerging world of Internet security.

Pomerance's quadratic sieve works by using Fermat's Factorisation

Method but continually changing the clock calculator being used to try to crack a number. The method is similar to the Sieve of Eratosthenes, the technique invented by the Alexandrian librarian, which sifts out primes by taking the primes in turn and then striking out all numbers which are multiples of that prime. Thus, by dropping numbers through different-sized sieves, non-primes are eliminated without having to consider them individually. In Pomerance's attack the prime sieves are replaced by varying the number of hours on the clock calculators. The calculations performed on each separate clock calculator provided Pomerance with more information about possible factors. The more clocks that could be used, the closer he could get to cracking a number into its prime constituents.

The ultimate test of this idea was to set it to work on the challenge of RSA 129. But in the 1980s this number still looked well out of reach of Pomerance's factorisation machine. In the early 1990s help arrived in the shape of the Internet. Two mathematicians, Arjen Lenstra and Mark Manasse, realised that the Internet would be the perfect ally for the quadratic sieve in an attack on RSA 129. The beauty of Pomerance's method was that the workload could be spread over many different computers. The Internet had been employed to find Mersenne primes by assigning different tasks to different desktops. Manasse and Lenstra realised that they could now use the Internet to co-ordinate an attack on RSA 129. Each computer could be assigned different clocks to sieve with. Suddenly, the Internet – which was meant to be protected by these codes – was being asked to help crack the RSA 129 challenge.

Lenstra and Manasse put Pomerance's quadratic sieve on the Internet and called for volunteers. In April 1994 came the announcement that RSA 129 had crumbled. By combining several hundred desktop computers across twenty-four countries, RSA 129 was cracked after eight months of real time in a project led by Derek Atkins at MIT, Michael Graff at Iowa State University, Paul Leyland of Oxford University and Arjen Lenstra. Even two fax machines had joined in the search – while they weren't handling messages, they were helping to look for the two prime numbers with 65 and 64 digits. The project used 524,339 different prime clocks.

In the late 1990s Rivest, Shamir and Adleman issued a new set of challenges. By the end of 2002 the smallest of their challenges to remain uncracked was a number with 160 digits. The trio's finances have improved since 1977, so you can now win $10,000 for cracking an RSA challenge number. Rivest threw away the primes he used to build these challenge numbers, so no one actually knows the answers until the numbers are cracked. RSA Security regards $10,000 as a small price to pay to keep ahead of the current band of number-crackers. And each time a new record is set, RSA simply advises businesses to increase the size of the primes.

Pomerance's quadratic sieve has been superseded by a new sieve called the number field sieve. This new sieve is responsible for the current record, set by cracking RSA 155. This was achieved by a network of mathematicians operating under the messianic name of Kabalah. RSA 155 was a significant psychological breakthrough. In the mid-1980s, when security agencies were still toying with the idea of using RSA, computer security with this level of complexity had been considered sufficient. As Ansgar Heuser of the BSI, the German security agency, admitted at a cryptography conference in Essen, had they gone ahead 'we would be in the middle of a disaster'. RSA Security is now recommending clocks for which *N*, the number of hours, has at least 230 digits. But the likes of the BSI, who need long-term security to protect their agents, are currently recommending clocks with over 600 digits.

Head in the sand

The number field sieve makes a brief appearance in the Hollywood film *Sneakers*. Robert Redford sits listening to a young mathematician lecturing about cracking very big numbers: 'The number field sieve is the best method currently available. There exists an intriguing possibility for a far more elegant approach . . . But maybe, just maybe, there is a short cut . . .' Sure enough, this whizz-kid, played by Donal Logue, has discovered such a method, 'a breakthrough of Gaussian proportions', and has wired it into a small box which unsurprisingly falls into the evil hands of the film's villain, played by Ben Kingsley. The plot is so outlandish that most viewers must imagine that this could never happen in the real world. Yet, as the credits for the film roll, up pops 'Mathematical Advisor: Len Adleman', the A in RSA. As Adleman admits, this is not a scenario that we can guarantee won't happen. Larry Lascar, who wrote *Sneakers*, *Awakening* and *War Games*, came to Adleman to make sure he got the maths right. 'I liked Larry and his desire for verisimilitude, so I agreed. Larry offered money, but I countered with Robert Redford – I would do the scene if my wife Lori could meet Redford.'

How prepared are businesses for such an academic breakthrough? Some more so than others, but on the whole most have their head in the sand. If you ask business and government security agencies, their answers are a little worrying. These are all comments I've recorded on the cryptographic circuit:

> 'We met the government standards, that's all we're worried about.'
> 'If we go down then at least a lot of other people will be going down with us.'
> 'Hopefully I will have retired anyway by the time such a mathematical breakthrough will have happened, so it won't be my problem.'
> 'We work on the principle of hope – for the time being nobody expects a dramatic breakthrough.'
> 'Nobody is able to give guarantees. We simply don't expect it.'

When I give presentations to businesses about Internet security I like to offer my own little RSA challenge: a bottle of champagne for the first person to uncover two prime numbers whose product is 126,619. The difference in the response to this challenge I gave in three banking seminars in different corners of the globe revealed the interesting cultural differences in the financial world's attitude to security. In Venice the challenge and the mathematics behind these codes washed straight over the heads of the European bankers, and I resorted to a plant in the audience to offer the solution. In contrast to the bankers of Europe, trained in the humanities as most of them are, the Far Eastern banking community has a far greater scientific pedigree. By the end of the presentation, in Bali, one man rose to his feet with the two primes and claimed the champagne. They showed that they appreciated the mathematics and its implementation in e-business much better than their European counterparts.

But the presentation to the Americas gave me the most striking insight. Within fifteen minutes of returning to my room after the presentation I received three phone calls with correct solutions. Two of the US bankers had logged on to the Internet, downloaded code-cracking programs and run 126,619 through them. The third was very cagey about his method, and he was strongly suspected of having eavesdropped on the other two.

Business has put its trust in a piece of mathematics that few have taken the time to examine for themselves. True, the immediate threat to the security of everyday Internet business is more likely to come from sloppy management leaving vital information unencrypted on a website. Like any cryptographic system, RSA is susceptible to human imperfection. During the Second World War the Allies benefited from a host of textbook errors made by German operators which helped them to crack Enigma. RSA can equally be weakened by operators choosing numbers which are too easily crackable. If you want to break codes, buying up second-hand computers is probably a better investment than enlisting for a Ph.D. in a pure mathematics department. The amount of sensitive information left on outmoded machines is frightening. Offering a simple bribe to someone guarding secret keys could get you far more for your money than sponsoring a team of mathematicians to crack numbers. As Bruce Schneier comments in his book *Applied Cryptography*, 'It's a whole lot easier to find flaws in people than it is to find them in crypto-systems.'

However, such security breaches, though serious for the company involved, pose no threat to the whole fabric of Internet business. This is what gives a film like *Sneakers* an edge. Although the probability of a breakthrough in cracking numbers is small, the risk is still there and the result would be globally devastating. It could become the Y2K of e-business, bringing the whole house of emails crumbling to the ground. We *think* that cracking numbers is inherently difficult, but we can't prove it. It would be a weight off a lot of executives' minds if we could assure them that it is impossible to find a fast program that can factorise numbers. Obviously it is difficult to prove that such a thing does not exist.

Cracking numbers is a complex task not because the mathematics is particularly difficult, but because there is such a huge haystack from which to pluck the two needles. There are many other problems related to this 'haystack' quality. For example, although every map can be coloured with four colours, how can you tell, given a particular map, whether you can actually get away with just using three colours? The only way to tell would seem to be by laboriously working through all possible combinations until, with luck, you hit upon a map that requires only three colours.

One of Landon T. Clay's Millennium Problems, known as *P versus NP*, raises a rather interesting question about such problems. If the complexity of a problem such as factorising numbers or colouring maps arises from the very large size of the haystack one has to search through, might there always be an efficient method to find the needle? Our hunch is that the answer to the P versus NP question is 'no'. There are problems which have an inherent complexity that can't be got around even with the hacking skills of a modern-day Gauss. If, however, the answer turns out to be 'yes', then as Rivest says, 'it would be a catastrophe for the cryptographic community'. Most cryptographic systems, RSA included, concern problems about large haystacks. A positive answer to this Millennium Problem would mean that there really is a fast way to crack numbers – it's just that we haven't found it yet!

Not too surprisingly, business is not overly concerned with mathematicians' obsession for building our mathematical edifice on foundations which are 100 per cent secure. Cracking numbers has remained difficult for the last few millennia, so the business world is happy to build the Internet shopping mall on a foundation which is 99.99 per cent secure. Most mathematicians think that there is something inherently computationally difficult about cracking numbers. But no one can predict what advances the next decades may bring. After all, RSA 129 looked secure some twenty years ago.

One of the main reasons why factorising numbers is so difficult is the randomness of the primes. Since the Riemann Hypothesis seeks to understand the source of the wild behaviour of the primes, a proof could provide new insights. In 1900 Hilbert had stressed in his description of the Riemann Hypothesis that its solution had the potential to unlock many other secrets about numbers. Given the central role of the Riemann Hypothesis in understanding the primes, mathematicians began to speculate that the proof, if it is ever found, might yield new ways to crack numbers. This is why businesses are now beginning to keep an eye on the abstruse world of prime number research. There is another reason why business is particularly interested in the Riemann Hypothesis. Before they can use the RSA code, Internet companies must first be able to find two prime numbers with sixty digits. If the Riemann Hypothesis is true, then there is a fast way to discover the primes used to build the RSA codes on which the security of e-business currently relies.

Hunting for big primes

Given the increasing pace of the Internet and the consequent demand for bigger and bigger prime numbers, Euclid's proof that the primes will never run dry suddenly takes on an unexpected commercial significance. If the primes are such an unruly bunch, how are businesses going to find these big prime numbers? There may be an infinite number of prime numbers, but as we count higher they get thinner on the ground. If there are fewer primes the further we count, then are there enough primes with around 60 digits for everyone in the world to have two of them in order to make their own private key? And even if there are enough, perhaps there are only *just* enough, in which case there is a high chance that two people will get the same pair.

Fortunately, Nature has been kind to the world of e-commerce. Gauss's Prime Number Theorem implies that the number of primes with 60 digits is approximately 10^{60} divided by the logarithm of 10^{60}. This means that there are enough primes with 60 digits for every atom in the Earth to have its own two primes. Not only that, the chance of you winning the National Lottery is higher than the likelihood of two different atoms being assigned the same pair of primes.

So, given that there are enough prime numbers to go round, how can we be sure that a number is prime? As we have seen, it is difficult enough to find the prime constituents of a non-prime number. If a candidate number is prime, then isn't it doubly difficult to discover this fact? After all, it means showing that no smaller number divides the candidate.

It turns out that to determine whether a number is prime isn't as tall an order as you might expect. There is a fast test to discover whether a number is not prime, even if you can't find any prime constituents. That is why Cole knew, as had the rest of the mathematical world for some twenty-seven years before he announced his calculation, that the number he was trying to crack wasn't prime. This test isn't much help in predicting the distribution of the primes, the heart of the Riemann Hypothesis. But because it tells us whether any particular number is prime, it lets us hear individual notes in the music, though it doesn't help us appreciate the overall melody encapsulated in the Riemann Hypothesis.

The origin of this test is Fermat's Little Theorem, which Rivest had exploited that night after the Passover wine when he had discovered RSA. Fermat found that if he took a number on a clock calculator with a prime number of hours, p, and raised it to the power p, he always got the number he started with. Euler realised that Fermat's Little Theorem could be used to prove that a number *isn't* prime. For example, on a clock with 6 hours, multiplying 2 together 6 times lands us at 4 o'clock. If 6 were prime, we would have arrived back at 2 again. So Fermat's Little Theorem tells us that 6 can't be a prime number – else it would be a counter-example to the theorem.

If we want to know whether a number p is prime, we take a clock calculator with p hours. We start testing different times to see whether raising the hour to the power of p gets us to the time we started from. Whenever it doesn't, we can throw the number out, confident that it is not a prime number. Each time we find an hour that does satisfy Fermat's test, we won't have proved that p is prime, but that hour on the clock is, if you like, bearing witness to p's claim to be prime.

Why is testing times on the clock any better than testing whether each number less than p divides p? The point is that if p fails the Fermat test, it fails it very badly. Over half the numbers on the clock face will fail this test and testify to the non-primality of p. That there are so many ways to prove that this number is not prime is therefore an important breakthrough. This method contrasts strongly with the step-by-step division test, checking off every number to see whether it is a factor of p. If p is the product of, say, two primes, then with the division test only those two primes can prove that p is not prime. None of the other numbers will be of any help. One has to get an exact hit for the division test to work.

In one of his multitude of collaborations, Erdős estimated (though did not rigorously prove) that to test whether a number less than 10^{150} is prime, finding just one time on the clock which passes Fermat's test already means that the odds of that number not being prime are as little as 1 in 10^{43}. The author of *The Book of Prime Number Records*, Paulo Ribenboim, points out that using this test, any business selling prime numbers could realistically peddle their wares under the banner 'satisfaction guaranteed or your money back', without too much fear of going bust.

Over the centuries mathematicians have refined Fermat's test. In the 1980s two mathematicians, Gary Miller and Michael Rabin, finally came up with a variation that would guarantee after just a few tests that a number is prime. But the Miller–Rabin test comes with a bit of mathematical small print: it works for really big numbers only if you can prove the Riemann Hypothesis. (To be precise, you need a slight generalisation of the Riemann Hypothesis.) This is probably one of the most important things that we know is hiding behind Mount Riemann. If you can prove the Riemann Hypothesis and its generalisation, then, as well as earning a million dollars, you will have guaranteed that the Miller–Rabin test really is a fast and efficient method for proving whether a number is prime or not.

In August 2002, three Indian mathematicians, Manindra Agrawal, Neeraj Kayal and Nitin Saxena, at the Indian Institute of Technology in Kanpur devised an alternative to the Miller–Rabin test. It was very slightly slower but avoided having to assume the Riemann Hypothesis. This came as a complete surprise to the prime number community. Within twenty-four hours of the announcement from Kanpur, 30,000 people across the world – Carl Pomerance among them – had downloaded the paper. The test was sufficiently straightforward for Pomerance to present the details to his colleagues in a seminar that same afternoon. He described their method as 'wonderfully elegant'. The spirit of Ramanujan still burns strong in India, and these three mathematicians were not afraid to challenge the received wisdom of how one should check whether a number is prime. Their story adds to the belief that one day some unknown mathematician will emerge with the idea that will finally solve the Riemann Hypothesis, the ultimate prime number problem.

It is amazing how kind Nature has been to the cryptographic community. She has provided a fast and easy way to produce the primes from which to build Internet cryptography, but has kept hidden from view any fast way to crack numbers into the primes from which they are built. But for how much longer will Nature be on the cryptographer's side?

The future's bright, the future's elliptic

The application of the theory of prime numbers to such a hard-core business problem has significantly raised the status of mathematics. When anyone questions the usefulness of funding such an esoteric area of research as number theory, pointing to the role of the primes in RSA has become a very powerful retort. Fields Medallist Timothy Gowers used this very example to justify the usefulness of mathematics in his talk 'The importance of mathematics' at the announcement of the Clay Millennium prizes.

In the days before the new cryptography, most mathematicians would have been hard pushed to come up with such a high-profile application of abstract mathematics, and one which so immediately captures people's attention. It has provided a lucky and timely break for the subject. You can almost guarantee that in any grant application for research in number theory the catchphrase 'and there may be cryptographic implications' will appear somewhere. In fairness, the mathematics behind RSA cryptography is not that deep. Most mathematicians would not compare the challenge of cracking numbers to the prospect of uncovering some long-standing mystery such as the Riemann Hypothesis.

Although solutions of the Riemann Hypothesis and the P versus NP problem might both have implications for RSA, it was another of the Millennium Problems that almost caused the Y2K of e-business. Early in 1999 rumours were rapidly circulating that something called the Birch–Swinnerton-Dyer Conjecture, a problem about things called elliptic curves, might expose the Achilles' heel of Internet security.

In January 1999, *The Times* ran a front-page article under the headline TEENAGER CRACKS EMAIL CODE. This achievement had earned the Irish teenager Sarah Flannery first prize in a science competition, but it promised far more lucrative riches. A picture showed her standing in front of an impressive blackboard of mathematics; the caption read, 'Sarah Flannery, 16, who baffled the judges with her grasp of cryptography. They described her work as "brilliant".' Given the Internet's dependence on 'email codes', this article was sure to capture the media and the public's attention. Further reading revealed that the 'cracking' referred to in the headline was not a new attack on the security of RSA, but the solution of a practical problem that plagues the implementation of RSA.

To encrypt and decrypt a credit card number using RSA, the number is multiplied together many times on a clock calculator whose number of hours runs into several hundred digits. Computations with such big numbers actually take quite a long time to do on a computer. Most websites ask you for more information than simply the details of your credit card, and they use RSA to decide on a private key that will be used by your computer and the website to encode all those details. Private keys, shared by the sender and the recipient, are much quicker at encrypting than the public RSA keys.

If you are shopping on the Internet from the comfort of your own home, using a personal computer with lots of memory and a fast processor, you won't even notice how long it takes to encrypt your credit card number. Increasingly, though, it is not just the comfort of home from which we access the Internet. Mobile phones, palmtops and other hand-held devices that will appear in the coming years are also capable of surfing the web. The so-called 3G (third-generation) technology provides the means for these machines to communicate with the Internet. But when it comes to encrypting a credit card number on a palmtop after a morning of Internet shopping, the power of the little hand-held computer is pushed to the extreme.

Mobile phones and palmtops are not built for performing big calculations. They have much less memory and slower processors than the computer sitting on your desk does. Not only that, but the bandwidth on which mobile devices transmit information is much narrower than is available over telephone lines or cables. It is therefore important to keep the amount of data that is broadcast to a minimum. The increasingly large numbers that RSA needs if it is to stay ahead of the ever-faster computers being used to crack codes makes it unsuited to the limited capabilities of mobile devices.

For some time, cryptographers have been after a new public-key cryptographic system with all the security and capability of RSA, but smaller and faster. In 1999, *The Times* and the rest of the media leapt at the possibility that sixteen-year-old Sarah Flannery had come up with just such a system. Flannery's code was much faster, but within six months of its announcement someone spotted a weakness which made the code insecure. The story is a salutary warning to the business world, some of whom had hoped to cash in on Flannery's new code. To her credit, she had never claimed that it was secure. Security can be demonstrated only by time and testing – neither of which the media appreciates too well. In the end, what had speeded up the code revealed too much of the hidden hand.

There has been a rival to RSA that is beginning to meet the challenges of the world of mobile, wireless communication, or m-commerce. It's not prime numbers behind these new codes, but something more exotic: *elliptic curves*. These curves are defined by special types of equations, and lay at the heart of Andrew Wiles's proof of Fermat's Last Theorem. They had already found their way into the cryptographic world as part of a new method to crack numbers into primes quickly. It seems to be an unwritten rule that the code-breakers pay back the code-makers with an even better code. Neal Koblitz of the University of Washington in Seattle was studying this method for cracking codes when he twigged that elliptic curves could equally well be used to make codes. Koblitz put forward his idea for elliptic curve cryptography in the mid-1980s. At the same time, Victor Miller at Ramapo College in New Jersey also discovered how to build a code from elliptic curves. Although they are more complicated than RSA, codes based on elliptic curves don't need such big numerical keys – and that's what makes them perfect for m-commerce.

Despite being sucked into the world of business by his creation of a cryptography suitable for mobile devices, Koblitz's heart is still in Hardy's world of pure number theory. One of the senior mathematicians on the number theory circuit, he retains his childhood enthusiasm for mathematics, which was sparked by a serendipitous sequence of events:

> When I was six years old, my family spent a year in Baroda, India. The math standards there were higher than in American schools. The next year, when I returned to the US, I was so far ahead of my classmates that my teacher erroneously believed that I had a special gift for mathematics. Like other mistaken notions that teachers get into their heads, this sort of erroneous belief has a way of becoming a self-fulfilling prophecy. As a result of all the encouragement after my return from India, I was on my way toward becoming a mathematician.

Koblitz's early years in India not only contributed to his mathematical development, but were also responsible for awakening his sense of the social injustice in the world. As an adult he has participated in mathematical missions to Vietnam and Central America. One of his numerous books on number theory and cryptography is dedicated 'to the memory of the students of Vietnam, Nicaragua, and El Salvador who lost their lives in the struggle against US aggression'. The profits from the book are used to supply books to the people of these three countries.

Closer to home, Koblitz resents the stranglehold that the NSA, the US Government's National Security Agency, has on his area of mathematics. It is now necessary to get clearance from the NSA before certain types of work on number theory can appear in print, even in the most esoteric of mathematical journals. Thanks to Koblitz's new ideas, elliptic curves have joined the primes on the 'restricted list' of research the authorities want to keep an eye on.

Rivest, Shamir and Adleman had used Gauss's clock calculators to scramble credit card numbers. Koblitz was now proposing to lose your credit card somewhere along these strange curves. Instead of multiplying hours on the clock, Koblitz wanted to exploit a strange multiplication that could be defined on points on these curves.

The joys of Chaldean poetry

At first, RSA felt quite threatened by the arrival of this new code on the block. It was a challenge to their monopoly on Internet cryptography. RSA's anxiety peaked in 1997, when they opened a website called ECC Central. Posted on the site were quotations from eminent mathematicians and cryptographers that cast doubt on claims that elliptic curves are secure. Some argued that the factorising of numbers has a longer heritage, dating back to Gauss, and if Gauss couldn't do it, then your security is assured. Others argued that the structure of elliptic curves is so rich that it would allow hackers to gain a foothold from which to attack the problem. The cryptography was too new for us to tell whether our current knowledge of elliptic curves would be sufficient to crack a code with such small key sizes. After all, Sarah Flannery's code only survived six months of scrutiny.

The RSA team also pointed out that if you are talking to bankers about what underpins the security of their billion-dollar transactions, explaining the problem of factorising numbers is not so difficult. But start writing $y^2 = x^3 + \ldots$, and very quickly eyes begin to glaze over. Certicom, the leading proponent of elliptic curve cryptography, counters this criticism, saying that by the end of the courses they run on financial security, bankers are happily playing with points on elliptic curves.

But what riled the elliptic curve camp most of all was a remark by Ron

Rivest, the 'R' of RSA: 'Trying to get an evaluation of the security of an elliptic curve crypto-system is a bit like trying to get an evaluation of some recently discovered Chaldean poetry.'

Neal Koblitz was lecturing on elliptic curves in Berkeley when the ECC Central website opened. He had never heard of Chaldean poetry, so he hurried across to the university library to look it up. There he found that the Chaldeans were an ancient Semitic people who ruled over southern Babylonia between 625 and 539 BC. 'Their poetry was really great stuff,' he says. So he got T-shirts made with a picture of an elliptic curve and I LOVE CHALDEAN POETRY emblazoned across the top, and handed them out at his lectures.

Elliptic curve cryptography has so far stood the test of time and has been enshrined in government standards. Mobile phones, palmtops and smart cards are happily implementing this new cryptography. Your credit card number is sped around these elliptic curves, covering its tracks as it goes. Although originally designed for small mobile devices, elliptic curve cryptography is even becoming the security of choice for larger systems. The BSI, the German security agency, is openly admitting that the lives of their agents have now been entrusted to the security of elliptic curves. Even our own lives will soon be placed in the hands of these curves, every time we fly. Elliptic curves are set to guard the security of air traffic control systems around the world. RSA subsequently closed its ECC Central website and is now doing its own research with the aim of implementing elliptic curve cryptography alongside its own RSA system.

Yet in the summer of 1998 the fears that the extra structure possessed by elliptic curves might be their cryptographic undoing came to haunt those who had invested in the security promised by them. A few months earlier, Neal Koblitz had stated that the Birch–Swinnerton-Dyer Conjecture, one of the greatest outstanding problems about elliptic curves, would surely never have implications for the use of elliptic curves in cryptography. But like Hardy's quote that number theory would never be useful, Koblitz's prediction came home to roost with a vengeance. Indeed, it may have been Koblitz's provocative statement that prompted Joseph Silverman of Brown University to propose an attack based on the heuristics of the Birch–Swinnerton-Dyer Conjecture.

The Birch–Swinnerton-Dyer Conjecture is one of the seven Millennium Problems. It proposes a way of establishing whether the equation of an elliptic curve has a finite or an infinite number of solutions. In 1960, two English mathematicians, Bryan Birch and Sir Peter Swinnerton-Dyer, conjectured that the answer lies hidden in an imaginary landscape like the one discovered by Riemann. Thanks to their conjecture, Birch and Swinnerton-Dyer are two names that (to mathematicians) are as inextricably linked as the names of Laurel and Hardy, although many have been tricked into believing that there are in fact three mathematicians behind the conjecture – Birch, Swinnerton and Dyer. Birch, with his rather bumbling manner, plays Stan Laurel to Swinnerton-Dyer's rather dour Oliver Hardy.

Riemann had discovered the wormhole that took you from the primes into the zeta landscape. Another Göttingen mathematician, Helmut Hasse, suggested that each elliptic curve had its own individual imaginary landscape. Hasse was rather a controversial figure in German mathematical history. During Hitler's destruction of the mathematics department in Göttingen, Hasse was appointed by the Nazis to take over the running of the department. Hasse's Nazi sympathies, combined with his mathematical abilities, made him a suitable candidate in the eyes of the authorities and the mathematicians in Germany who hoped to preserve Göttingen's tradition.

There are very mixed feelings about Hasse in the mathematical community. Few can forgive the political choices he made. He even wrote to the authorities, in 1937, asking for one of his Jewish ancestors to be deleted from the records so that he could join the Nazi Party. Carl Ludwig Siegel recalls returning from a trip in 1938 to find 'Hasse for the first time wearing Nazi-party insignia! It is incomprehensible to me how an intelligent and conscientious man can do such a thing.' Despite his politics, his mathematical insight proved much sounder. His name is immortalised in the Hasse zeta functions, which are responsible for building the landscapes that hold the secrets of finding solutions to the equations of elliptic curves.

Whereas Riemann was able to show how to build the complete landscape covering the map of imaginary numbers, Hasse could not do the same for these elliptic landscapes. For each elliptic curve, he could construct part of the associated landscape, but beyond a certain point he found himself up against a ridge running north–south which he did not have the techniques to pass. It was in fact Wiles's solution of Fermat's Last Theorem that eventually showed how to cross this boundary and map the rest of the landscape.

However, years before we even knew whether there was any landscape beyond this ridge, Birch and Swinnerton-Dyer were already making conjectures about what this hypothetical landscape might be able to tell us. They predicted that there should be one point in each landscape which held the secret to whether the particular elliptic curve that was being used to build the landscape had infinitely many solutions or not. The trick was to measure the height of the landscape at the point above the number 1. If the landscape is actually at sea level here, then the elliptic curve would have infinitely many fractional solutions. If, on the other hand, the landscape was not at sea level, there must be finitely many fractional solutions. If the Birch–Swinnerton-Dyer Conjecture is true, and this point in each landscape really does hold the secret to finding solutions on elliptic curves, then it is another striking example of the power of these imaginary landscapes.

Although Birch and Swinnerton-Dyer were motivated by theoretical considerations, their conjecture was very much a result of experimenting with particular elliptic curves. Birch recalls the eureka moment when everything fell into place. He was playing around with the numbers that came out of his calculations. 'It was whilst I was staying in a lovely hotel in the Black Forest in Germany. I plotted the numbers I got, and lo and behold there were a dozen dots arranged in four parallel lines . . . Wonderful!' The pattern of those lines implied that there was some strong relationship that was forcing the dots to line up. 'From that point on it was absolutely clear that there was something there. I went back to Peter and said, "Oh, look at this!"' And as if this were another fine mess that Birch had got them into, 'Peter replied "I told you so" – as he does.'

There has been significant progress on this conjecture since it was proposed in the 1960s. Both Wiles and Zagier have made important contributions, but there is still a long way to go. The importance of the conjecture was marked by its selection as one of the Millennium Problems. It is the only Millennium Problem on which there has been continuous work towards a solution. Birch, however, believes it will be a long time before anyone will claim Clay's prize. Nonetheless, the Birch–Swinnerton-Dyer Conjecture almost became the passport, not to Clay's million, but to the millions of dollars that rely on the security of the Internet codes.

The codes that depend on elliptic curves rely on the difficulty of finding solutions to certain arithmetic problems. Joseph Silverman saw that the heuristics of the Birch–Swinnerton-Dyer Conjecture might provide him with a way to twist the cryptographic problem to reveal clues as to where to look for the solutions. It was certainly a long shot, and he admits himself that he doubted whether it would be a very efficient means of attack. But none of the experts could easily rule out the possibility that it might turn out to be one of the fast programs that the hackers were after.

Silverman could have gone public with his proposed attack; the media would have gone into a frenzy; RSA would be gloating; Certicom's stock would have collapsed; and elliptic curves would never have recovered from the perceived insecurity even if the attack was repulsed. Instead, Silverman chose to proceed in a more scholarly fashion. He emailed Koblitz with his proposal, three weeks before the conference where he was to deliver a paper on his ideas.

Koblitz was due to fly off to Waterloo in Canada, the home of Certicom, at the end of the week. The directors of Certicom were very quickly faxing him, hoping for a quick fix or an explanation of why the attack would fail. 'At the beginning, I couldn't see any reason why Silverman's proposal couldn't work.' Koblitz likes to rise early on the day of a flight, and he knew he had to do something to comfort his friends in Waterloo. By the time he boarded the plane he had convinced himself that if Silverman's attack was successful, it could be used to bring down RSA too. So if they were going down, RSA was coming with them.

'It was a terrifying moment,' recalls Koblitz. 'I emailed Silverman to say that it's at times like this that one is glad to be a mathematician and not a businessman. You begin to realise that life is far more exciting than the movies.' Silverman, however, may not have been too upset to see RSA go down as well. He was part of a team developing a new cryptography which goes under the name of NTRU. They are a little cagey about revealing what NTRU stands for, but it is generally believed to be Number Theorists 'R' Us. Unlike the other codes, theirs would be unaffected by his attack. A nice turn of events for the stocks of NTRU.

Within two weeks Koblitz had identified enough of the special structure of elliptic curves to show that Silverman's proposal was still computationally unfeasible. Elliptic curve cryptography was saved by a technicality called the height function, something Koblitz now refers to as the 'golden shield'. It appears to protect the codes not just from Silverman's approach but from a battery of other attacks. Whilst there was an initial panic, a scholarly calm has returned to the proceedings and Koblitz still enjoys delivering his lecture on this whole saga, entitled 'How pure mathematics almost brought down e-business'. The story highlights how advances in the most obscure or abstract corners of the mathematical world now have the potential to bring business to its knees.

It is exactly for reasons such as this that the likes of AT&T and national security agencies are keeping a watchful eye on Hardy's 'clean and gentle world' of number theory. During the 1980s and 1990s, AT&T director Andrew Odlyzko started to point the company's supercomputers at regions of Riemann's landscape that had never been contemplated before. You may ask what the point of these calculations is. If you are not expecting to find a counter-example to the Riemann Hypothesis, then why spend so much energy and so much of AT&T's money on calculating zeros? Odlyzko's interest had been piqued by hearing of some strange theoretical predictions made by the American mathematician Hugh Montgomery about the zeros far up Riemann's ley line. Odlyzko recognised that if the predictions were correct, then one of the strangest and most unexpected twists in the story of the primes was about to unravel.

CHAPTER ELEVEN

From Orderly Zeros to Quantum Chaos

The real voyage of discovery consists not in seeking new landscapes but in having new eyes. Marcel Proust, *Remembrance of Things Past*

How do the points at sea level in the zeta landscape arrange themselves along Riemann's magic ley line? It seemed a crazy question, but Hugh Montgomery hadn't intended to ask it. Most thought it foolhardy to ask such a question when no one could prove that they were really on this line in the first place. Yet the surprising patterns Montgomery discovered once he had asked the question are the best evidence to date for where we should be looking for a solution to the Riemann Hypothesis. Montgomery was asking himself this question in the first place because it helped him to understand a completely unrelated question, one that had intrigued him as a graduate student. He'd been wandering around in a seemingly unconnected part of the mathematical world, trying to make his mark when, like Alice, he tumbled unsuspectingly through a secret passage and found himself in a mysterious landscape that turned out to be none other than Riemann's.

In contrast to the cohort of sandal-wearing mathematicians clad in T-shirt and jeans, Montgomery is a dapper dresser, invariably turned out in suit and tie. His dress reflects the reserve and control that he brings to his life as a mathematician. Although originally from America, he chose to do his doctorate in Cambridge, England, where he warmed to the pomp of college living. Montgomery had blossomed as a young mathematician thanks to a 1960s educational experiment in teaching mathematics to schoolchildren. The aim was not to teach an accepted canon with no explanation of how mathematicians had arrived at their discoveries, but to capture the true spirit of the practising mathematician. Montgomery and his contemporaries were told the basic axioms and encouraged to deduce the canon for themselves. Equipped only with the rules of deduction, they were left to reconstruct the mathematical edifice on their own, rather than just be shown the monument as if they were tourists. This gave Montgomery his start:

> I was really lucky because it turned me on to mathematics. I understood even at high school what it is to be a mathematician. The trouble with the

course was that you had to retrain all the mathematics teachers to be able to teach it. I was lucky to be taught by one of the creators of the system. Although it was only taught to relatively few students it actually produced a surprising number of professional mathematicians.

At school, Montgomery particularly enjoyed exploring properties of numbers, especially the primes. But he also found how little was known about the primes. Were there infinitely many twin primes, such as 17 and 19, or 1,000,037 and 1,000,039? Was every even number the sum of two primes, as Goldbach had conjectured? Montgomery had to wait until he was a graduate student in Cambridge before hearing about the greatest prime problem of all: the Riemann Hypothesis. However, it was another problem that seized his attention when he first fell under the spell of Cambridge's great mathematical tradition.

Montgomery arrived at Cambridge in the late 1960s to a party atmosphere. The mathematics department was celebrating a breakthrough on a problem that had been posed by the great Gauss. Alan Baker, a Fellow of Trinity College, had made significant progress on the difficult question of how well imaginary numbers factorised. This was a problem Gauss had written about extensively in his *Disquisitiones Arithmeticae*. For an ordinary number, say 140, there is a single set of prime building blocks, in this case 2, 2, 5 and 7. There isn't another choice of primes we can multiply together to get 140. Imaginary numbers, though, are not so well behaved. Gauss was quite shocked to find that there was sometimes more than one way to construct an imaginary number from prime building blocks.

Montgomery was keen to tap into the excitement that Baker's solution of one of Gauss's problems had generated. He thought he could make his mathematical mark by extending Baker's ideas to some of the other problems that Gauss had posed. It would be difficult to push Baker's contribution further, but Montgomery was undaunted. He began reading widely, taking in as much number theory as he could. He couldn't have asked for a better environment. Cambridge, its long tradition strengthened by Hardy and Littlewood, was a great place to lap up new ideas. He learnt that Hardy and Littlewood had made some fascinating conjectures about the frequency of twin primes, which had so intrigued him at school.

He also learnt about the rather disconcerting theorems of Gödel. At school, Montgomery had discovered how the mathematical edifice was built by deducing theorems from an accepted set of axioms. According to Gödel, though, for some problems that technique wouldn't work. Apparently there would always be conjectures about numbers that could never be proved from the axioms Montgomery had spent his school years learning. What if one of the problems about primes he was intending to tackle turned out to have no proof? He might spend his life chasing a shadow.

To broaden his horizons beyond the spires and quadrangles of Cambridge, Montgomery decided to spend a year at the Institute for Advanced Study in Princeton. There, he got his chance to express his worries about trying to prove the unprovable. By custom, every visitor to the Institute, however junior or senior, is invited to lunch with the Director. When the Director asked him what he was working on, Montgomery said that he'd been interested in the Twin Primes Conjecture for quite a while, but had to admit that Gödel's theorems perturbed him. The Director unnerved the young Montgomery with his reply, 'Well, why don't we ask Gödel?', and Gödel was duly ushered over to give his opinion. Sadly for Montgomery, Gödel couldn't reassure him that something like the Twin Primes Conjecture would be provable from the current axioms of number theory.

Gödel himself had voiced such concerns in relation to the Riemann Hypothesis: perhaps the axioms that formed the foundations of the mathematical edifice were not broad enough to carry the required proof, in which case you might continue building upwards and never find a connection to the Hypothesis. However, he did offer some consolation. Gödel believed that any conjecture of genuine interest cannot be forever out of reach. It was just a matter of finding a new foundation stone to extend the base of the edifice. Only by going back to the subject's foundations and seeking to broaden them would you be able to build up to the missing proof. If the conjecture was something you really cared about – if the conjectured result was a natural extension of what had already been proved – then, thought Gödel, you would always be able to find a stone that fits equally naturally into the existing foundations, from which you could then prove your conjecture. Gödel had proved that this would always leave other conjectures unaccounted for, but the continuing evolution of the axiomatic foundations of mathematics held the prospect of capturing more and more of these unsolved problems.

Montgomery returned to Cambridge reassured that his dream of understanding the mysteries of the universe of numbers was not completely futile. He returned to Gauss's problem about factorising imaginary numbers. He knew from what he had read that the properties of Riemann's landscape were not unrelated to Gauss's endeavours. In particular, at the beginning of the twentieth century the Riemann Hypothesis played a rather paradoxical role in proving one of Gauss's conjectures about factorising imaginary numbers. This was the so-called Class Number Conjecture.

In 1916 a German mathematician, Erich Hecke, succeeded in proving that if the Riemann Hypothesis was true, then Gauss's Class Number Conjecture was also true. This was one of many conditional proofs that would emerge over the century that depended on having to climb Riemann's peak if the treasures it was hiding were to be claimed. None of them could truly be called 'proofs' until the Riemann Hypothesis was proved. The rather paradoxical twist Montgomery learnt about Gauss's Class Number Conjecture came a few years later. Three mathematicians, Max Deuring, Louis Mordell and Hans Heilbronn, succeeded in showing that if the Riemann Hypothesis was false, then this also could be used to prove that Gauss's conjecture about the factorisation of imaginary numbers was correct. Here was a 'no lose' situation. Either way, it meant that Gauss had been right about his hunch on factorisation. The unconditional proof of Gauss's Class Number Conjecture, which combined Hecke's proof with that of Deuring, Mordell and Heilbronn, was one of the strangest applications of the Riemann Hypothesis.

Montgomery now knew how important the Riemann zeros would be in addressing some of Gauss's unsolved problems about factorising imaginary numbers. He was sure he could make some headway with extending Baker's acclaimed work if he could show that zeros liked to bunch together along Riemann's ley line. His belief that one zero would be followed very quickly by another was inspired by the Twin Primes Conjecture, which had fascinated him for so long. Could he show that points at sea level can be very close together, like the twin primes we expect to see infinitely often? Close bunching of points at sea level would have important implications for the problem of factorising imaginary numbers. Would this be Montgomery's first scalp, the kind of prize that every graduate student dreams will make their mark in the cut-throat world of academe?

Montgomery was putting his money on the zeros being randomly scattered along Riemann's ley line, somehow reflecting the apparently random distribution of the primes along the number line. After all, if the primes looked as though they'd been chosen by the toss of a coin, it was a fair bet that the zeros of the zeta function would be randomly distributed too. Randomness invariably creates clusters which is why buses always come in threes and winning lottery numbers are often close to each other. Montgomery was hoping to find, within this random distribution, clusters of zeros close together. On his northward march along the critical line,

Montgomery expected to see a succession of batches of zeros, which he could then use to prove things about factorising imaginary numbers.

The trouble was, there was little evidence to go on. Not enough zeros had been calculated for any clustering of zeros to be directly visible, so Montgomery had to take a lateral approach. In the absence of experimental evidence, was there some aspect of the theory that predicted this bunching? What he did was to engineer an intriguing reversal of the role usually played by the zeros. The explicit formula that Riemann had discovered using the zeta landscape expressed a direct connection between primes and zeros. It was intended as a way to understand the primes by investigating zeros. What Montgomery did was to turn the equation on its head. He would use knowledge about the primes to deduce the behaviour of the zeros along Riemann's ley line. Montgomery remembered that Hardy and Littlewood had guessed how often twin primes should crop up as one counted through the primes. Perhaps he could extend this guess to the behaviour of the zeros. But when he fed Hardy and Littlewood's guess into Riemann's explicit formula, much to his surprise and disappointment, it forecast that the zeros would not bunch together at all.

Montgomery began to explore the forecast in more detail. It seemed to be saying that, as one ventured north along Riemann's line, zeros – unlike primes – would actually repel one another. Montgomery soon came to realise that zeros didn't like to be close together at all. Unlike the behaviour of primes, zeros would not sometimes be accompanied by several other zeros close behind them. In fact, Montgomery's prediction was suggesting that the zeros might well be arranging themselves in a completely uniform manner along Riemann's line, in contrast to the random distribution he had been expecting (see figure overleaf).

Montgomery looked for a way to describe his prediction of how the gaps between points at sea level behaved. He prepared what is known as a pair-correlation graph to show the expected range of separations between zeros (see figure on page 261). The curve was unlike any other that Montgomery had seen. It looked nothing like the sort of graph you would get if you were plotting differences in height, say, of a randomly selected group of people, a graph related to the classical Gaussian bell-shaped curve.

Montgomery's graph records how many pairs of zeros there should be for each possible separation of pairs. The early part of the graph shows that zeros don't like to be close together, because its height stays very low. Montgomery believed that to the right on the graph a wiggly shape should creep in, signifying an unusual and distinctive set of statistics. He couldn't

The gaps between random raindrops, primes and Riemann zeros.

prove that this was how the distances between zeros would pan out, nor did he have enough calculations of the positions of zeros to test experimentally whether his prediction was correct. His proposal for this strange graph was based purely on Hardy and Littlewood's conjecture about how often to expect twin primes. The graph, however, turned out not to be as new as Montgomery had first thought.

Since he had been hoping to find that zeros were close together, Montgomery saw his work as something of a failure. He had planned to use clusters of zeros on Riemann's critical line to resolve some of Gauss's unanswered questions about the factorisation of imaginary numbers. But the opposite had happened. If Montgomery was right about his new conjecture, and zeros like to repel one another, it shed little light on his initial ideas. Still, when you set out on a journey, you don't always know where you'll end up. As Littlewood had advised Montgomery during his time in Cambridge, 'Don't be afraid to work on hard problems because you might solve something interesting along the way.' This dictum Littlewood had learnt the hard way when, as a graduate student, his supervisor had unwittingly given him the Riemann Hypothesis to solve.

Montgomery had stumbled upon this unexpected distribution of gaps between zeros in the autumn of 1971. By March 1972 he had defended his Ph.D. thesis and accepted a job at the University of Michigan, where he is now a professor. He believed that his perspective was genuinely new and interesting, but a serious doubt remained in his mind. He knew that Atle Selberg had become something of a latter-day Gauss. 'Selberg had a lot of unpublished work, and there was always the danger that he would say, "Oh yes, I've known this for several years."' Just as Legendre's

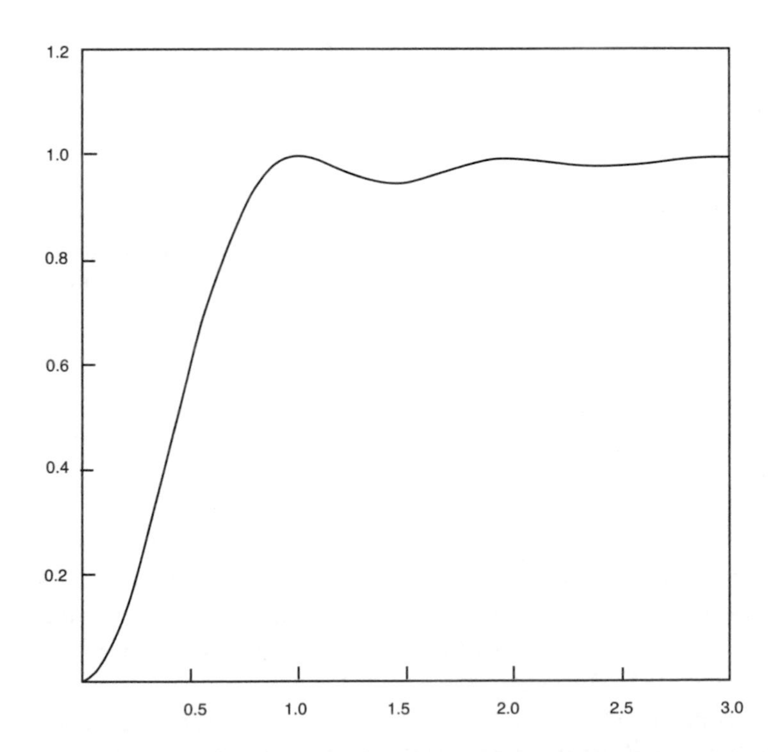

Montgomery's graph. The horizontal axis keeps track of the distance between pairs of zeros, whilst the vertical axis measures how many pairs there are at each given distance.

announcements of new discoveries would turn out to be old results that Gauss had recorded in unpublished manuscripts years before, modern-day mathematicians often found that Selberg had beaten them to it. Having been burnt by his interaction with Erdős over the elementary proof of the Prime Number Theorem, Selberg worked very much alone on his ideas in number theory, and many of them were never published.

So, on his way to a number theory conference in the spring of 1972, Montgomery took a detour via Princeton to show his discovery to Selberg. Something else was bugging him: 'I was a little bit troubled because I thought there was a message in what I had done and I didn't know what the message was.' However, it wasn't Selberg who was to help Montgomery to interpret the message, but another member of the Princeton mafia.

Dyson, the frog prince of physics

The British physicist Freeman Dyson made his name just after the war by championing the maverick scientist Richard Feynman. After finishing his degree in Cambridge, Dyson won a fellowship to study physics at Cornell University. It was there that he met the young Feynman, who was developing a very unique and personal take on quantum physics. At first, many dismissed what Feynman had to say because they couldn't understand the very personal language he was using. Dyson appreciated the power of Feynman's perspective and helped him to articulate more clearly his revolutionary ideas. The tools Feynman developed are now the key to most calculations made by particle physicists. If it hadn't been for Dyson's interpretative skills, those tools might have been lost for ever.

Physics wasn't the first thing to capture Dyson's imagination. He came from a family with a strong musical tradition but little interest in the sciences. At school, however, the intoxicating melodies of mathematics bewitched the young Dyson. He became fascinated with Ramanujan's theory of partitions after winning a copy of one of Hardy's books on number theory. 'In the forty years since that happy day, I have intermittently been coming back to Ramanujan's garden. Every time I come back, I find fresh flowers blooming. That was the most remarkable thing about Ramanujan. He discovered so much and yet he left so much in his garden for other people to discover.'

According to Dyson, although scientists all explore the same terrain, they fall into two distinct groups: the birds and the frogs. The birds soar high above their field, able to see grand connections across the landscape;

the frogs spend their time splashing around in the mud and swimming in a small pond with which they become very familiar. Mathematics was very much a subject for the birds, but Dyson saw himself then as a frog, which prompted his move to the practical concerns of physics.

His successful promotion of Feynman's quantum physics brought him to the attention of the head of the Institute for Advanced Study, Robert Oppenheimer, the physicist who had headed the US nuclear programme during the Second World War. Dyson accepted Oppenheimer's offer of a permanent position at the Institute in 1953. Although he is a softly spoken, unassuming character, Dyson's forthright opinions helped him make a name for himself outside academic circles. He became known for his speculations on the possibility of extraterrestrial civilisations. His cult status amongst a public fascinated by outer space was enhanced by his work in the late 1950s and early 1960s on Project Orion, a proposal to build spacecraft capable of taking humans to Mars and Saturn.

Although Montgomery had spent the academic year 1970/71 at the Institute, which was when he first met Gödel, he had had little contact with the physicists. There were more than enough number theorists at Princeton to keep him busy. Nevertheless, as he recalls, 'I knew Dyson by sight, and we'd had a nodding, smiling acquaintance although I doubt he knew who I was. I knew who he was because he'd worked during the Second World War on number theory things in London.'

On his pre-conference detour, Montgomery spent the day at Princeton explaining his ideas to Selberg and to some of the other number theorists who were visiting the Institute. When the time came, they broke for the ritual observed in most mathematics departments, afternoon tea. Tea-time has always been an important occasion at the Institute because it allows members from different disciplines to exchange ideas. Montgomery was chatting with one of the number theorists who had attended his talk, Sarvadaman Chowla. Chowla, an Indian student of Littlewood's, had fled to America when his native Lahore was made part of Pakistan when the new states of Pakistan and India were founded in 1947. He became a regular visitor at the Institute, where his lively personality and good humour endeared him to the permanent members. As Chowla and Montgomery chatted away, the Indian mathematician spotted Dyson across the room.

'Chowla said, "Have you met Dyson?" and I said no. "I'll introduce you", I said no.' Chowla, though, has a reputation for not taking 'no' for an answer – he's the only person ever to have bullied Selberg into writing a joint paper. 'Chowla was very insistent, and dragged me over to introduce me. I was embarrassed to bother Dyson, but he was very cordial and he asked me what I was working on.' Montgomery started talking about how he thought the gaps between pairs of zeros might behave, and as soon as he mentioned his graph of the distribution of gaps, Dyson's eyes lit up. 'But that's just the same as the behaviour of the difference between pairs of eigenvalues of random Hermitian matrices!'

Dyson quickly explained to Montgomery that this strange-sounding mathematics was being used by quantum physicists to predict the energy levels in the nucleus of a heavy atom when it is bombarded with low-energy neutrons. Dyson, who was at the forefront of this work, pointed Montgomery towards some of the experiments that had been done to record these energy levels. Sure enough, when Montgomery looked at the gaps between the energy levels in the nucleus of erbium, the 68th element in the Periodic Table, there was a striking familiarity. If he took a strip of zeros from Riemann's ley line and placed it alongside these experimentally recorded energy levels, he could see at once that the two were uncannily similar. Both the zeros and the energy levels were spaced in a much more orderly fashion than if they had been chosen randomly.

Montgomery couldn't quite believe it. The patterns he was predicting for the way zeros were distributed were the patterns quantum physicists were finding among the energy levels in the nuclei of heavy atoms. They were such distinctive patterns that the strong resemblance could not be a coincidence. Here was the message that Montgomery was looking for: perhaps the mathematics behind the quantum energy levels in heavy nuclei is the mathematics that determines the locations of Riemann's zeros.

The mathematics responsible for explaining these energy levels goes back to the revelation that sparked the twentieth-century development of quantum physics. Elementary particles such as electrons and photons have two seemingly contradictory characteristics. From one perspective they behave remarkably like tiny billiard balls. Yet experiment had also revealed a different character that can be explained only if these fundamental 'particles' are regarded as waves. Quantum physics was born out of science's attempts to explain this subatomic split personality, known as wave–particle duality.

Quantum drums

In the early twentieth century, a picture emerged of the atom as a minute solar system made out of indivisible particles. The Sun at the centre of this mini-solar system was called the nucleus; physicists would later discover that this nucleus was built from particles called protons and neutrons. Orbiting the nucleus were the electrons, the planets of the atomic structure. Advances in theory and experiment soon forced physicists to rethink this model. They began to realise that the atom behaves less like a planetary system, more like a drum. The vibrations you create when you bang a drum are built from certain basic wave patterns, each with its own frequency. There are in theory infinitely many possible frequencies, and the sound of the drum is thus a combination of these various frequencies. Unlike the harmonics of the violin string, the noise of the drum is a much more complicated mixture of frequencies determined by the shape of the drum, the tension in the drumskin, the external air pressure and other influences. The complexity of the different wave patterns produced by a drum explains why many of the percussion instruments in the orchestra don't produce an identifiable note.

There is a way to see the complexity of the vibrations that make up the sound of a drum. The eighteenth-century scientist Ernst Chladni devised an experiment which he used to perform at the courts of Europe. (Napoleon was particularly fascinated by his demonstration, and made him a gift of 6,000 francs.) Chladni took a square metal plate to represent the drum. When he banged the plate, it made a horrible clanging sound, but by skilfully vibrating the plate with a violin bow Chladni could pick out each of its individual frequencies. By covering the plate with a thin layer of sand, he showed his audience the different types of vibrations set up in the plate at each basic frequency. The sand would collect along the parts of the plate that weren't vibrating, and strange patterns would emerge. Each time Chladni set the plate vibrating with a new stroke of his violin bow, a new shape would appear in the sand, indicating a new frequency (see figure overleaf).

Physicists in the 1920s realised that the mathematics that describes the frequencies in the sound of a drum could also be used to predict the characteristic energy levels at which electrons in an atom like to vibrate. The confines of the atom are like the boundaries of the drum: forces in the atom control the vibrations of subatomic particles, just as the tension in the drumskin or the surrounding air pressure govern the vibrations that go to make the sound of the drum. Each atom was like one of Chladni's plates. The electrons in the atom vibrate only in very set patterns, like those made visible by Chladni. Exciting an electron makes it begin to vibrate at a new frequency, just as Chladni had been able to create new patterns in the sand on his plates by using a violin bow. Each different atom in the Periodic Table has its own distinctive set of frequencies at

Some of the exotic vibrations of a metal plate with which Chladni entertained Napoleon.

which the electrons inside it prefer to vibrate. These frequencies are the atoms' individual signatures, exploited by spectroscopists to identify the atomic species present in substances under investigation.

A mathematical theory had been developed to account for the patterns – or waveforms – that appear on the surface of a drum. The theory dates back to Euler's *wave equation*. Feed into it the physical properties of the drum – its shape, the tension in the drumskin, the surrounding air pressure, and so on – and its solutions give you the possible waveforms. The physics of the atom differs from the physics of a drum in that it involves imaginary numbers. To solve the equations that dictate the behaviour of the atom, physicists found themselves obliged to enter the intangible world of imaginary numbers. And it is imaginary numbers that give quantum physics its strange probabilistic character.

In our everyday, macroscopic world, we can make measurements without affecting what we are measuring. When we use a stopwatch to time athletes, we don't slow them down; when we measure where a javelin falls, we don't alter the distance of the throw. As an observer, we are independent of the system we are measuring. But things are different in the microscopic world. When we observe an electron we interact with it, invariably changing its behaviour.

Quantum physics attempts to explain what is happening to the particle before the observer gets involved. For as long as the quantum world remains unobserved by us in our macroscopic world, it exists only in the world of imaginary numbers. It is these imaginary numbers that explain the apparently inexplicable observations from our macroscopic perspective. For example, it seems that an unobserved electron can be in two different places at the same time, or can be vibrating at several different frequencies or energy levels. When we observe an event in the quantum world, it is as though we are seeing not the event itself in its natural domain, but a shadow of the event projected into our 'real' world of ordinary numbers. The act of observation causes the two-dimensional imaginary world to collapse into the one-dimensional line of ordinary numbers. Before we observe an electron, it will be vibrating, like a drum, in a combination of different frequencies. But when we observe it, it's not like us listening to a drum and hearing all the frequencies at the same time – all we hear is the electron vibrating at a single frequency.

Two of the key figures in mapping the new world of the quantum were Göttingen physicists Werner Heisenberg and Max Born. Looking down from his office, Hilbert would often see Heisenberg and Born strolling up and down the lawns outside the mathematics department, deep in discussion, putting together the twentieth-century model of the atom. Hilbert began to wonder whether the locations of the zeros in Riemann's landscape could be explained by the mathematics of vibrations that Heisenberg was developing to explain the energy levels in the atom. But there had been little to go on at the time. Montgomery's discoveries relaunched Hilbert's idea that the best chance of understanding Riemann's zeros would come from the mathematics of the quantum drums which Born and Heisenberg were then creating to explain energy levels. The mix of imaginary numbers and waves gave rise to a characteristic set of frequencies unique to drums with their source in quantum physics rather than a classical orchestra. But as Montgomery learnt from Dyson during their meeting in the common room at Princeton, the characteristic frequencies that would ultimately be most in tune with the location of the Riemann zeros came from some of the most complicated atoms in the quantum orchestra.

Fascinating rhythm

The first atom that quantum physicists were able to analyse was hydrogen. A hydrogen atom is a very simple sort of drum: there is one electron orbiting one proton. The equations determining the frequencies or energy levels of this electron and proton are simple enough to be solved precisely. The frequencies of the electron have much in common with the harmonics produced by a violin string. Although quantum physicists were successful with hydrogen, as soon as they moved further into the Periodic Table, they found it impossible to describe the mathematical drum precisely. The more neutrons and protons in the nucleus, and the more orbiting electrons, the more difficult the task grew. By the time they reached the 92 protons and 146 neutrons that form the nucleus of uranium-238, the physicists were lost. The most difficult problem was to determine the possible energy levels in the nucleus, the sun at the heart of the atomic solar system. Working out the shape of the mathematical drum that determined these nuclear energy levels was just too complicated. Even if physicists could determine which mathematical drums were responsible for the energy levels, the drums would be so complex that it would be impossible to determine their frequencies.

It was not until the 1950s that a way was found to analyse such a complicated set-up. Rather than trying to find the precise values of all the different energy levels, Eugene Wigner and Lev Landau decided instead to look at the statistics of these energy levels. They did for energy levels what Gauss had done for primes. Gauss had changed the focus from trying to predict precisely when a prime would occur to estimating on average how many primes one would see as one counted higher. In the same way, Wigner and Landau were advocating a more tractable approach to understanding atomic energy levels. The statistics would reveal the chance of finding, in some small region of the spectrum of all frequencies, the energy levels of a particular nucleus.

The nucleus of uranium was so complicated that there was a whole host of possible equations that would determine the energy levels according to what state the uranium was in. So there was little hope in assessing the statistics of these energy levels if the statistics changed dramatically with a change in the state of the nucleus. Since the energy levels were determined by analysing quantum drums, Wigner and Landau decided to see whether the statistics of the frequencies varied wildly if you changed the shape of the drums. Fortunately, for most drums this turned out not to be the case. Wigner and Landau discovered that when they took quantum drums at random, the specific frequencies might change but the statistics of the frequencies did not. The statistics of most quantum drums were the same, but did the nucleus of a heavy atom behave like the average quantum drum? Wigner and Landau believed that there was nothing special about the particular drums describing the uranium nucleus, for example, that made them any different to the majority of quantum drums.

Wigner and Landau's hunch was spot on. When they compared the statistics of the energy levels of a random quantum drum with the statistics of energy levels observed in experiment, the fit was excellent. In particular, when they looked at the gaps between the energy levels in a uranium nucleus, it seemed as though the energy levels were repelling one another. That was why Freeman Dyson had got so excited during his meeting with Montgomery at Princeton – the graph Montgomery had shown him bore the characteristic stamp of the statistics of energy levels. But Montgomery had unearthed this strange pattern in an apparently unrelated area of science.

The next question, then, was why and how did these two areas – energy levels and Riemann zeros – have anything to do with each other. Montgomery must have felt as astonished as an archaeologist who'd discovered identical Palaeolithic paintings in caves at the opposite ends of the world. There just had to be a link between the two. Montgomery admits that his conversation with Dyson had to be one of the most fortuitous coincidences in scientific history: 'It was really serendipity that I happened to be in just the right place.' Since Galileo and Newton, physics and mathematics have often ranged over similar territory, but no one had expected Riemann's number theory and quantum physics to be so intimately connected. Montgomery's attempt to understand the factorisation of imaginary numbers had come to nothing, but he had stumbled across something far more interesting. 'As failed research projects go, this was better than most,' Montgomery smiles.

What did these revelations over tea at Princeton mean for the Riemann Hypothesis? If the points at sea level in Riemann's landscape can be explained by the mathematics of energy levels in physics, then there was the exciting prospect of actually proving why the points at sea level lie in a straight line. A zero off the line would be like having an imaginary energy level, something not permitted by the equations of quantum physics. Here was the best hope yet for providing some sort of explanation for Riemann's Hypothesis.

Although experiments had been done to confirm Wigner and Landau's model for energy levels in large atoms, Montgomery still had no experimental confirmation that the points at sea level in Riemann's landscape were doing what he believed the theory predicted they should. No one had tested to see whether the zeros really did repel one another, as he was suggesting. The trouble was that the regions of Riemann's landscape where these statistics were likely to appear lay far beyond Montgomery's computational reach.

In Cambridge, Montgomery had learnt of Littlewood's discoveries about how long one has to count prime numbers before they show their true colours. Despite Littlewood's theoretical proof that Gauss's guess at the number of primes would sometimes be an underestimate, no one had succeeded in demonstrating this experimentally. Montgomery was resigning himself to suffering the same fate. It had taken some time for experimental physicists to build particle accelerators that generated enough energy for Wigner and Landau's predictions to be confirmed. Montgomery feared that mathematicians would never be able to calculate numbers high enough to test whether zeros far up the critical line were doing what he was predicting they should.

But Montgomery hadn't counted on the computational powers of Andrew Odlyzko and his Cray supercomputer sitting in AT&T's research laboratory at the heart of New Jersey. Odlyzko had heard of Montgomery's prediction about the gaps between the zeros and the parallel with the random drums behind energy levels in heavy nuclei. Here was just the kind of challenge that appealed to him. He began casting for zeros at a distance of 10^{12} units up Riemann's ley line. This was to be a phenomenal feat of computation. If we imagine that the map of the Riemann landscape is centred on New Jersey, and represent each unit along the critical line by a step of one centimetre, then Odlyzko was looking at sections of Riemann's ley line that were twenty-five times as distant as the Moon. Once the Cray supercomputer had churned out about a hundred thousand zeros, Odlyzko could begin to look at the statistics of the gaps between them. By the middle of the 1980s, he was ready to publish the results of his calculations. The zeros in Riemann's landscape were indeed showing some similarities of spacing to the energy levels in heavy atoms, but it was clear that the fit wasn't perfect. No statistician would be satisfied with the match. Was Montgomery wrong, or did he have to probe even farther north?

Odlyzko, undaunted by the magnitude of the task, pressed on to 10^{20} steps north. In terms of our imaginary map centred on New Jersey, Odlyzko was now exploring 100 light years from home, a distance well beyond Vega, the star-system from which the prime number message was broadcast in Carl Sagan's *Contact*. In 1989 Odlyzko plotted the distances between the zeros and compared it to Montgomery's prediction. This time the fit was staggering. Here was convincing evidence of a new aspect of the zeros. From as far away as 10^{20} they were sending out the very clear message that they were being produced by some complicated mathematical drum.

Mathematical magic

How significant was the statistical match that Andrew Odlyzko had discovered? Perhaps statistics like this could be replicated by using some completely unrelated piece of mathematics. Were Montgomery and Odlyzko pointing us in the right direction, or sending us on a wild-goose chase?

For the answers to these questions, we can do no better than ask the Stanford statistician Persi Diaconis, master debunker of psychic phenomena and the man who helped expose the 'Bible code' – which claimed to have found hidden messages in the ancient Hebrew text – for the hoax it was. Presented with the Riemann data, Diaconis admits you would be hard pushed to find a better statistical match. 'I've been a statistician all my life, and I've never seen such a good fit of data.' Diaconis knows better than most that what looks good from one angle needs to be examined from all other perspectives to make sure that some telling flaw hasn't been obscured. Diaconis is a master of this kind of trickery – it was magic, not mathematics, that first captured his imagination.

As a child in New York, Diaconis would spend all his time skipping school to hang out in magic shops. His sleight of hand caught the eye of one of America's greatest magicians, Dai Vernon. Diaconis recalls that Vernon, then aged sixty-eight, offered him the chance to come on the road as his assistant: 'I'm off to Delaware tomorrow, do you want to come along?' The fourteen-year-old Diaconis packed a bag and left, without telling his parents. They spent the next two years travelling the country:

> We were like Oliver Twist and Fagin. The magic community is a very supportive community. It's not grungy carnivals, or anything like that, it's upper middle class amateurs. Magicians are fascinated by gamblers. Vernon and I would seek out crooked gamblers, and if we heard there was an Eskimo who could deal the second card with snow shoes, we'd be off to Alaska – it was that type of adventure. We just did it for two years, just follow the wind. Hanging out with gamblers, there was always talk of odds, and I got fascinated by probability and I wanted to learn more about it.

On his travels, Diaconis started to read about the mathematics of probability. Again it was the fateful intervention of one particular book that helped spark the career of one of the most fascinating mathematicians of our generation. He was given William Feller's book on probability, *An Introduction to Probability Theory and Its Applications*, one of the standard university texts on the subject. Knowing no calculus, he couldn't get anywhere with it. Diaconis decided the only way forward was to enrol for night school at City College in New York. The bug took hold, and within two and a half years he'd graduated and was keen to apply for graduate college. Harvard took a chance on this unconventional student, who has never looked back.

Diaconis remains true to his magical roots and admits there is a lot in common between the two arts:

> The way I do mathematics is very similar to magic. In both subjects you have a problem you're trying to solve with constraints. In mathematics, it's the limitations of a reasoned argument with the tools you have available, and with magic it's to use your tools and sleight of hand to bring about a certain effect without the audience knowing what you're doing. The intellectual process of solving problems in the two areas is almost the same. One difference in magic and mathematics is the competition. The competition in mathematics is a lot stiffer than in magic.

As a statistician, Diaconis is interested in whether something is random or not. He made it on to the front page of the *New York Times* for his analysis of card-shuffling. According to Diaconis, it takes seven shuffles for the average card player to put the cards in random order. But that's an average card player and an average shuffle. Things are rather different if the shuffler has Diaconis's magical hands. Many of his tricks depend on his skill in performing the perfect shuffle. He knows that eight perfect shuffles in a row bring the cards back to their original order, though audiences will be convinced that the cards are in random order. He is highly tuned to whether a shuffled pack of cards has been 'spiked'. Diaconis has made such a name for himself in detecting patterns where others see only chaos that he is employed in Las Vegas to check that electronic shuffling machines don't give anything away to the eagle-eyed punter.

Diaconis was particularly intrigued when number theorists started to broadcast Montgomery and Odlyzko's claims that the zeros in Riemann's landscape looked like the frequencies of some random drum. If anyone was equipped to sniff out a rat, it was him. 'So I called Andrew and I said I wanted some zeros. So he gave me about 50,000 zeros of my very own,

Persi Diaconis, professor at Stanford University.

starting at about 10^{20}.' Then he tried a new test he'd discovered when he was at AT&T working on encrypting telephone calls. 'I tested the hell out of it, and they matched the predictions perfectly,' he says. Here was further evidence that the zeros are from the beats of a random mathematical drum whose frequencies behave like the energy levels of quantum physics. For Diaconis, the connections between primes and energy levels is not Nature practising some malicious deceit, but genuine magic.

Once these new statistics had been discovered, they began to surface everywhere: heavy nuclei, zeros of the Riemann zeta function, DNA sequencing, the properties of glass. Most curious, perhaps, is Diaconis's discovery that these statistics might help answer another unsolved problem: how often can you expect to win a game of patience?

In one of the commonest patience games, seven piles of cards are dealt, one card in the first pile, two in the second and seven in the last. The top card of each pile is turned over. The remaining cards are turned over in groups of three. Acceptable moves are to place one exposed card on another if the card you are moving is of a different colour and one lower in value than the card it is being put on top of. So, for example, a red seven can go on a black eight, and a black jack on a red queen. Aces are placed to one side as they appear, on which sequences of each suit are built until all the cards are cleared.

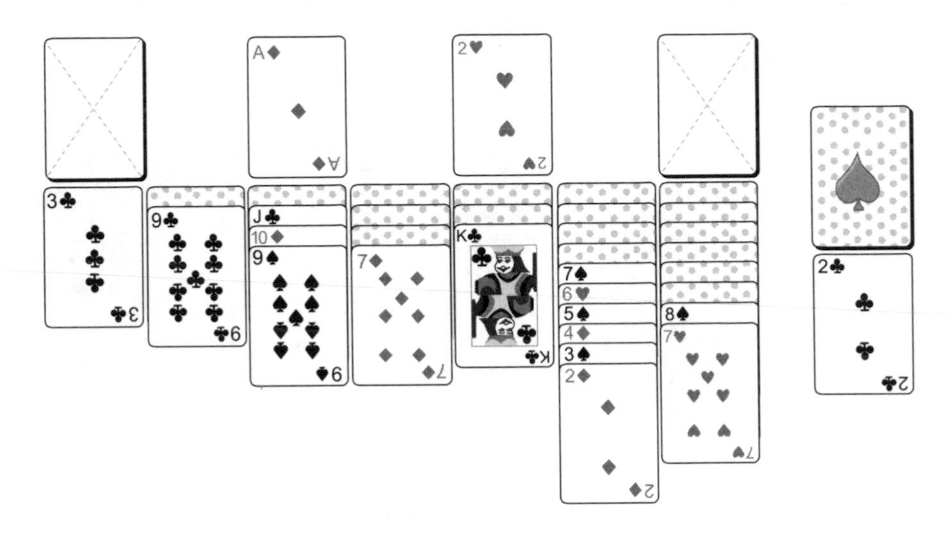

Klondike, or Idiot's Delight – one of the most popular games of patience, yet a mystery to mathematicians.

The game goes by various names, including Klondike and Idiot's Delight. There are variants of the game. In Las Vegas you can buy a deck of cards for $52 and instead of continually recycling through the remaining pack seeing every third card, you are allowed to see every card but only once. For every card you move onto the suit stacks, the casino pays you $5.

Even though the game has been played since around 1780 and is familiar to almost every owner of a desktop computer, no one knows the average success rate for clearing the deck. Considering you can make $5 a card in Vegas, it would be worth knowing the odds against you. Even such a simple-looking game has enough complications to fox Diaconis in his attempts to calculate an average success rate. But from the data he's collected over the years, it looks as though you clear the whole deck around 15 per cent of the time. But Diaconis would love to have a proof.

A common strategy for solving a mathematical problem is to start with an easier problem. Diaconis has analysed a much simpler version of Klondike, called patience sorting. He was thrilled to find that the frequency of winning this simplified game of patience has its heart in the theory of the frequencies of those random mathematical drums. Despite his progress, he believes we are still a long way from a full analysis of Klondike itself. He promises his students they'll make it to the front page of the *New York Times* if they can make the breakthrough. Despite their tantalising connections with random mathematical drums, solutions to both Klondike and the Riemann Hypothesis remain elusive.

Quantum billiards

Number theorists were trying to come to terms with the strange turn their subject had taken since Montgomery's cup of tea with Dyson. Although Montgomery's analysis seemed to indicate that it was the physics of quantum drums that might be the source of the Riemann zeros, there was little else to illuminate this new avenue. Where was the magical drum hiding? From the statistics and evidence gathered so far, this drum looked remarkably like any drum chosen at random. That was not going to be much help in finding the specific drum responsible for the Riemann zeros. As this strange link was investigated further, it became apparent that the connection with quantum physics wasn't the only surprising twist in the story of the Riemann zeros. A new connection emerged to help mathematicians in their search for the drum.

Diaconis and other statisticians have developed a sophisticated array of weapons with which they can test any given proposition. The Bible code looked statistically significant because its proponents kept making you look at the data from one angle only. It was under the pressure of other tests that the Bible code eventually buckled. Even though Diaconis's studies had not blown Montgomery's predictions out of the water, back in New Jersey Odlyzko was beginning to worry about some of his own new calculations. He had started to look at a different statistical test to see whether the connection between the Riemann zeros and quantum physics was justified. He'd noticed that some rather troublesome discrepancies were beginning to creep into the data for the Riemann zeros.

Odlyzko was looking at the graph of a statistical measure called the number variance. He plotted the graph corresponding to the Riemann zeros and compared it to the corresponding graph arising from the frequencies of a random quantum drum. As he looked at how the graphs evolved he could see that although the match was very good to begin with, the data from the Riemann zeros was suddenly dipping away from the graph predicted by the random quantum drums. The early part of the graph was still testing the statistics of the distance between neighbouring zeros. But when Odlyzko analysed the continuing development of his graph, he found that discrepancies were beginning to creep in. As the graph evolved, it was tracking not the statistics of the distances between close zeros, as the early part of the graph was doing, but something more like the statistics of the gaps between the Nth zero and the $(N + 1000)$th zero. Odlyzko thought at first that he must have made some computational errors to cause this deviation. As it turned out, he was witnessing the first evidence of the influence on Riemann's landscape of another major theme of twentieth-century science: *chaos theory*.

Like quantum physics, chaos theory has managed to establish itself in the mainstream of popular culture. No nineties rave was complete without a fractal projected onto the walls. Despite their apparent complexity, fractals are generated by very innocent-looking rules. Chaos theory, the mathematics behind these pictures, helps explain why, although the rules of Nature might be simple, reality looks infinitely complex. The term 'chaos' is used when a dynamical system is very sensitive to initial conditions. If a slight change in how an experiment is set up results in a dramatic difference in the outcome, that is the signature of chaos.

One of the manifestations of the mathematics of chaos is in the game of billiards. If you shoot a ball across the billiard table, it will follow a path determined by the angles at which it hits the cushions of the table. The interest comes in altering very slightly the initial direction in which you shoot off the ball. Does the path suddenly diverge dramatically from the path it took before? The answer lies in the shape of the table. On a conventional rectangular table there is no chaotic behaviour in the path the ball takes (despite what most amateur players may think). The path is very predictable, and a slight change in the initial direction of the ball does not dramatically alter the ball's path. However, on a billiard table shaped like a stadium, the paths of balls take on quite a different character. If we now fire off two balls in only slightly different initial directions, we discover that they follow wildly different paths which appear to have nothing to do with each other. As the figure on the opposite page indicates, the physics of the stadium-shaped table is chaotic, in contrast to the very predictable paths on a rectangular table.

As the mathematics of chaos emerged during the 1970s, a number of quantum physicists became interested in the implications of this new theory for their own subject. In particular, they wondered what would happen if you played this game of billiards on the atomic scale. After all, from one perspective electrons were behaving remarkably like microscopic billiard balls.

Using semiconducting material, from which computer chips are made,

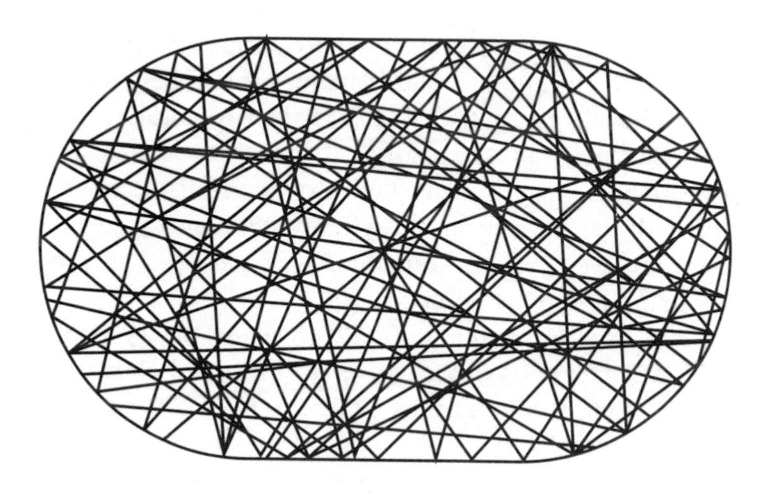

Chaotic motion: the paths mapped out by balls on a billiard table shaped like a stadium.

it is possible to carve out a billiard table so small that hundreds could fit onto the tip of a pin. Physicists began to explore the motion of an electron as it bounced around the tiny table. The electron is no longer bound to an atom, but is free to move through the semiconductor. It is this movement that is responsible for the transfer of data through the computer chip. But the electron's path is not completely unrestricted. Although it is no longer orbiting the nucleus of an atom, the electron is now constrained by the boundaries of the table. Physicists were interested in what effect different shaped tables would have on the electron's wave-like behaviour as well as its particle-like, billiard ball motion. Just as an electron confined to an atom vibrates at certain characteristic frequencies, so does a free electron as it chases out a path on the tiny billiard table.

When physicists analysed the statistics of the energy levels, they found that the statistics varied according to whether the billiard table gave rise to chaotic paths or to regular paths. If the electrons were confined to bouncing around a rectangular region, tracing out regular, non-chaotic paths, then their energy levels were quite randomly distributed. In particular, energy levels were often bunched close together. However, the statistics were quite different when the electrons were confined to a stadium-shaped region, in which the paths are chaotic. The energy levels were no longer random, but arranged themselves in a much more uniform pattern where no two energy levels were close to each other.

Here was yet another manifestation of the strange repulsion of energy levels. Chaotic quantum billiards was creating the same patterns that had been observed in the energy levels in heavy atoms and by Montgomery and Odlyzko in the location of the Riemann zeros. These different levels fitted very well with the statistics of a random quantum drum. But it turned out that not all the statistical measures matched perfectly. Physicists were beginning to understand that the statistics of the distance between the *N*th energy level and the ($N + 1000$)th energy level depended on whether you were playing quantum billiards or simply measuring frequencies in a random quantum drum.

One of the experts on this cocktail of chaos theory and quantum physics is Sir Michael Berry of the University of Bristol. Berry was the first to understand that the deviations Odlyzko had noticed between the number variance graphs of the Riemann zeros and of random quantum drums were signifying that a chaotic quantum system might offer the best physical model for the behaviour of the primes. Berry is a charismatic figure on the scientific circuit. He brings an air of sophistication to his subject that is sometimes lacking in those immersed in the world of science. He is a Renaissance man, happy to quote from the giants of literature as well as science to persuade others of his view of the world. He is an expert in finding the perfect image to see through the complexities of mathematical formulas. Mathematicians have been very lucky to recruit this English knight in their attempts on the Riemann Hypothesis.

He became fascinated by prime numbers in the 1980s after reading an article in the *Mathematical Intelligencer* entitled 'The first 50 million prime numbers'. The article was by Don Zagier, the mathematical musketeer at the Max Planck Institute who had battled with Bombieri over the Riemann Hypothesis. Instead of tediously listing millions of numbers, Zagier described how the zeros in Riemann's landscape could be used to create waves which magically reproduced how many primes one can expect to find as one counts higher. 'It was a beautiful article. I thought the Riemann zeros were wonderful things.' Berry was taken with the very physical interpretation of Riemann's discovery – that there is music in the primes.

As a physicist, Berry brings to the subject of prime numbers a physical intuition which is lacking in most mathematicians. Mathematicians can spend so long in a world of mental constructs that they forget about all the links between the abstract mathematical world and the physical world around us. Riemann had turned the primes into wave functions; for a physicist such as Berry, these waves are not just abstract music, but can be translated into physical sounds anyone can listen to. His presentations on the Riemann Hypothesis used to feature a recording of Riemann's music –

a low, rumbling white noise. Berry describes it as 'rather a post-modern sort of music, but thanks to Riemann's work, we can say what Bernard Shaw said about Wagner: this music is better than it sounds'.

Berry's interest in the primes coincided with his growing understanding of the differences between the statistics of energy levels in electrons playing quantum billiards and the energy levels in a random quantum drum. 'I thought it might be interesting to look again at the story of the Riemann zeros and Dyson's ideas in the light of the new connections with quantum chaos.' Would the special statistics that Berry had discovered in the energy levels of quantum billiards be reflected in the statistics of the zeros of the Riemann zeta landscape? 'I thought it would be very nice to see if the zeros actually behaved in this way, and I did some rough calculations.' But he didn't have enough data. 'Then I heard of Odlyzko, who'd done these epic calculations. I wrote to him and he was wonderfully helpful. He explained to me that he'd been a little worried because his calculations beyond a certain point had started to show some deviations. He thought he must have made a mistake in his computations.'

But Odlyzko did not have the insights of a physicist. When Berry compared the zeros to the energy levels of chaotic quantum billiards, he found a perfect match. The discrepancies that Odlyzko had observed turned out to be the first sign of the difference between the statistics of frequencies in a random quantum drum and the energy levels of chaotic quantum billiards. He had not been aware of this new chaotic quantum system, but Berry recognised it straight away:

> This was a great moment because it was obviously right. That was to me absolute convincing circumstantial evidence that if you think the Riemann Hypothesis is true, then the Riemann zeros would have underlying them not just a quantum system, but a quantum system with a classical counterpart, moderately simple but chaotic. It was a lovely moment. That was, if you like, something that quantum mechanics provided for the theory of the Riemann zeros.

Curiously, if the secret of the prime numbers really is a game of chaotic quantum billiards, then the primes are represented by very special paths around the billiard table. Some of the paths return the ball to its starting point after a certain number of trips around the table, after which the pattern repeats. It seems that these special paths are representations of the primes: each path corresponds to one of the primes, and the longer the path before it repeats itself, the bigger the corresponding prime.

Berry's new twist could end up uniting three of the great themes of science: quantum physics (the physics of the very small), chaos (the mathematics of unpredictability) and prime numbers (the atoms of arithmetic). Perhaps, after all, the order that Riemann had hoped to uncover in the primes is described by quantum chaos. Once again, the primes are asserting their enigmatic character. The apparent link between the statistics of zeros and of energy levels has persuaded many physicists to enlist in the search for a proof of the Riemann Hypothesis. The source of the zeros may turn out to be the frequencies of a mathematical drum; if so, quantum physicists more than anyone are best equipped to locate those drums. Their lives reverberate to the sound of drums.

We have all this evidence that the Riemann zeros are vibrations, but we don't know what's doing the vibrating. It may be that the source is very mathematical, with no physical model. The mathematics explaining the zeros might be the same as the mathematics of quantum chaos, but that does not imply there will necessarily be a physical manifestation of a solution. But Berry doesn't think so. He believes that once the mathematics is in place, there will be a corresponding physical model whose energy levels will reflect the Riemann zeros. 'I've no doubt that when someone has found the source of the zeros that someone will make it.' Could it possibly exist already, hiding somewhere out there in the universe waiting for us to discover it? Maybe the cosmic primes that Ellie Arroway detected in Carl Sagan's book *Contact* weren't a sign of alien life after all, but just the frequencies emitted by some vibrating neutron star. As Berry explains, 'There is this well-known totalitarian principle that whatever the laws of physics permit can be found somewhere naturally. I'm sceptical that this can be applied to this case, but certainly you might be able to concoct it somehow.'

Where Odlyzko had AT&T, Berry and his research group benefited for a number of years from the support of another leading commercial player. At its major UK facility in Bristol, Hewlett-Packard enlisted the help of members of Berry's research group to harness the power of quantum physics. The company knew that any progress made on the Riemann Hypothesis had the potential to improve our understanding of the game of quantum billiards. And because the rules of quantum billiards help determine the behaviour of computer circuitry, as electrons career around the grooves carved into computer chips, they knew how important it was to keep up to date with progress made by the expert quantum billiard players on their doorstep.

42 – the answer to the Ultimate Question

Although the big guns like AT&T and Hewlett-Packard have been forced to reduce their investment in the primes by the economic downturn in the fortunes of the computer industry, there is still one commercial player who is happy to continue promoting research into this seemingly abstract game. Fry Electronics is a chain of around twenty huge electronics stores spread across the Western USA that supplies the nation with computer gadgets and electronic gizmos. The company can't match the sponsorship of giants like AT&T or Hewlett-Packard. But if you visit Fry Electronics' headquarters in Palo Alto, California, you'll find, next to the main entrance to the store, a grungy metal door with the sign 'American Institute of Mathematics'.

The Institute is the brainchild of one of the directors of the company, John Fry. He and Brian Conrey studied mathematics together at Santa Clara University. Whilst Conrey went on to earn his place in the record books for proving the largest percentage of zeros so far to be found on Riemann's line, Fry embarked on a more commercial venture, but he never lost his interest in mathematics. As the electronics industry boomed, he wondered whether there was some way he could support the subject. He'd already sponsored a five-a-side football team, so he turned his ideas to sponsoring a mathematics team.

Fry contacted Conrey, and together they concocted a plan to try to co-ordinate efforts to prove the Riemann Hypothesis. To launch their venture they sponsored a meeting in Seattle to be held in 1996 on the anniversary of the proof of the Prime Number Theorem. It wasn't just a matter of putting up the money – they were looking to foster a new ethos of collaboration. The Riemann prize was now so coveted that many were reluctant to share even the sketchiest of ideas in case it provided the vital missing piece for someone else's jigsaw. Conrey and Fry wanted to break this cycle, which they felt was leading nowhere. The emphasis at conferences and meetings should be on sharing ideas that might not necessarily have come to fruition. They even went to the extent of sitting mathematicians round a table as if they were thrashing out a business plan.

Out of the Seattle meeting came some of the best evidence yet that the Riemann Hypothesis has something to do with quantum chaos. The evidence materialised after some of the mathematicians present expressed their concern about basing the connection purely on the observation that two graphs looked indistinguishable. One who voiced some healthy scepticism was Peter Sarnak. Although he was very impressed by all the analogies between quantum chaos and the zeros of the Riemann zeta function, he was yet to be convinced of a genuine link.

Sarnak is one of the leading lights at Princeton and was a confidant of Andrew Wiles during Wiles's secretly mounted attack on Fermat's Last Theorem. Sarnak's interest in the Riemann Hypothesis began in the mid-1970s, when he arrived from South Africa to work with Paul Cohen, just up the road from Fry Electronics at Stanford University. As a student, Sarnak had approached Cohen because he had become interested in mathematical logic. Cohen had shocked the world in 1963, ten years earlier, by employing a cunning sequence of logical arguments to settle the first of Hilbert's twenty-three problems. Contrary to Hilbert's expectation that his question had a 'yes' or 'no' answer, Cohen proved that you could *choose* which answer you wanted to be true.

The South African arrived in Stanford thinking he'd be put to work on some equally fiendish logic conundrum. But Cohen had set his sights on another of Hilbert's problems, the eighth. Solving Hilbert's first problem was a hard act to follow, and Cohen felt that only the Riemann Hypothesis would give him more of the buzz he'd experienced. He shared his ideas about the problem with Sarnak, and thus was sparked Sarnak's lifelong addiction to number theory.

Sarnak's enthusiasm for his subject is contagious. There is an energy and excitement about him when he talks of mathematics. Selberg, who now admits to being old and hard of hearing, says that Sarnak is one of the few mathematicians at Princeton he can still hear properly. His South African accent rings out across the department as he enthuses about some new development in the subject. There was excitement over the entry of quantum physics into the sacrosanct halls of number theory, but Sarnak wanted more: was there any evidence that making a connection between energy levels and zeros had led to any real progress?

It may have suggested where to look for an explanation, but this crossover had told us nothing we didn't know already. The connection seemed to be based on the close match between various statistics. The fact that two pictures look very similar was not the sort of thing that number theorists acknowledged as convincing evidence for a link. After all, mathematicians were still sceptical of the power of pictures to reveal the truth despite Riemann having brought geometry into the mainstream.

Sarnak arrived for the meeting in Seattle doubting that anything other than deep mathematical insight could reveal something significant about Riemann's landscape. After hearing talks about analogies between the Riemann zeros and energy levels in quantum chaotic billiards, and listening to Berry's performance of the music of the primes, Sarnak could bear it no longer. It was all very fascinating seeing the same pictures cropping up in both areas, but who could point to some genuine contribution to prime number theory that these connections had made possible? He offered the quantum physicists a challenge: use the analogy between quantum chaos and prime numbers to tell us something we don't already know about Riemann's landscape – something specific that couldn't be hidden behind statistics. There was a bottle of good wine as an incentive.

One of Berry's former students, Jon Keating, would win Sarnak's bottle thanks to the important part played by one special number, 42. If you are well versed in popular fiction, you will know that the number 42 has held a special significance. In Douglas Adams's *The Hitch Hiker's Guide to the Galaxy*, Zaphod Beeblebrox discovers that 42 is the Answer to the Ultimate Question of Life, the Universe and Everything (even though it's not too clear what the question was). The number 42 was also dear to Lewis Carroll, himself a mathematician in Oxford in the second half of the nineteenth century. At the trial of the Knave of Hearts in Carroll's *Alice in Wonderland*, the King declares, 'Rule Forty-two. ALL PERSONS MORE THAN A MILE HIGH TO LEAVE THE COURT.' Carroll picks up the number again and again in his writing. The Beaver in *The Hunting of the Snark* arrives with 'forty-two boxes, all carefully packed / With his name painted clearly on each'. Strangely enough, the number was about to enter the story of the Riemann Hypothesis, helping to convince doubting number theorists that quantum chaos was another side of the same prime number coin.

Hearing of Sarnak's offer of a bottle of good wine, Conrey presented the physicists with a very specific challenge as a test case. The challenge was close to Conrey's heart because it had to do with something he'd been working on for many years, albeit with very little success. There are certain attributes of the Riemann zeta function, called its *moments*, which it was known should give rise to a sequence of numbers. The trouble was that mathematicians had very little clue as to how to calculate the sequence itself. Hardy and Littlewood had shown that the first number in the sequence was 1. Albert Ingham, one of Littlewood's students, proved in the 1920s that the next number was 2. This didn't go very far to establishing a pattern that could help with further exploration.

Before the Seattle meeting, Conrey had done a huge amount of work on the problem of the next number in collaboration with a colleague, Amit Ghosh, which suggested that the third number in the sequence was a big jump away, at 42. For Conrey, that this should be the number next in the sequence 'was kind of surprising. It gave some indication that there was some level of complexity.' There were no guesses at how the sequence would go on from there. Conrey challenged the physicists to explain 42 in terms of the analogy with quantum physics. As Conrey pointed out, '42 is a number. You've either got it or you haven't. It's not just seeing how close the fit of a curve is.'

Undaunted, Jon Keating left Seattle and started beavering away. The meeting had been such a success that Fry and Conrey decided to organise a second one. It was held two years later at the Schrödinger Institute in Vienna, perhaps an appropriate venue considering the new partnership that was emerging between number theory and the quantum physics that Schrödinger had helped to found.

In the meantime, Conrey had joined forces with another mathematician, Steve Gonek. With a huge effort, squeezing all they could from their knowledge of number theory, they came up with a guess for the fourth number in the sequence – 24,024. 'So we had this sequence: 1, 2, 42, 24,024, . . . We tried like the Dickens to guess what the sequence was. We knew our method couldn't go any further because it was giving a negative answer for the next number in the sequence.' It was known that all the numbers in the sequence were bigger than zero. Conrey arrived at Vienna prepared to talk about why they thought the next number in the sequence was 24,024.

'Keating arrived a little late. On the afternoon he was going to give his lecture I saw him, and I'd seen his title and I had begun to wonder whether he had got it. As soon as he showed up I went and immediately asked him, "Did you figure it out?" He said yes, he'd got the 42.' In fact, with his graduate student, Nina Snaith, Keating had created a formula that would generate every number in the sequence. 'Then I told him about the 24,024.' This was the real test. Would Keating and Snaith's formula match Conrey and Gonek's guess of 24,024? After all, Keating had known that he was meant to be getting 42, so he might have cooked his formula to get this number. This new number, 24,024, was completely new to Keating and not one he could fake.

'Just before Jon was due to give his lecture we went up to one of the blackboards in the Schrödinger Institute and calculated what his formula was predicting for the fourth number in the sequence.' They kept on making numerical mistakes – mathematicians are sometimes not the best at mental arithmetic after years of abstract thought, which rarely calls upon the multiplication tables we learnt as kids. Then finally they got the arithmetic to work out. 'When it came out at 24,024 it was just incredible,' recalls Conrey. Seconds later, Keating rushed on to give his lecture where he and Snaith's formula was first publicly announced, still buzzing with the excitement that it checked out with Conrey and Gonek's guess. Keating described the experience at the blackboard as 'the most exciting few seconds of my scientific life'.

Keating had been nervous at the thought of addressing the cream of number theorists – here was he, a physicist, about to lecture them on something they had been working for years to understand. But the thrill of getting 24,024 gave him the confidence he needed. In the audience sat Selberg, by now the grandfather of the subject. At the end of his lecture the audience was invited to ask questions. Selberg has a reputation not for asking questions but for making pronouncements at the end of lectures along the lines of 'I proved this in the fifties,' or 'I tried the same approach thirty years ago but it doesn't work.' Keating braced himself for the inevitable. But instead, Selberg began asking question after question, clearly captivated by this new idea. Only after Keating had bravely answered all Selberg's questions did Selberg make a pronouncement: 'It must be right.' Keating had risen to Sarnak's challenge, and told mathematicians something they didn't know. Sarnak duly handed over the promised bottle of wine.

The power of the analogy between the Riemann zeros and quantum physics is twofold. First, it tells us where we should be looking for a solution to the Riemann Hypothesis. And second, as Keating had now proved, it can predict other properties of Riemann's landscape. As Berry says, 'The analogy doesn't have a firm mathematical foundation. It has to be judged by how useful it is in suggesting things for the mathematicians to prove. I'm unashamed about that – as a physicist I hold Feynman's dictum dear: "A great deal more is known than has been proved."' Even if the physicists can't come up with a physical model that generates zeros, mathematicians admit that it could well be a physicist who finally proves the Riemann Hypothesis. And that is why Bombieri's April Fool, with which we opened our story, was so believable.

Riemann's final twist

The physicists believe that the reason Riemann's zeros will be in a straight line is that they will turn out to be frequencies of some mathematical drum. A zero off the line would correspond to an imaginary frequency which was prohibited by the theory. It was not the first time that such an argument had been used to answer a problem. Keating, Berry and other physicists all learnt as students about a classical problem in hydrodynamics whose solution depends on similar reasoning. The problem concerns a spinning ball of fluid held together by the mutual gravitational interactions of the particles inside it. For example, a star is a ball of spinning gas kept together by its own gravity. The question is, what happens to the spinning ball of fluid if you give it a small kick? Will the fluid wobble briefly and remain intact, or will the small kick destroy the ball completely? The answer depends on showing why certain imaginary numbers lie in a straight line. If they do, the spinning ball of fluid will remain intact. The reason why these imaginary numbers do indeed line up is related very closely to the quantum physicists' ideas about proving the Riemann Hypothesis. Who discovered this solution? Who used the mathematics of vibrations to force these imaginary numbers onto a straight line? None other than Bernhard Riemann.

Shortly after his triumph at the Schrödinger Institute, Keating was due to visit Göttingen to lecture on using connections with quantum physics to shed light on the Riemann Hypothesis. Most mathematicians passing through Göttingen take the time to visit the library to examine Riemann's famous unpublished scribblings, his *Nachlass*. Not only is it a moving experience to feel a bond with such an important figure in the history of mathematics, but the *Nachlass* still contains many unsolved mysteries, locked inside Riemann's illegible scribbles. It has become the Rosetta Stone of mathematics.

Before Keating set off for Göttingen, one of his colleagues in the mathematics department, Philip Drazin, recommended looking at the part of the *Nachlass* in which Riemann tackles the classical problem of hydrodynamics. Although Riemann's housekeeper destroyed so many of his papers, the *Nachlass* still contains a wealth of material, and has been divided into many sections covering different periods of Riemann's life and his many varied interests.

At the library in Göttingen, Keating ordered the two different parts of the *Nachlass* that he wanted to consult: one on Riemann's ideas about the zeros in his zeta landscape, and the second on his work on hydrodynamics. When only one pile of papers appeared from the vaults, Keating mentioned that he had asked to see two parts. Both 'parts' were on the same sheets of paper, the librarian told him. As Keating explored the pages, he found to his amazement that Riemann had been concocting his proof about rotating balls of fluid at the very same time that he'd been thinking about the points at sea level in his zeta landscape. The very method by which modern-day physicists were proposing to force Riemann's zeros to line up had been used by Riemann to answer the hydrodynamics problem.

There, in front of Keating on the same pieces of paper, were Riemann's thoughts on both problems.

Yet again, the *Nachlass* had revealed how far Riemann was ahead of his time. He could not have failed to recognise the significance of his solution of the problem in fluid dynamics. His method had shown why certain imaginary numbers that emerged from his analysis of the ball of fluid were all in a straight line. Yet at the same time, and on the same paper, he was trying to prove why the zeros in his zeta landscape all lay on a straight line. In the year following his discoveries about primes and hydrodynamics, he was recording his new ideas in the little black book which, infuriatingly, disappeared from the archives. With it have disappeared Riemann's thoughts on uniting these two themes from number theory and physics.

During the few decades following Riemann's death, mathematics and physics began to diverge. Although Riemann was happy to combine the two disciplines, increasingly the scientists that came after him were less interested in the crossover between the two. Only in the twentieth century did physics and mathematics get back to working side by side, and it is this reunion that might lead to that elusive breakthrough that Riemann had dreamed of.

Although these connections with physics were exciting, many mathematicians still believed in the power of their own subject to solve the riddle of the primes. Many agreed with Sarnak that the solution to the Riemann Hypothesis lies deep in the heart of mathematics. The justification for believing that mathematics alone can deliver the answer can be traced back to 1940 and the activities of a rather special French prisoner.

CHAPTER TWELVE

The Missing Piece of the Jigsaw

It is said that the history of mathematics should proceed in the same way as the musical analysis of a symphony. There are a number of themes. You can more or less see when a given theme appears for the first time. Then it gets mixed up with the other themes and the art of the composer consists in handling them all simultaneously. Sometimes the violin plays one theme, the flute another, then they exchange and this goes on. The history of mathematics is just the same. André Weil, *Two Lectures on Number Theory, Past and Present*

Despite all the excitement about a game of quantum billiards that might explain the Riemann Hypothesis, many mathematicians remained sceptical of the intrusion by physicists into the world of pure number theory. Most of them continued to trust in the strength of their own subject to explain why the primes behave as we believe they do. It was plausible that the same style of mathematics might be responsible for both the quantum phenomenon and the primes, but many mathematicians thought it unlikely that physical intuition would help to prove the Riemann Hypothesis. When word started to get round that one of the most successful architects of pure mathematical theory had turned his attention to the Riemann Hypothesis, mathematicians' self-confidence seemed to be justified. Alain Connes had started giving lectures on his ideas for a solution in the mid-1990s. Many felt that the Riemann Hypothesis' time had finally come.

The fact that Connes was tackling the Riemann Hypothesis head on was noteworthy in itself. Selberg, for example, admits that he has never really tried to prove the Riemann Hypothesis. There's no point going off to fight a battle, he says, if you haven't got a weapon to fight with. Connes writes of his decision to take on the fight that, 'According to my first teacher, Gustave Choquet, one does, by openly facing a well known unsolved problem, run the risk of being remembered more by one's failure than anything else. After reaching a certain age, I realised that waiting "safely" until one reaches the end-point of one's life is an equally self-defeating alternative.'

Connes appeared to have access to a potent armoury of techniques which he had used to lay bare mysteries in other corners of mathematics. His creation of a subject called non-commutative geometry has been heralded as a modern version of Riemann's vision of geometry which had had such a significant impact on the course of mathematics in the nineteenth century. Just as Riemann's work paved the way for Einstein's breakthroughs in the theory of relativity, Connes's non-commutative geometry has proved a powerful language for understanding the complexities of the world of quantum physics.

The new mathematics that Connes had created is regarded as one of the milestones of twentieth-century mathematics, and it earned him a Fields Medal in 1983. Yet Connes's new language had not appeared out of the blue: it was part of a resurgence of French mathematics which began during the Second World War. Whilst the Institute in Princeton had flourished thanks to the influx of intellectuals fleeing persecution in Europe, Connes was a professor at an institute in France created in the 1950s which would help return Paris to the centre of the mathematical stage, a position it had lost to Göttingen during Napoleon's reign.

Connes's ideas are part of a mathematical movement towards a very sophisticated and abstract view of the subject. Over the last fifty years, the very language in which mathematics is expressed has undergone an evolutionary shift. It is still evolving, and many believe that until this process is complete we will not have a language which is sufficiently advanced to articulate an explanation for why the primes behave as Riemann's Hypothesis predicts. This new French mathematical revolution has its origins in a prison cell in France during the Second World War. From this cell there emerged a new language of mathematics, one which soon showed its power to explore landscapes like the one Riemann constructed to understand the primes.

Speaking in many tongues

In 1940, Elie Cartan received an envelope addressed to him as editor of the prestigious French journal *Comptes Rendus*. Ever since the beginning of the nineteenth century, when Cauchy had published his epic papers on the mathematics of imaginary numbers, *Comptes Rendus* had become one of the premier journals for making announcements of exciting new results. What intrigued Cartan about the package was the address from which it had been sent: the Bonne-Nouvelle military prison in Rouen. If it weren't for the fact that Cartan recognised the handwriting on the envelope, he might well have cast it aside, convinced it was from yet another crank announcing a proof of Fermat's Last Theorem. The handwriting belonged to a mathematician named André Weil. Weil had already established a reputation as one of France's leading young mathematical stars. Cartan knew that anything of Weil's would be worth reading.

When he slit open the envelope, his shock at getting a paper from a military prison was surpassed by his surprise at its contents. Weil had found a way to show why points at sea level in certain landscapes like to lie on a straight line. Although the technique didn't work for Riemann's landscape, the fact that it worked for other landscapes was proof enough for Cartan that here was something significant. Weil's theorem subsequently became a guiding light for mathematicians in search of a proof of the Riemann Hypothesis. Connes's own approach owes much to the ideas that Weil forged in the quiet of his prison cell in Rouen.

Weil's ability to navigate some of these landscapes where others had failed can be traced back to his early love of ancient languages, especially Sanskrit. He believed that the development of new mathematical ideas went hand in hand with the development of sophisticated forms of language. For Weil it was no surprise that in India the invention of grammar had preceded that of decimal notation and negative numbers, and that the algebra of the Arabs was born out of the sophisticated development of the Arabic language in medieval times.

His strong linguistic skills contributed to Weil's remarkable ability to create a new mathematical language which allowed him to articulate hitherto inexpressible subtleties. But it was his obsession with languages and in particular his love of the ancient Sanskrit text, the *Mahabharata*, that at the beginning of 1940 landed the eminent young mathematician in jail.

Weil's mathematical talents had been evident from an early age. His first teacher said of her six-year old pupil, 'No matter what I tell him on that subject, he seems to know it already.' His mother was sure that her son couldn't be getting sufficient intellectual stimulation if he was top of the class all the time. She went to the headmaster and insisted that her young André be moved up several classes. The astonished headmaster replied, 'Madame, this is the first time a mother has ever complained to me that her son's class rank is too high.' Thanks to his pushy mother, André found himself in the class of Monsieur Monbeig.

Monbeig had an unconventional approach to learning which Weil credits for his development as a mathematician. For example, rather than learning grammar by rote, his teacher had developed an elaborate personal system of algebraic notation to lay bare the patterns beneath. Later in life, when Weil encountered Noam Chomsky's revolutionary ideas on linguistics, nothing in Chomsky seemed new to him. Weil recognised that 'this early practice with a non-trivial symbolism must have been of great educational value, particularly for a future mathematician'.

Mathematics became Weil's passion and his addiction. 'Once when I took a painful fall, my sister Simone could think of nothing for it but to run and fetch my algebra book, to comfort me.' Weil's talents came to the attention of one of the great legends of French mathematics. Jacques Hadamard, who had made his name at the turn of the century by proving Gauss's Prime Number Theorem, encouraged Weil to pursue mathematics, and at the age of sixteen Weil entered the École Normale Supérieure, one of the Paris academies created during the French Revolution, to begin his professional mathematical training.

As well as studying mathematics at the École, Weil indulged his passion for ancient languages. Out of this love affair would later come a new mathematical world, but at the time Weil simply wanted to be able read the epic poetry of ancient Greece and India in the original ancient tongues. One epic in particular was to become a close companion throughout his life: the *Bhagavad-Gita*, the Song of God from the *Mahabharata*. In Paris, he spent as much time learning Sanskrit as he did doing mathematics.

Weil believed that the only way to see the full beauty of any text, not just epic poetry, was to read the original. Weil believed that in mathematics too one should go back and read the original papers of the masters, not rely on secondary accounts of their work. 'I had become convinced that what really counts in the history of humanity are the truly great minds, and that the only way to get to know these minds was through direct contact with their works,' he would write in his autobiography, *The Apprenticeship of a Mathematician*. This was how he came to study Riemann's work. 'Starting out thus was a stroke of luck for which I have always been grateful.' Riemann's Hypothesis on the nature of the primes was to infuse Weil's mathematical life.

When he finished his exams at the École, Weil was still below the age of compulsory military service, and instead set off for a grand mathematical tour of Europe. He criss-crossed the continent – Milan, Copenhagen, Berlin, Stockholm – sitting in on lectures and talking to the mathematical pioneers of the age. At Göttingen, as yet untouched by Hitler's academic purging, ideas came together in Weil's mind that would form the basis for his doctoral thesis. In the hometown of three of Europe's most eminent mathematicians – Gauss, Riemann and Hilbert – it was obvious to Weil that Paris had lost the mathematical reputation that it had enjoyed in the heady days of Fourier and Cauchy. This was partly because many budding French mathematicians who might have become senior figures during the 1930s had lost their lives in the First World War. There was a missing generation. In the aftermath of the war, few of the great German names came to Paris to present their work, starving the city of new ideas. What would happen to the great French mathematical heritage, going back to Fermat? Weil and a number of other young mathematicians decided to take things into their own hands.

There was no father figure around which the ambitious young students could gather, so they invented one: Nicolas Bourbaki. Under this pseudonym they would collectively compile an account of the contemporary state of mathematics. Their guiding ethos went back to what makes mathematics so unique amongst the sciences. Mathematics is an edifice, built upon axioms, in which a theorem proved in ancient Greece is still a theorem in twenty-first century mathematics. The Bourbaki group set out to survey the current state of the edifice and provide a comprehensive report written in the language of modern mathematics. Taking inspiration from Euclid's great treatise which had launched Western mathematics two thousand years before, they gave their work the title *Éléments de Mathématique*. Despite this Greek heritage, it was a uniquely French affair. The emphasis was on the broadest possible context for any result. If this meant losing sight of the specific question that the mathematics had originally been developed to answer, then so be it.

The origins of the choice of 'Nicolas Bourbaki' – in fact the name of a little-known French general – to lead their mathematical assault lie in a ritual observed at the École Normale at the beginning of the twentieth century. First-year students faced an initiation ceremony in which a senior student posing as a distinguished foreign visitor presented an elaborate lecture on well-known mathematical theorems. The lecturer would slip deliberate mistakes into some of the proofs he presented, which the first-year students had to spot. The clue was that the theorems with the mistakes were mis-attributed to obscure French generals rather than their rightful authors.

The meetings of these young French writers were anarchic, chaotic affairs. One of its founders, Jean Dieudonné, described how 'certain foreigners, invited as spectators to Bourbaki meetings, always come out with the impression that it is a gathering of madmen. They could not imagine how these people, shouting – sometimes three or four at the same time – could ever come up with something intelligent.' The members of Bourbaki believed that this anarchic character was essential to the functioning of the project. As they battled to unify the current state of mathematics, the new language that Weil would later develop began to emerge.

André Weil (1906–98) in Aligarh, India with Vijayaraghavan (next to Weil) and two students, pictured in 1931.

Weil's love of ancient languages and Sanskrit literature led to his first appointment, in 1930, as a professor at the Aligarh Muslim University not far from Delhi. The university had intended him to teach French Civilisation, but at the last minute decided that he would teach mathematics instead. During his time in India, Weil met Gandhi. Weil's exposure to Gandhi's philosophy, together with his reading of the *Gita*, were to have fateful consequences for Weil when he returned to a Europe gearing up for war. Krishna's advice to Arjuna in the *Gita* is to act according to his *dharma*, his personal code of behaviour. For Arjuna, who was of warrior caste, this meant fighting despite the inevitable devastation. Weil felt that his *dharma* was telling him the opposite – to be true to his pacifist beliefs. He resolved that if war broke out he would avoid the call-up for the French Army by fleeing to a neutral country.

During the summer of 1939, he travelled with his wife to Finland. His hope that Finland would be a suitable stepping stone for an escape to America turned out to be a great mistake. On the night of August 23, 1939, Stalin signed the Nazi–Soviet Pact which allied the USSR with Germany. In exchange for Soviet neutrality, Hitler promised Stalin a free hand in Estonia, Latvia, eastern Poland and Finland. When war broke out in September 1939, the Finnish government knew that it would not be long before their country too was at war. Any Soviet links were therefore regarded with the utmost suspicion. So when the authorities came across letters being mailed to Soviet addresses by a French visitor and filled with incomprehensible equations, they quickly jumped to the conclusion that the visitor was acting for the enemy. By December 1939 the Frenchman had been arrested and charged with spying for Moscow. The evening before he was due to be executed, the chief of police was attending a state dinner and found himself sitting next to a mathematician from the University in Helsinki, Rolf Nevanlinna.

Over coffee, the chief of police turned to Nevanlinna. 'Tomorrow we are executing a spy who claims to know you. Ordinarily I wouldn't have troubled you with such trivia, but since we're both here anyway, I'm glad to have the opportunity to consult you.' 'What is his name?' asked the academic. 'André Weil,' the chief responded. Nevanlinna was flabbergasted. He had entertained Weil and his wife during the summer at his lakeside country home. 'Is it really necessary to execute him?' he pleaded. 'Couldn't you just escort him to the border and deport him?' 'Well, there's an idea; I hadn't thought of it.' And with that fortuitous meeting, Weil was spared a bullet and mathematics was spared losing one of its greatest practitioners of the twentieth century.

By February 1940 Weil was back in France, but languishing in a prison cell in Rouen awaiting trial on charges of desertion. One of the joys of mathematics is that it requires little equipment beyond pen, paper and imagination. The prison provided the first two of these. Weil already had plenty of the third. Selberg, in his native Norway, had found the isolation imposed by the war years perfect for doing mathematics. As a clerk in India, Ramanujan had thrived without any access to formal training. One of Hardy's students, Vijayaraghavan, who had been a colleague of Weil's in India, had often joked to Weil that 'if I could spend six months or a year in prison I would most certainly be able to prove the Riemann Hypothesis'. Weil suddenly had the chance to put Vijayaraghavan's theory to the test.

Riemann had constructed a landscape in which the points at sea level held the secrets of the behaviour of the primes. To prove the Riemann Hypothesis, Weil needed to show why these points at sea level were lining up in a straight line. He made several attempts to navigate Riemann's landscape, but without success. But since Riemann's discovery of the wormhole connecting primes to his zeta landscape, mathematicians had come across similar landscapes which helped to explain other problems in number theory. Such was the power of these diverse landscapes, each defined by a variant of the zeta function, that they were beginning to gain almost a cult status. They would become so ubiquitous a way of answering problems in number theory that Selberg once declared he thought there should be a test-ban treaty on the further proliferation of zeta functions.

It was while exploring some of these related landscapes that Weil discovered a method that would explain why points at sea level in them like to be in a straight line. The landscapes where Weil was successful did not have to do with prime numbers, but held the key to counting how many solutions an equation such as $y^2 = x^3 - x$ will have if you are working on one of Gauss's clock calculators. For example, take the above equation and a five-hour calculator. If we put x equal to 2 on the right-hand side of the equation, we get $2^3 - 2 = 8 - 2 = 6$, which is 1 o'clock on the five-hour clock calculator. Similarly, taking y equal to 4 on the other side gives us 16, which is also 1 o'clock on the five-hour clock calculator. This result, which we can write as $(x, y) = (2, 4)$, is called a solution of the equation because both sides of the equation match up when calculated on the five-hour clock calculator. There are in fact seven different choices for the pair of numbers (x, y) which make the equation true:

$$(x, y) = (0, 0),\ (1, 0),\ (2, 1),\ (2, 4),\ (3, 2),\ (3, 3),\ (4, 0)$$

What if we chose a different prime number clock with p hours on the clock face? The number of choices which will satisfy the equation is approximately p. But not exactly. Just as Gauss's logarithmic guess for the number of primes fluctuates either side of the true number of primes, so does the number p under- or over-estimate the true number of solutions to the equation. In fact, it was Gauss – in the very last entry of his mathematical diary – who first proved for this particular equation that the error in the estimate would be no more than twice the square root of p. Gauss had used ad hoc methods which would not work on other equations. The beauty of Weil's proof was that it applied to any equation built from the variables x and y. By proving that points at sea level in the zeta landscape of each equation are in a straight line, Weil had generalised Gauss's discovery that the error in the estimate would be no more than essentially the square root of p.

Although it had no direct relationship to the Riemann Hypothesis for primes, Weil's proof was nonetheless a psychologically important breakthrough. He had found a way to show that points at sea level in a landscape built from equations such as $y^2 = x^3 - x$ all lie in a straight line. The reason Cartan was so excited when he opened Weil's parcel and saw the proof was that he could visualise these new techniques helping to make sense of Riemann's original landscape.

Weil had taken the first steps towards a whole new language for understanding solutions to equations. A school of Italian mathematicians led by Francesco Severi and Guido Castelnuovo based in Rome had already been doing something like this, and Weil would have learnt of their work on his European tour. But these Italian foundations were distinctly shaky and weren't up to supporting the mathematics that Weil required. Weil's ideas became the foundations of the subject we call algebraic geometry, which is at the heart of the proof of Fermat's Last Theorem.

Working with this new language, Weil was able to construct for each equation a very special sort of mathematical drum. It had a finite number of frequencies, in contrast to the infinitely many frequencies of physical drums and the infinitely many energy levels in quantum physics. The frequencies of Weil's drum marked precisely the coordinates of the points at sea level in the landscape corresponding to the equation. But he still had to work hard to make them line up in a straight line. These were no longer frequencies that mirrored energy levels in quantum physics, where a zero off the line would imply an imaginary energy level, something forbidden by the physical theory. He needed something different to force the zeros to lie on a straight line.

As Weil sat in his cell listening to the drum he had constructed, it suddenly dawned on him that he already had the last piece of the jigsaw that would explain why the frequencies of this drum lay on a straight line. During his tour of Europe as a graduate student he had learnt about a theorem derived by the Italian mathematician Guido Castelnuovo that now proved crucial in forcing the zeros of these equation-counting landscapes to line up in an orderly fashion. Without the lucky break that Castelnuovo's result provided, these landscapes might have remained as inaccessible as Riemann's. As Peter Sarnak at Princeton admits, 'the fact that Weil managed to make his proof work was something of a miracle'.

Weil had partially succeeded in realising Vijayaraghavan's dream. He may not have cracked the Riemann Hypothesis for primes, but he had found a way to show that points at sea level in related landscapes like to lie on a straight line. He wrote to his wife, Eveline, on April 7, 1940, telling her, 'My mathematics work is proceeding beyond my wildest hopes, and I am even a bit worried – if it is only in prison that I work so well, will I have to arrange to spend two or three months locked up every year?' Usually Weil would have waited before publishing anything, but the future was far too uncertain to risk delay. Weil prepared his note for *Comptes Rendus* and mailed it to Elie Cartan.

Weil wrote to his wife about his paper, saying, 'I am very pleased with it, especially because of where it was written (it must be the first in the history of mathematics) and because it is a fine way of letting all my friends around the world know that I exist. And I am thrilled by the beauty of my theorems.' On reading the note, Elie Cartan's son Henri, a friend and mathematical contemporary of Weil's, wrote back enviously: 'We're not all lucky enough to sit and work undisturbed like you . . .'

Cartan senior was only too happy to publish the paper. On May 3, 1940, Weil's period of productive incarceration was brought to an end. Cartan testified at the trial, which Weil described as a 'rather badly acted comedy'. Weil was sentenced to five years in prison for failing to report to duty, but the sentence would be suspended if he agreed to serve in a combat unit. Despite the mathematics that had flowed during his time in Rouen prison, Weil agreed to go into the army. It turned out to be a wise move. A month later, when the Germans advanced, all the prisoners in Rouen were shot by the French, allegedly to speed the warden's retreat.

Using a forged medical certificate that he had obtained in England, Weil was discharged in 1941 for having pneumonia. He managed to get visas for himself and his family to travel to America, where he met up with Siegel at the Institute for Advanced Study in Princeton. The two had become friends during Weil's tour of Europe. When Siegel went to pick up Riemann's unpublished notes, and discovered Riemann's secret formula for calculating zeros, Weil had accompanied him. Siegel was obviously keen to understand whether Weil's successful navigation of a related landscape could be extended to understand the original Riemann landscape.

Many believed, like Siegel, that whatever it was that made Weil's proof work on one landscape should provide an essential clue in the search for the true Grail of the Riemann Hypothesis. Weil tried for years to find that elusive link with the landscape created by Riemann. Unfortunately, as a free man he was never to match the success he had achieved as a prisoner in Rouen. One can hear the melancholy as Weil described in later life that desire to relive the rush of that first discovery: 'Every mathematician worthy of the name has experienced . . . the state of lucid exaltation in which one thought succeeds another as if miraculously . . . this feeling may last for hours at a time, even for days. Once you have experienced it, you are eager to repeat it but unable to do it at will, unless perhaps by dogged work . . .'

In an interview for *La Science* in 1979, Weil was asked which theorem he most wished he had proved. He replied that 'In the past it sometimes occurred to me that if I could prove the Riemann Hypothesis, which was formulated in 1859, I would keep it secret in order to be able to reveal it only on the occasion of its centenary in 1959.' But despite a concerted effort, nothing gave. 'Since 1959, I have felt that I am quite far from it; I have gradually given up, not without regret.'

Weil was in close contact throughout his life with Goro Shimura, one of the Japanese mathematicians who formulated the conjecture that Andrew Wiles solved on his way to Fermat's Last Theorem. Shimura recalls how Weil admitted to him once in later life, 'I'd like to see the Riemann Hypothesis settled before I die, but that is unlikely.' Shimura remembers a conversation they once had about Charlie Chaplin. In his youth Chaplin had visited a fortune-teller who had predicted accurately what the future held in store for him. Weil wistfully joked, 'Well, in my autobiography I might write that in my youth I was told by a fortune-teller that I would never be able to solve the Riemann Hypothesis.'

Even though Weil's dream of proving the Riemann Hypothesis, or at least of seeing it proved, never came true, there is no doubt that his work is hugely significant. Weil's proof gave mathematicians some faith that it could be conquered. It also helped them believe that Riemann's hunch is probably right. If the points at sea level were lining up in one zeta landscape, there was hope that they would also line up in the landscape of the primes. Not only that, Weil had even used a strange mathematical drum to navigate his landscape long before the connections with quantum chaos had told us that this was a good way to look for a solution. As Peter Sarnak expressed it, 'Weil's result has become our guiding light in our search for the Riemann Hypothesis.'

Weil's new mathematical language, algebraic geometry, had enabled him to articulate subtleties about solutions to equations that hitherto had been impossible. But if there was any hope of extending Weil's ideas to prove the Riemann Hypothesis, it was clear they would need to be developed beyond the foundations he had laid in his prison cell in Rouen. It would be another mathematician from Paris who would bring the bones of Weil's new language to life. The master architect who performed this task was one of the strangest and most revolutionary mathematicians of the twentieth century – Alexandre Grothendieck.

A new French Revolution

Napoleon had forged his academic revolution with the creation of institutions such as the École Polytechnique and the École Normale Supérieure. But too strong an emphasis on mathematics serving the needs of the state had seen Paris lose its place as the focus of mathematical activity to the medieval town of Göttingen, where the more abstract approach of Gauss and Riemann was allowed to flourish. In the second half of the twentieth century there was a new optimism in France that Paris could regain its position as a key player in the world of mathematics.

Through the initiative of the Russian émigré industrialist Léon Motchane, who had a passion for science, and with the academic guidance of key figures in the Bourbaki group, it was decided to create a new institute that would be modelled on the successful Institute for Advanced Study in Princeton. Unlike Napoleon's academies, this new institute would be outside state control. Funded by private enterprise, the Institut des Hautes Études Scientifiques was established in 1958. Its buildings are tucked away in the woods of Bois-Marie, not far outside Paris. Over the years it has successfully realised the dreams of its creators. Marcel Boiteux, a former president of the Institut, described it as 'a radiant hearth, a vibrant hive, and a monastery, where deep-sown seeds germinate and grow to maturity at their own pace'. One of the first professors to be appointed at the Institut was a young mathematical star by the name of Alexandre Grothendieck. This first seed would blossom in most spectacular fashion.

Grothendieck is an austere mathematician. His office in the Institut was bare except for an oil painting of his father. The painting was by a fellow inmate at one of the camps where his father was confined before he was sent to Auschwitz, where he perished in 1942. Grothendieck shares with his father the fiery expression in the eyes that burn out of the shaven head in the portrait.

Although he didn't know his father, the adulation with which his mother spoke of him had a profound effect on Grothendieck. As he once commented, his father's career read like a veritable who's who of European revolutions from 1900 to 1940: from being a leader in the Bolshevik revolution in October 1917, to armed clashes on the streets of Berlin with the Nazis, to serving in the anarchist militia during the Spanish Civil War. The Nazis finally caught up with him in France, where he was handed over as a Jew by the Vichy Government.

Grothendieck's own revolution would be brought about not on the political battlefield but on the mathematical stage. From Weil's tentative

Alexandre Grothendieck, professor at the Institut des Hautes Études Scientifiques until 1970.

beginnings he developed a new language for mathematics. Just as Riemann's new insights marked a turning point in mathematics, Grothendieck's new language of geometry and algebra saw the creation of a whole new dialectic which allowed mathematicians to articulate ideas which were previously inexpressible. This new perspective can be compared to the new vistas that had opened up at the end of the eighteenth century, when mathematicians finally accepted the concept of imaginary numbers. But the new language was not an easy one to learn. Even Weil was rather disconcerted by Grothendieck's new abstract world.

The Institut des Hautes Études Scientifiques became the natural postwar home of the Bourbaki project, which was still busy producing further volumes of its encyclopaedic survey of modern mathematics. Grothendieck became one of its major contributors. As senior members of the group reached the age of fifty, they retired from Bourbaki and new young French blood was sought to fill their places. More than anything else, Bourbaki's publications helped to re-establish France as the centre of the mathematical world. In the minds of many mathematicians Bourbaki was a person in his own right, and he even applied to become a member of the American Mathematical Society.

Many outside France have criticised Bourbaki's effect on mathematics, claiming that he has been selective in what he has chosen to document. They believe that he has sterilised mathematical research by presenting mathematics as a finished product rather than an evolving organism. His emphasis on the broad sweep ignores the quirky and often special aspects of the subject. But Bourbaki believes that his project has been misinterpreted. The tomes that bear his name are there to confirm the solidity of the position we now occupy. They are meant as a new *Elements*, a modern equivalent of the springboard that Euclid had provided two millennia ago.

The old guard, those mathematicians active before the Second World War, began to complain that they no longer recognised the subject that they had worked in for so many years. Siegel had this to say about an account of his own work that was cast in the new language:

> I was disgusted with the way in which my own contribution to the subject had been disfigured and made unintelligible. The whole style . . . contradicts the sense for simplicity and honesty which we admire in the works of the masters in number theory – Lagrange, Gauss, or on a smaller scale, Hardy, Landau. I see a pig broken into a beautiful garden and rooting up all flowers and trees.

He was pessimistic about the future of mathematics in the face of such abstraction: 'I am afraid that mathematics will perish before the end of this century if the present trend for senseless abstraction – as I call it: theory of the empty set – cannot be blocked up.'

It was a view shared by many. Selberg described his impressions after listening to a lecture outlining an abstract framework in which the Riemann Hypothesis might be proved. 'My thought was that such lectures were never given in earlier times. I said to someone after the lecture a thought which had come into my mind: if wishes were horses, then beggars can ride.' In the lecture a whole framework of abstract hypotheses had been proposed. If the language could be made to fit the theory of the primes, then the lecturer would have been able to prove the Riemann

Hypothesis. But as Selberg complains, 'He had none of these hypotheses that he wanted. This is probably not the right way to think about mathematics. One should try to start with something you can get hold of and get a grip of. The talk contained very interesting things but it is an example of a trend that I think is very dangerous.'

But for Grothendieck this was not abstraction for abstraction's sake. In his view this was a revolution that was necessitated by the questions that mathematics was trying to answer. He wrote volume after volume describing this new language. Grothendieck's vision was messianic, and he began to attract a following of faithful young disciples. His output was huge, covering some ten thousand pages. When a visitor complained at the poor state of the library at the Institut, he replied, 'We don't read books here, we write them.'

Gödel had spoken of the need to expand the foundations of the subject before the Riemann Hypothesis could be truly mastered. Grothendieck's revolutionary new language was a first step in an attempt to do just that, yet despite all his efforts the Riemann Hypothesis remained frustratingly out of his reach. His revolution answered many other problems, including important conjectures made by Weil in connection with counting solutions to equations, but not this one.

In fact, it was his father's political past that was ultimately responsible for Grothendieck's failure to climb to the summit of Mount Riemann. Grothendieck did his best to live up to his father's political ideals. He became a staunch pacifist, campaigning vociferously against the military build-up in the 1960s. He objected strongly to the worsening political situation in Russia – so much so that when he was awarded his Fields Medal in 1966, in recognition of his advances in algebraic geometry, he refused to travel to Moscow to collect the prize in protest at Soviet military escalation.

All that time spent exploring the mathematical world had left Grothendieck politically rather naive. When he was shown a poster advertising a conference sponsored by NATO at which he was to be the main speaker, Grothendieck asked innocently what NATO stood for. As soon as he learnt that it was a military organisation, he wrote to the conference organisers threatening to withdraw. (The organisers actually waived the money in order to keep him.) In 1967, Grothendieck gave a short lecture course on abstract algebraic geometry to a bemused audience in the jungle of North Vietnam, where the University of Hanoi had been evacuated during the bombing. He saw his talk, full of abstruse ideas, as a protest against the war which was raging within earshot.

Grothendieck lecturing at the Institut des Hautes Études Scientifiques.

Things came to a head in 1970, when Grothendieck found that some of the private finance for the Institut was coming from military sources. He went straight to the director, Léon Motchane, threatening to resign. Motchane, the force behind the establishment of the Institut, was not as flexible as those conference organisers a few years before. Grothendieck stuck to his principles and left. Those close to him believe that he may have used the issue of military funding as an excuse to escape from the golden cage that the Institut had become. It seemed to Grothendieck that he was no more than a mathematical mandarin for the establishment. He was happier as an outcast – he hated being on the cosy inside of things. He was also forty-two. The myth that a mathematician has done his best work by the age of forty was beginning to worry Grothendieck. What if the rest of his mathematical life was devoid of creativity? He wasn't one to live on past laurels. He was also becoming increasingly disillusioned by his failure to make any more headway on charting the points at sea level. In the comfort of the Institut, Grothendieck had got no further than Weil had in his prison cell. When he left the Institut des Hautes Études Scientifiques, he pretty much left mathematics.

Grothendieck started to drift. He set up a group called Survive, dedicated to anti-military and ecological issues. He began practising Buddhism with a fervour that his Hasidic ancestors would have recognised. The bitterness he felt at being unable to complete his mathematical vision boiled over in an extraordinary thousand-page autobiography in which he violently attacked what had been made of his mathematical legacy. He couldn't accept that his mathematical disciples were now the new leaders of the revolution he had inspired and were putting their own individual stamps on the subject.

Some thirty years after walking out of the Institut, Grothendieck now lives in a remote hamlet in the Pyrenees. According to a couple of mathematicians who visited him a few years ago, 'he is obsessed with the Devil, which he sees at work everywhere in the world, destroying the divine harmony'. He held the Devil responsible for, among other things, changing the speed of light from the nice round value of 300,000 km/s to the 'ugly' 299,887 km/s. All mathematicians need to have a little bit of madness if they are to feel at home in the mathematical world. The sheer number of hours Grothendieck had spent exploring at the edges of the world of mathematics left him unable to chart his way home.

Grothendieck is not the only mathematician who has gone crazy trying to prove the Riemann Hypothesis. In the late 1950s, after his early successes, John Forbes Nash became captivated by the prospect of proving the Riemann Hypothesis. According to Sylvia Nasar's biography of Nash, *A Beautiful Mind*, people were 'gossiping that Nash was in love with Cohen', who was also wrestling with the Riemann Hypothesis. Nash talked extensively with Paul Cohen about his own ideas, but Cohen could see they were going nowhere. Some believe that Cohen's rejection of Nash – both emotionally and mathematically – contributed to Nash's subsequent mental decline. In 1959 he was invited to present his ideas for a solution to the Riemann Hypothesis to a meeting of the American Mathematical Society in Columbia University, New York. It was a disaster. The audience sat in stunned silence as he went off the rails before their very eyes, offering a series of nonsensical arguments purporting to prove the Riemann Hypothesis. Grothendieck and Nash illustrate the dangers of mathematical obsession. (Unlike Grothendieck, Nash made it back from the brink and went on to share the 1994 Nobel Prize for Economics for his mathematical development of game theory.)

In contrast to the psychological collapse that Grothendieck has suffered, his mathematical structure remains erect. Many believe that crucial ideas we are still missing will extend Grothendieck's revolution and will finally unveil the mysteries of the primes. In the mid-1990s, the mathematical community began to buzz with the news that maybe Grothendieck's successor was at hand.

The last laugh

When the word started to spread that Alain Connes was working on the Riemann Hypothesis, many eyebrows were raised. Connes, a professor at the Institut des Hautes Études Scientifiques and at the Collège de France, is a heavyweight with a reputation to match Grothendieck's. His invention of non-commutative geometry does indeed go beyond the geometry of Weil and Grothendieck. Connes, like Grothendieck, is someone who is able to see structure where others see only a mess.

In mathematics, 'non-commutative' means that it matters in which order you do something. For example, take a square photograph of someone's face and place it face down. First, flip the photo over from right to left and then rotate it 90 degrees clockwise. Now repeat the experiment, but do the rotation before the flip (again, making sure you flip from right to left), and you'll find that the face is pointing in the opposite direction. It matters which operation you do first. The same principle is at the heart of many of the mysteries of quantum physics. Heisenberg's uncertainty principle says that we can never know precisely the position and the momentum of a particle. The mathematical reason for this uncertainty is that it matters in which order you measure the position and the momentum.

Connes has taken the algebraic geometry of Weil and Grothendieck into regions of mathematics where such symmetries break down, revealing a completely new mathematical world. Whilst most mathematicians spend their lives gaining a better understanding of the mathematical landscape they see around them, every few generations comes an explorer who can strike out and find undiscovered continents. Connes is such an explorer.

These explorations are an all-consuming passion for Connes. His love of the subject goes back to when he first began to contemplate elementary mathematical problems at the age of seven. 'I very clearly remember the intense pleasure that I had plunging into the special state of concentration that one needs in order to do mathematics.' It seems he has never emerged from this trance. And for all his intimidating theory and abstraction, Connes retains something of that boyish playfulness that he had when he was seven. For Connes, mathematics more than anything else can bring you closest to a concept of ultimate truth. And the joyful pursuit of that goal has been part of his dedication to the subject ever since he was a boy. As he puts it, since 'mathematical reality cannot be located in space or time, it affords – when one is fortunate enough to uncover the minutest portion of it – a sensation of extraordinary pleasure through the feeling of timelessness that it produces'.

He describes a mathematician as someone who is always active, always on the lookout for new territory to move into. Whilst others will sail close to the shores of the land that they recognise, Connes will leave the familiar mathematical landscape behind him and voyage across uncharted waters that lie well beyond our current mathematical horizon. The fact that he was able to see the connections between the primes and the harsh abstract world of non-commutative geometry owes much to his talent for borrowing from the different mathematical cultures he visits on his mathematical travels. Some mathematicians prefer to explore in pairs or groups. Their pooled skills can help them cross mathematical oceans where a single-handed attempt might fail. But Connes is one of those travellers who revel in their solitude: 'If one really wants to find something it is necessary to be alone.'

The new geometry that Connes discovered grew out of Weil and Grothendieck's development of algebraic geometry. What Weil and Grothendieck had provided was a new dictionary which could be used to translate geometry into algebra. The dictionary comes into its own when a problem expressed in the language of geometry is obscure and shrouded in mystery, but when translated into algebra becomes immediately transparent. This is how Weil had managed to count solutions to equations and prove that the zeros in the associated landscape are on a straight line. He would have got nowhere by trying to understand the geometrical shapes carved out by these equations. But once he had his algebraic–geometric dictionary, he had the means to understand.

Where Weil's geometry answered questions of pure number theory, Connes's ideas provided a mathematical description of a geometry that the string theorists and quantum physicists have been desperate to construct. Physicists at the end of the twentieth century were crying out for a new geometry to underpin string theory, which had been introduced in the 1970s as a possible solution to the incompatibility of quantum physics and relativity. Connes was intrigued and began to search for the geometry that the physicists thought should exist. He realised that, even without a clear picture of the physical side of the geometry, he could still construct its abstract algebraic side. It was a discovery that only someone reared in the abstract world of mathematics could have made – physical intuition would not have turned it up.

The strange behaviour of the subatomic world forced Connes to completely set aside the standard ways by which to understand conventional geometry. If Riemann's geometrical revolution provided Einstein with the language in which to describe the physics of the very big, then Connes's geometry gives mathematicians the chance to reach into the strange geometry of the very small. Thanks to him, we might at last be able to decipher the fine structure of space.

Hugh Montgomery and Michael Berry had highlighted the possible connection between the primes and quantum chaos. The fact that Connes's language was perfectly suited to quantum physics contributed to the growing sense of optimism about his assault on the Riemann Hypothesis. Given his emergence from a French mathematical renaissance which had already created new techniques for navigating zeta landscapes, the mathematical community were perhaps justified in believing that the end was near. Finally, all the strands appeared to be coming together.

What Connes believes he has identified is a very complicated geometrical space, called the non-commutative space of Adele classes, built in the world of algebra. To build his space he used strange numbers discovered at the beginning of the twentieth century called *p*-adic numbers. There is one family of *p*-adic numbers for each prime, *p*. Connes believes that by gluing all these numbers together and looking at how multiplication works in this highly singular space, the Riemann zeros should appear naturally as resonances in this space. His approach is an exotic cocktail of many of the ingredients that had emerged over the centuries of studying the primes. It wasn't surprising then that mathematicians were hopeful that he might indeed succeed.

Connes is not only a masterful mathematician but a charismatic performer. Many have been mesmerised by his presentations on the Riemann Hypothesis. I have sat listening to him talk, convinced that from the work he was describing a proof would inevitably follow, that he has done all the hard work and all that's left is for others to add the finishing touches. But though it might appear that he has the big idea that everyone has been hankering for, Connes himself knows that there's a lot of ground still to cover. 'The process of verification can be very painful: one's terribly afraid of being wrong . . . it involves the most anxiety, for one never knows if one's intuition is right – a bit as in dreams, where intuition very often proves mistaken.'

In the spring of 1997, Connes went to Princeton to explain his new ideas to the big guns: Bombieri, Selberg and Sarnak. Princeton was still the undisputed Mecca of the Riemann Hypothesis despite Paris's push to reclaim its dominance. Selberg had become godfather to the problem – nothing could pass muster before being vetted by a man who had spent half a century doing battle with the primes. Sarnak was the young gun whose rapier-like intellect would cut through anything that was found slightly wanting. He'd recently joined forces with Nick Katz, also at Princeton, one of the undisputed masters of the mathematics developed by Weil and Grothendieck. Together they had proved that the strange statistics of random drums that we believe describe the zeros in Riemann's landscape are definitely present in the landscapes considered by Weil and Grothendieck. Katz's eyes were particularly sharp, and little escaped his penetrating stare. It was Katz who, some years before, had found the mistake in Wiles's first erroneous proof of Fermat's Last Theorem.

And finally there was Bombieri, sitting in state as the undisputed master of the Riemann Hypothesis. He had earned his Fields Medal for the most significant result to date about the error between the true number of primes and Gauss's guess – a proof of something mathematicians call the 'Riemann Hypothesis on average'. In the quiet of his office overlooking the woods that surround the Institute, Bombieri has been marshalling all his insights of previous years for a final push for the complete solution. Bombieri, like Katz, has a fine eye for detail. A keen philatelist, he once had the chance to purchase a very rare stamp to add to his collection. After scrutinising it carefully he discovered three flaws. He returned the stamp to the dealer, pointing out two of them. The third subtle flaw he kept to himself – in case he is offered an improved forgery at a future date. Any aspiring proof of the Riemann Hypothesis is subjected to an equally painstaking examination.

Selberg, Sarnak, Katz, Bombieri: a formidable line-up, but Connes was not at all intimidated. The force of his argument and personality were easily a match for Princeton's big guns. He knew that he had no proof yet, but he was convinced that his perspective offered the best prospect to date for a solution to the Riemann Hypothesis. It united many of the ideas that had emerged from the quantum physicists and from the mathematical insights of Weil and Grothendieck.

The Princeton mafia agreed that some progress had been made, but they could also see that problems remained. Sarnak recognised that Connes had successfully developed ideas he had first heard for himself from his supervisor, Paul Cohen, shortly after he had arrived in Stanford. The difference was that Connes was now equipped with a sophisticated new language and techniques, which helped crystallise Cohen's ideas. But there was a problem with Connes's approach: he seemed to have arranged things so that any points which did happen to lie off Riemann's line couldn't be seen. Like a magician, Connes was making you look only at points on the line, whilst any points off the line disappeared up his mathematical sleeve.

'Connes hypnotises the audience,' says Sarnak. 'He's a really convincing guy. He's charming. You point out a difficulty with his approach, then next time you see him he says, "You were right." That's how he buys you over so easily.' And then, Sarnak explains, Connes quickly introduces the next new twist to his argument. Sarnak believes, though, that Connes is still missing the kind of magic that allowed Weil to make his breakthrough while sitting in prison in 1940. Bombieri agrees: 'I still think that some major new idea is needed here.'

Soon after Connes's lecture, Bombieri received an email from a friend, Doron Zeilberger of the University of Temple, who seemed to be claiming to have discovered the most fantastic new properties of π. Bombieri, though, was wise enough to pick up on the date: April the First. To show he had got the joke, Bombieri replied in kind. He mischievously tapped into the fever that was brewing over Connes's contributions to finding patterns in the primes: 'There are fantastic developments to Alain Connes's lecture at IAS last Wednesday . . .' A young physicist in the audience had seen in a flash how to complete Connes's project. The Riemann Hypothesis was true. 'Please give this the highest diffusion.'

Zeilberger obliged, and a week later the announcement had been splashed across the electronic bulletin board of the forthcoming International Congress for the world's mathematicians to read. It took some time to damp down the excitement that Bombieri had whipped up. Connes arrived back in Paris to discover people talking about the news. Even though the joke was really on the physicists, he was apparently quite upset.

Bombieri's April Fool joke seems to have marked the end of the excitement about Connes's work on the Riemann Hypothesis. Now the dust has settled it looks as though much of the hope that Connes's ideas might solve the secret of the primes has evaporated. Even in his sophisticated world of non-commutative geometry, the primes remain elusive. Several years after Connes burst upon the scene, Fortress Riemann is still holding out. Of course, it's still possible that Connes's approach will bear fruit. It has many things going for it. But the feeling that he has opened up easy access to the proof has disappeared. The walls that shield the Riemann Hypothesis may now look a little different, but they are as impenetrable as they were before.

Connes himself is philosophical about the impasse. As he said when the million-dollar prize for a solution of the Riemann Hypothesis was announced, 'For me, mathematics has always been the greatest school of humility. Mathematics is mainly valuable because of the immensely difficult problems which are like the Himalayas of Mathematics. To reach the peak will be extremely difficult and we might even have to pay the price. But what is true is that if we reach the peak, the view from there will be fantastic.' He has not given up on his quest and is still battling away, hoping for the last big idea that will complete the journey. He longs for that wonderful moment in every mathematician's life when everything comes together. 'The moment illumination occurs, it engages the emotions in such a way that it's impossible to remain passive or indifferent. On those rare occasions when I've actually experienced it, I couldn't keep tears from coming to my eyes.'

Still, then, we go on listening to that mysterious prime number beat: 2, 3, 5, 7, 11, 13, 17, 19, . . . The primes stretch out into the far reaches of the universe of numbers, never running dry. They are central to mathematics, the building blocks from which everything else follows. Do we really have to accept that, despite our desire for order and explanation, these fundamental numbers might forever remain out of reach?

Euclid proved that the primes go on for ever. Gauss guessed that the primes were picked out at random, as if determined by the tossing of a coin. Riemann was sucked down a wormhole into an imaginary landscape where the primes turned into music. In this landscape, each point at sea level sounded a note. The search was on to interpret Riemann's treasure map, and discover the location of each point at sea level. Armed with a formula he kept secret from the world, Riemann discovered that whereas primes appeared chaotic, the points in his map were full of order. Instead of being randomly dotted around they were lining up in a straight line. He couldn't see far enough across his landscape to tell whether this would always be true, but he believed it would be. The Riemann Hypothesis was born.

If Riemann's Hypothesis is true, then none of the notes will sound louder than the others; the orchestra playing the music of the primes will be in perfect balance. It will explain why we see no strong patterns in the primes. A pattern would correspond to one instrument playing louder than the others. It is as if each instrument plays its own pattern, but by combining together so perfectly, the patterns cancel themselves out, leaving just the formless ebb and flow of the primes.

If true, Riemann's Hypothesis will help us to see why the primes look as though they've been selected by tossing a coin. But maybe Riemann's hunch about these points at sea level is just wishful thinking. Perhaps, as the music evolves, one particular instrument in this orchestra of the primes will start to dominate the music. Maybe there are patterns to be discovered in the far reaches of the numbers. Maybe the prime number coin started to show a bias as Nature tossed it again and again in the process of creating the mathematical universe we inhabit. As we have discovered, prime numbers can be a malicious bunch, hiding their true colours from view.

So began the quest to confirm Riemann's belief that the points in his treasure map of the primes were all in a straight line. We have criss-crossed the historical and physical world: Napoleon's Revolutionary France; the neo-humanistic revolution in Germany, from grand Berlin to the medieval streets of Göttingen; the strange alliance between Cambridge and India; the isolation of war-torn Norway; the New World, and a new academy founded in Princeton for those brave seekers of Riemann's Grail expelled from Europe by the ravages of war; and finally to modern Paris and a new language, first spoken in a prison cell and which unhinged the mind of one of its key developers.

The story of the primes spreads well beyond the mathematical world. Technological advances have changed the way we do mathematics. The computer, born in Bletchley Park, has provided us with the ability to see numbers that were previously confined to the unobservable universe. The language of quantum physics has allowed mathematicians to articulate patterns and connections that might never have been discovered without this cross-fertilisation of scientific cultures. Even the corporate world of AT&T, Hewlett-Packard and an electronics warehouse in California have had a part to play. The central role of the primes in computer security has forced these numbers into the limelight. The primes now affect all our lives as they protect the world's electronic secrets from the prying eyes of Internet hackers.

Despite these twists and turns, prime numbers have remained elusive. Each time we chase them into new territory, be it Connes's non-commutative world or Berry's quantum chaos, they find new places to hide.

Many of the mathematicians who have contributed to our understanding of the primes have been rewarded with long lives. Having proved the Prime Number Theorem in 1896, Jacques Hadamard and Charles de la Vallée-Poussin both lived into their nineties. People had begun to believe that their having proved the Prime Number Theorem had made them immortal. The belief in a connection between longevity and the primes has been further fuelled by Atle Selberg and Paul Erdős, whose alternative elementary proof of the Prime Number Theorem in the 1940s saw both of them live into their eighties. Mathematicians joke about a new conjecture: anyone who proves the Riemann Hypothesis will indeed become immortal. Another joke is that someone somewhere has actually already disproved the Riemann Hypothesis, but no one has heard about it because that unlucky mathematician was struck dead immediately.

Opinions differ on how far we are from a solution. Andrew Odlyzko, who has calculated so many points at sea level in Riemann's treasure map, believes we just can't predict when: 'It could be next week, it could be a century from now. The problem seems to be hard. I doubt whether it is going to be something very easy, just because so many very good people have looked at it for so long and so hard. But it could be that someone will have a brilliant idea next week.' Others reckon we are still at least two good ideas away from a solution.

Hugh Montgomery believes that, given the outcome of his conversation with the quantum physicist Freeman Dyson over tea at Princeton, we have completed a good part of the climb to the top of Mount Riemann. But there is a sobering footnote to his optimism: 'We have a proof of the Riemann Hypothesis, except for a gap. Unfortunately that gap appears right at the beginning.' As Montgomery points out, that's a bad place to have a gap. Any gap is fatal. A gap in the middle would at least mean we'd made some progress on our journey. But a gap at the beginning means that unless we find a way through the first gate, the rest of the path we have laid out to the top of Mount Riemann is useless. 'It's producing a logjam in the theory that we can't get this first theorem proved.'

Many mathematicians are still too frightened to go near this notoriously difficult problem, despite the incentive of a million dollars for a solution. So many great names have tried and failed: Riemann, Hilbert, Hardy, Selberg, Connes . . . But there are still those brave enough to try, and names to look out for in the future include Christopher Deninger in Germany and Shai Haran in Israel.

Many predict that Riemann's Hypothesis will survive its bicentenary. Some believe its time has come, and with so much evidence for where we should be looking for a solution, it can't last out. Some believe that its fate lies in Gödel's hands: it will turn out to be true but unprovable. Some believe it is false. Some believe they have already proved it and the mathematical establishment dare not let go of its enigma. Some have gone mad in search of a solution.

Maybe we have become so hung up on looking at the primes from Gauss's and Riemann's perspective that what we are missing is simply a different way to understand these enigmatic numbers. Gauss gave an estimate for the number of primes, Riemann predicted that the guess is at worst the square root of N off its mark, Littlewood showed that you can't do better than this. Maybe there is an alternative viewpoint that no one has found because we have become so culturally attached to the house that Gauss built.

Like characters in a murder mystery, we've been working our way through the mathematical suspects. Who or what put the zeros on Riemann's critical line? The scene is strewn with evidence, fingerprints everywhere, we have a photo-fit of the solution – yet the answer eludes us. The consolation is that, even if the primes never yield up their secrets, they are leading us on the most extraordinary intellectual odyssey. They have acquired an importance which extends well beyond their fundamental role as the atoms of arithmetic. As we have discovered, they have opened doors between hitherto unrelated areas of mathematics. Number theory, geometry, analysis, logic, probability theory, quantum physics – all have been drawn together in our search for the Riemann Hypothesis. And that search has put mathematics in a new light. We marvel at its extraordinary interconnectedness: mathematics has gone from a subject of patterns to a subject of connections.

These linkages don't exist only within the mathematical world. The primes were once regarded as the ultimate abstract concept, devoid of any significance beyond the ivory tower. Mathematicians, and G.H. Hardy was perhaps the best example, once relished the thought of being able to examine the objects of their study in isolation, undistracted by concerns of relevance in the outside world. But no longer do the primes provide an escape from the pressures of the real world, as they did for Riemann and others. The primes are central to the security of the modern electronic world, and their resonances with quantum physics may have something to tell us about the nature of the physical world.

Even if we do succeed in proving the Riemann Hypothesis, there are many more questions and conjectures champing at the bit, many new exciting pieces of mathematics just waiting for the Hypothesis to be proved before they can be launched. The solution will be just a beginning, an opening up of uncharted virgin territory. In Andrew Wiles's words, the proof of the Riemann Hypothesis will allow us the possibility to navigate this world in the same way that the solution to the problem of longitude helped eighteenth-century explorers to navigate the physical world.

Until then, we shall listen enthralled by this unpredictable mathematical music, unable to master its twists and turns. The primes have been a constant companion in our exploration of the mathematical world yet they remain the most enigmatic of all numbers. Despite the best efforts of the greatest mathematical minds to explain the modulation and transformation of this mystical music, the primes remain an unanswered riddle. We still await the person whose name will live for ever as the mathematician who made the primes sing.

Acknowledgements

Many of my colleagues were very generous with their time and support. I would particularly like to thank the following, who were happy to sit and talk with me about their views and ideas: Leonard Adleman, Sir Michael Berry, Bryan Birch, Enrico Bombieri, Richard Brent, Paula Cohen, Brian Conrey, Persi Diaconis, Gerhard Frey, Timothy Gowers, Fritz Grunewald, Shai Haran, Roger Heath-Brown, Jon Keating, Neal Koblitz, Jeff Lagarias, Arjen Lenstra, Hendrik Lenstra, Alfred Menezes, Hugh Montgomery, Andrew Odlyzko, Samuel Patterson, Ron Rivest, Zeev Rudnick, Peter Sarnak, Dan Segal, Atle Selberg, Peter Shor, Herman te Riele, Scott Vanstone and Don Zagier.

I'd especially like to thank Sir Michael Berry, whom I first met on the stairs at No. 10 Downing Street while queuing to shake hands with a prime minister, and who first brought to my attention the music contained in the primes. The title of this book was inspired by that meeting.

I am indebted to a number of people who carefully read early versions of part or all of the manuscript: Sir Michael Berry, Jeremy Butterfield, Bernard du Sautoy, Jeremy Gray, Fritz Grunewald, Roger Heath-Brown, Andrew Hodges, Jon Keating, Angus Macintyre, Dan Segal, Jim Semple and Eric Weinstein. Any errors in the text that remain are of course my responsibility.

I have also benefited from the numerous books, articles and papers that have provided invaluable background material. Many of these sources are listed in Further Reading. I should particularly like to single out the *Notices of the American Mathematical Society*, which ceaselessly publishes articles full of wonderful insights into mathematics and the mathematical community.

Several institutions have been very co-operative in assisting me during the writing of this book, including the American Institute of Mathematics, Certicom, the University Library in Göttingen, AT&T Labs at Florham Park, the Institute for Advanced Study in Princeton, Hewlett-Packard Laboratories in Bristol and the Max Planck Institute for Mathematics in Bonn.

I am extremely glad to be able to acknowledge the debt I owe to those in publishing who made this book possible: my agent, Antony Topping at Greene & Heaton, who was there for me from the very first spark of an idea to the final publication; Judith Murray, who brought us together; my editors, Christopher Potter, Leo Hollis and Mitzi Angel at Fourth Estate, and Tim Duggan at HarperCollins; and my copy-editor, John Woodruff. I should particularly like to thank Leo, who spent so many hours bending his head round the fourth dimension.

I would not have been able to write this book without the support of the Royal Society. Being a Royal Society Research Fellow has provided me with the chance not only to follow my mathematical dreams, but also to communicate the excitement I've experienced along the way. The Royal Society is more than just a bank account – they care about who they fund. Their support of my activities to bring maths to the masses was invaluable.

I would also like to thank a number of people in the media who were brave enough to take the risk to publish and broadcast my first pieces about serious mathematics and who took the time to help a mathematician to write: Graham Patterson, Philippa Ingram and Anjana Ahuja at *The Times*; John Watkins and Peter Evans at the BBC; and Gerhart Friedlander at Science Spectra. I am grateful also to NCR and Milestone Pictures for the chance to bring mathematics to the banking community.

I became a mathematician because of one teacher at my secondary school, Mr Bailson, who first showed me some of the music behind the arithmetic of the schoolroom. I am indebted to his inspiration and to Gillotts Comprehensive School, King James's 6th Form College, and Wadham College, Oxford, for the exceptional education I received.

Thank you to Arsenal for winning the double while I was writing this book. Highbury provided an important venue to let off steam after wrestling with Riemann.

On a personal note, I want to thank my friends and family for their support: my father, who helped me understand the power of numbers; my mother, who helped me understand the power of words; my grandparents, especially Peter, who were an inspiration; and my partner, Shani, for tolerating a book in the house and for her belief that I could write it. My biggest thank-you goes to my son, Tomer, for playing at the end of a day of work and without whom I would not have survived writing this book.

Further Reading

Many of the following books and articles provided material which was important in the writing of this book. For those who have been stimulated to dig deeper into the subject, I can recommend anything in this list. I have not included here any highly technical material that requires a mathematics degree to appreciate unless it contains some interesting non-technical insights.

Albers, D.J., Interview with Persi Diaconis, in *Mathematical People: Profiles and Interviews*, ed. D.J. Albers and G.L. Alexanderson (Boston: Birkhäuser, 1985), pp. 66–79

Aldous, D., and Diaconis, P., 'Longest increasing subsequences: from patience sorting to the Baik–Deift–Johansson theorem', *Bulletin of the American Mathematical Society*, vol. 36, no. 4 (1999), pp. 413–32

Alexanderson, G.L., Interview with Paul Erdős, in *Mathematical People: Profiles and Interviews*, ed. D.J. Albers and G.L. Alexanderson (Boston: Birkhäuser, 1985), pp. 82–91

Babai, L., Pomerance, C., and Vértesi, P., 'The mathematics of Paul Erdős', *Notices of the American Mathematical Society*, vol. 45, no. 1 (1998), pp. 19–31

Babai, L., and Spencer, J., 'Paul Erdős (1913–1996)', *Notices of the American Mathematical Society*, vol. 45, no. 1 (1998), pp. 64–73

Barner, K., 'Paul Wolfskehl and the Wolfskehl Prize', *Notices of the American Mathematical Society*, vol. 44, no. 10 (1997), pp. 1294–1303

Beiler, A.H., *Recreations in the Theory of Numbers: The Queen of Mathematics Entertains* (New York: Dover Publications, 1964)

Bell, E.T., *Men of Mathematics* (New York: Simon & Schuster, 1937)

Berndt, B.C., and Rankin, R.A. (eds), *Ramanujan: Letters and Commentary*, History of Mathematics, vol. 9 (Providence, RI: American Mathematical Society, 1995)

Berndt, B.C., and Rankin, R.A. (eds), *Ramanujan: Essays And Surveys*, History of Mathematics, vol. 22 (Providence, RI: American Mathematical Society, 2001)

Berry, M., 'Quantum physics on the edge of chaos', *New Scientist*, November 19 (1987), pp. 44–7

Bollobás, B. (ed.), *Littlewood's Miscellany* (Cambridge: Cambridge University Press, 1986)

Bombieri, E., 'Prime territory: exploring the infinite landscape at the base of the number system', *The Sciences*, vol. 32, no. 5 (1992), pp. 30–36

Borel, A., 'Twenty-five years with Nicolas Bourbaki, 1949–1973', *Notices of the American Mathematical Society*, vol. 45, no. 3 (1998), pp. 373–80

Borel, A., Cartier, P., Chandrasekharan, K., Chern, S.-S., and Iyanaga, S., 'André

Weil (1906–1998)' *Notices of the American Mathematical Society*, vol. 46, no. 4 (1999), pp. 440–47

Bourbaki, N., *Elements of the History of Mathematics*, translated from the 1984 French original by John Meldrum (Berlin: Springer-Verlag, 1994)

Breuilly, J. (ed.), *Nineteenth-Century Germany: Politics, Culture and Society 1780–1918* (London: Arnold, 2001)

Calaprice, A. (ed.), *The Expanded Quotable Einstein* (Princeton, NJ: Princeton University Press, 2000)

Calinger, R., 'Leonhard Euler: the first St Petersburg years (1727–1741)', *Historia Mathematica*, vol. 23, no. 2 (1996), pp. 121–66

Campbell, D.M., and Higgins, J.C. (eds), *Mathematics: People, Problems, Results*, 2 vols (Belmont, CA: Wadsworth International, 1984) [Includes chapters on Bourbaki, Gauss, Littlewood, Hardy, Hasse, Cambridge mathematics, Hilbert and his problems, the nature of proof and Gödel's theorem]

Cartan, H., 'André Weil: memories of a long friendship', *Notices of the American Mathematical Society*, vol. 46, no. 6 (1999), pp. 633–6

Cartier, P., 'A mad day's work: from Grothendieck to Connes and Kontsevich. The evolution of concepts of space and symmetry', *Bulletin of the American Mathematical Society*, vol. 38, no. 4 (2001), pp. 389–408

Changeux, J.-P., and Connes, A., *Conversations on Mind, Matter, and Mathematics*, edited and translated from the 1989 French original by M.B. DeBevoise (Princeton, NJ: Princeton University Press, 1995)

Connes, A., Lichnerowicz, A., and Schützenberger, M.P., *Triangles of Thoughts*, translated from the 2000 French original by Jennifer Gage (Providence, RI: American Mathematical Society, 2001)

Connes, A., 'Noncommutative geometry and the Riemann zeta function', in *Mathematics: Frontiers and Perspectives*, edited by V. Arnold, M. Atiyah, P. Lax and B. Mazur (Providence, RI: American Mathematical Society, 2000), pp. 35–54

Courant, R., 'Reminiscences from Hilbert's Göttingen', *The Mathematical Intelligencer*, vol. 3, no. 4 (1981), pp. 154–64

Davenport, H., 'Reminiscences of conversations with Carl Ludwig Siegel. Edited by Mrs Harold Davenport', *The Mathematical Intelligencer*, vol. 7, no. 2 (1985), pp. 76–9

Davis, M., *The Universal Computer: The Road from Leibniz to Turing* (New York, NY: W.W. Norton, 2000)

Davis, M., 'Book review: *Logical Dilemmas: The Life and Work of Kurt Gödel* and *Gödel: A Life of Logic*', *Notices of the American Mathematical Society*, vol. 48, no. 8 (2001), pp. 807–13

Dyson, F., 'A walk through Ramanujan's garden', in *Ramanujan Revisited*, edited by G.E. Andrews, R.A. Askey, B.C. Berndt, K.G. Ramanathan and R.A. Rankin (Boston, MA: Academic Press, 1988), pp. 7–28

Edwards, H.M., *Riemann's Zeta Function*, Pure and Applied Mathematics, Vol.

58 (New York, NY: Academic Press, 1974) [Contains a translation of Riemann's ten-page paper on the primes, 'Über die Anzahl der Primzahlen unter einer gegebenen Grösse', as an appendix]

Flannery, S., with Flannery, D., *In Code: A Mathematical Journey* (London: Profile Books, 2000)

Gardner, J.H., and Wilson, R.J., 'Thomas Archer Hirst – Mathematician Xtravagant III. Göttingen and Berlin', *American Mathematical Monthly*, vol. 100, no. 7 (1993), pp. 619–25

Goldstein, L.J., 'A history of the prime number theorem', *American Mathematical Monthly*, vol. 80, no. 6 (1973), pp. 599–615

Gray, J.J., 'Mathematics in Cambridge and beyond', in *Cambridge Minds*, ed. R. Mason (Cambridge: Cambridge University Press, 1994), pp. 86–99

Gray, J.J., *The Hilbert Challenge* (Oxford: Oxford University Press, 2000)

Hardy, G.H., 'Mr S. Ramanujan's mathematical work in England', *Journal of the Indian Mathematical Society*, vol. 9 (1917), pp. 30–45

Hardy, G.H., 'Obituary notice: S. Ramanujan', *Proceedings of the London Mathematical Society*, vol. 19 (1921), pp. xl–lviii

Hardy, G.H., 'The theory of numbers', *Nature*, September 16 (1922), pp. 381–5

Hardy, G.H., 'The case against the Mathematical Tripos', *Mathematical Gazette*, vol. 13 (1926), pp. 61–71

Hardy, G.H., 'An introduction to the theory of numbers', *Bulletin of the American Mathematical Society*, vol. 35 (1929), pp. 778–818

Hardy, G.H., 'Mathematical proof', *Mind*, vol. 38 (1929), pp. 1–25

Hardy, G.H., 'The Indian mathematician Ramanujan', *American Mathematical Monthly*, vol. 44, no. 3 (1937), pp. 137–55

Hardy, G.H., 'Obituary notice: E. Landau', *Journal of the London Mathematical Society*, vol. 13 (1938), pp. 302–10

Hardy, G.H., *A Mathematician's Apology* (Cambridge: Cambridge University Press, 1940)

Hardy, G.H., *Ramanujan. Twelve Lectures on Subjects Suggested by His Life and Work* (Cambridge: Cambridge University Press, 1940)

Hodges, A., *Alan Turing: The Enigma* (New York, NY: Simon & Schuster, 1983)

Hoffman, P., *The Man Who Loved Only Numbers. The story of Paul Erdős and the Search for Mathematical Truth* (London: Fourth Estate, 1998)

Jackson, A., 'The IHÉS at forty', *Notices of the American Mathematical Society*, vol. 46, no. 3 (1999), pp. 329–37

Jackson, A., 'Interview with Henri Cartan', *Notices of the American Mathematical Society*, vol. 46, no. 7 (1999), pp. 782–8

Jackson, A., 'Million-dollar mathematics prizes announced', *Notices of the American Mathematical Society*, vol. 47, no. 8 (2000), pp. 877–9

Kanigel, R., *The Man Who Knew Infinity: A Life of the Genius Ramanujan* (New York, NY: Scribner's, 1991)

Koblitz, N., 'Mathematics under hardship conditions in the Third World', *Notices of the American Mathematical Society*, vol. 38, no. 9 (1991), pp. 1123–8

Knapp, A.W., 'André Weil: a prologue', *Notices of the American Mathematical Society*, vol. 46, no. 4 (1999), pp. 434–9

Lang, S., 'Mordell's review, Siegel's letter to Mordell, Diophantine geometry, and 20th century mathematics', *Notices of the American Mathematical Society*, vol. 42, no. 3 (1995), pp. 339–50

Laugwitz, D., *Bernhard Riemann, 1826–1866: Turning Points in the Conception of Mathematics*, translated from the 1996 German original by Abe Shenitzer (Boston, MA: Birkhäuser, 1999)

Lesniewski, A., 'Noncommutative geometry', *Notices of the American Mathematical Society*, vol. 44, no. 7 (1997), pp. 800–805

Littlewood, J.E., *A Mathematician's Miscellany* (London: Methuen, 1953)

Littlewood, J.E., 'The Riemann hypothesis', in *The Scientist Speculates: An Anthology of Partly-Baked Ideas*, edited by I.J. Good, A.J. Mayne and J. Maynard Smith (London: Heinemann, 1962), pp. 390–91

Mac Lane, S., 'Mathematics at Göttingen under the Nazis', *Notices of the American Mathematical Society*, vol. 42, no. 10 (1995), pp. 1134–8

Neuenschwander, E., 'A brief report on a number of recently discovered sets of notes on Riemann's lectures and on the transmission of the Riemann *Nachlass*', *Historia Mathematica*, vol. 15, no. 2 (1988), pp. 101–13

Pomerance, C., 'A tale of two sieves', *Notices of the American Mathematical Society*, vol. 43, no. 12 (1996), pp. 1473–85 [An article about factorising numbers]

Reid, C., *Hilbert* (New York, NY: Springer, 1970)

Reid, C., *Julia, A Life in Mathematics* (Washington, DC: Mathematical Association of America, 1996) [With contributions from Lisl Gaal, Martin Davis and Yuri Matijasevich]

Reid, C., 'Being Julia Robinson's sister', *Notices of the American Mathematical Society*, vol. 43, no. 12 (1996), pp. 1486–92

Reid, L.W., *The Elements of the Theory of Algebraic Numbers*, with an Introduction by David Hilbert (New York, NY: Macmillan, 1910)

Ribenboim, P., *The New Book of Prime Number Records* (New York, NY: Springer, 1996)

Sacks, O., *The Man Who Mistook His Wife for a Hat* (New York, NY: Simon & Schuster, 1985)

Sagan, C., *Contact* (New York: Simon & Schuster, 1985)

Schappacher, N., 'Edmund Landau's Göttingen: from the life and death of a great mathematical center', *The Mathematical Intelligencer*, vol. 13, no. 4 (1991), pp. 12–18

Schechter, B., *My Brain Is Open. The Mathematical Journeys of Paul Erdős* (New York, NY: Simon & Schuster, 1998)

Schneier, B., *Applied Cryptography*, second edition, (New York, NY: John Wiley, 1996)

Segal, S.L., 'Helmut Hasse in 1934', *Historia Mathematica*, vol. 7, no. 1 (1980), pp. 46–56

Selberg, A., 'Reflections around the Ramanujan centenary', in *Ramanujan: Essays and Surveys*, History of Mathematics, vol. 22, edited by B.C. Berndt and R.A. Rankin (Providence, RI: American Mathematical Society, 2001), pp. 203–13

Shimura, G., 'André Weil as I knew him', *Notices of the American Mathematical Society*, vol. 46, no. 4 (1999), pp. 428–33

Singh, S., *The Code Book* (London: Fourth Estate, 1999)

Struik, D.J., *A Concise History of Mathematics*, (New York, NY: Dover Publications, 1948)

Weil, A., 'Two lectures on number theory, past and present', *L'Enseignement Mathématique*, vol. 20, no. 2 (1974), pp. 87–110

Weil, A., *Number Theory: An Approach Through History from Hammurapi to Legendre* (Boston, MA: Birkhäuser, 1984)

Weil, A., *The Apprenticeship of a Mathematician*, translated from the 1991 French original by Jennifer Gage (Basel: Birkhäuser, 1992)

Wilson, R., *Four Colours Suffice: How the Map Problem Was Solved* (London: Allen Lane, 2002)

Zagier, D., 'The first 50,000,000 prime numbers', *Mathematical Intelligencer*, vol. 0 (1977), pp. 7–19 [To the mathematicians who founded this journal, it seemed fitting to give the number zero to the first issue]

Websites

All of the articles in the above list from the *Notices of the American Mathematical Society* and the *Bulletin of the American Mathematical Society* are available on-line at http://www.ams.org/notices/ and http://www.ams.org/bull/

http://www.musicoftheprimes.com
My own website, which will include an evolving resource to supplement the book.

http://www.claymath.org/
A description of all seven Clay prizes as well as streamed videos of presentations by Connes, Wiles and Clay himself.

http://www.msri.org
The website of the Mathematical Sciences Research Institute at Berkeley, which has a large resource of streamed videos including a number aimed at a general audience.

http://www.rsasecurity.com/rsalabs/faq/
http://www.rsasecurity.com/rsalabs/challenges/
Here you can find RSA's cryptographic challenges.

http://www.mersenne.org/prime.htm
Visit this site to join the Great Internet Mersenne Prime Search.

http://www.eff.org
Information on the Electronic Frontier Foundation's prizes for the discovery of big primes.

http://www.maths.ex.ac.uk/~mwatkins/
An interesting resource of quotes and other material related to prime numbers and the Riemann Hypothesis.

http://www.certicom.com/research/ecc_chal_contents.html
An explanation of elliptic curve cryptography including Certicom's cryptographic challenges.

http://www-groups.dcs.st-andrews.ac.uk/history
'The MacTutor History of Mathematics archive' – a wonderful resource maintained by the University of St Andrews of mathematical biographies.

http://www.phys.unsw.edu.au/music/
A fascinating site exploring the acoustic qualities of different musical instruments with connection to Ernst Chladni's plates.

http://www.utm.edu/research/primes/
A good resource for information about prime numbers.

http://www.naturalsciences.be/expo/ishango/en/index.html
A chance to see the Ishango bone.

http://www.turing.org.uk/
A website maintained by Andrew Hodges, Alan Turing's biographer.

http://www.salon.com/people/feature/1999/10/09/dyson
'Freeman Dyson: frog prince of physics', an article by Kristi Coale.

Illustration and Text Credits

p15 courtesy of Clay Mathematics Institute; © 2000 Clay Mathematics Institute, All Rights Reserved; pp21, 42 and 133 Science Photo Library; p37 SCALA, Florence; p73 Universitatsbiblioteheck Gottingen; p177 Cambridge University Library; p214 Photography by Ingrid von Kruse, Freibildnerische Photographie; p221 photo courtesy of Andrew Odlyzko; p229 Photo courtesy of Professor Leonard M. Adleman.

Material from *Contact* by Carl Sagan: Copyright © 1985, 1986, 1987 by Carl Sagan. Reprinted with the permission of Simon & Schuster Adult Publishing Group and Orbit, a Division of Time Warner Books, UK.

The quotes by Julia Robinson in section 'From the chaos of uncertainty to an equation for the primes' in Chapter Eight are taken from Reid, C., *Julia, A Life in Mathematics* (Washington, DC: Mathematical Association of America, 1996). The quotes by André Weil in section 'Speaking in many tongues' in Chapter Twelve are taken from Weil, A., *The Apprenticeship of a Mathematician* (Basel: Birkhäuser, 1992). The quotes by G.H. Hardy throughout the book are taken from Hardy, G.H., *A Mathematician's Apology* (Cambridge: Cambridge University Press, 1940) and from other articles by Hardy listed in Further Reading.

All reasonable efforts have been made by the author and the publisher to trace the copyright holders of the images and material quoted in this book. In the event that the author or publisher are contacted by any of the untraceable copyright holders after the publication of this book, the author and the publisher will endeavour to rectify the position accordingly.

Index

Figures in italics indicate captions.